Dehydrogenases
Requiring Nicotinamide Coenzymes

EXS
Experientia Supplementum
Vol. 36

Dehydrogenases

Requiring Nicotinamide Coenzymes

Edited by Jonathan Jeffery
Department of Biochemistry
University of Aberdeen, Aberdeen, Scotland, U.K.

1980
Birkhäuser Verlag
Basel · Boston · Stuttgart

Library of Congress Cataloging
in Publication Data

Main entry under title:
Dehydrogenases requiring nicotinamide
coenzymes.
(Experientia supplementum; 36)
1. Dehydrogenases. 2. Vitamin PP.
3. Coenzymes.
I. Jeffery, Jonathan, 1935 – II. Series.
[DNLM: 1. Oxidoreductases.
2. Nicotinamide. 3. Coenzymes.
W1 EX23 v. 36 / QU140 D323]
QP603.D4D43 574.1'9258 79-22456
ISBN 3-7643-1104-5

CIP-Kurztitelaufnahme
der Deutschen Bibliothek

Dehydrogenases requiring nicotinamide coenzymes / ed. by Jonathan Jeffery. –
Basel, Boston, Stuttgart: Birkhäuser, 1980.
(Experientia: Suppl.; 36)
ISBN 3-7643-1104-5
NE: Jeffery, Jonathan [Hrsg.]

ISBN 3-7643-1104-5

PREFACE

This book brings together important work on the structure and function of the dehydrogenases requiring nicotinamide coenzymes.

The structure and function of lactate dehydrogenase, alcohol dehydrogenase, and glyceraldehyde 3-phosphate dehydrogenase are compared. These are the relatively simple oligomeric dehydrogenases for which very extensive structural and mechanistic information is available.

Kinetic studies are dealt with in general terms, with selected examples.

Genetic and structural relationships revealed by primary structure are discussed.

General features of the specificity of dehydrogenase reactions are described.

The pyridine nucleotide - disulphide oxidoreductases exemplify a type of multicomponent enzyme complex. Lipoamide dehydrogenase, glutathione reductase, and thioredoxin reductase are described and compared.

The fatty acid synthesizing systems of some organisms are nonaggregated, whereas in other organisms they are multienzyme complexes containing, in animal tissues for example, two closely related multifunctional polypeptide chains. The dehydrogenase activities of the fatty acid synthesizing systems afford many points of interest, which are described and discussed.

Hydroxymethylglutaryl-CoA reductase occupies a key position in the synthesis of isoprenoid compounds, including cholesterol. It probably occurs in all forms of life; in animal cells it is a microsomal enzyme. Its properties are reviewed.

It is hoped that this book brings together much of the most interesting work on dehydrogenases, and presents it in a form that is easy to follow. The more general textbooks must omit much that is of interest and importance. The more specialized reviews do not make easy reading. It is hoped that this book strikes a balance that will meet a need.

July, 1979 — Jonathan Jeffery

CONTRIBUTORS

CARL-IVAR BRÄNDÉN
Sveriges Lantbruksuniversitet, Institutionen för Kemi,
S - 750 07 Uppsala, Sweden

WILLIS E. BROWN
Department of Biochemistry, Purdue University,
West Lafayette, Indiana 47907, U.S.A.

HANS EKLUND
Sveriges Lantbruksuniversitet, Institutionen för Kemi,
S - 750 07 Uppsala, Sweden

ARNE HOLMGREN
Karolinska Institutet, Kemiska Institutionen,
S - 104 01 Stockholm 60, Sweden

JONATHAN JEFFERY
Department of Biochemistry, University of Aberdeen,
Marischal College, Aberdeen AB9 1AS, Scotland, U.K.

HANS JÖRNVALL
Karolinska Institutet, Kemiska Institutionen,
S - 104 01 Stockholm 60, Sweden

SARVAGYA S. KATIYAR
Department of Chemistry, Indian Institute of Technology,
Kanpur 208016, India

JOHN W. PORTER
Lipid Metabolism Laboratory, William S. Middleton Memorial
Veterans Hospital, and the Department of Physiological
Chemistry, University of Wisconsin, Madison,
Wisconsin 53706, U.S.A.

VICTOR W. RODWELL
Department of Biochemistry, Purdue University,
West Lafayette, Indiana 47907, U.S.A.

CONTENTS

KINETIC ASPECTS OF SOLUBLE DEHYDROGENASES REQUIRING NICOTINAMIDE COENZYMES

Jonathan Jeffery

CONTENTS

SUMMARY

Methods for establishing the kinetic mechanisms of dehydrogenase reactions are dealt with in general terms. Examples ranging from relatively simple to obviously complex enzymes, and showing various mechanistic features of interest and importance are discussed.

1. INTRODUCTION

The overall reaction catalysed by many of the enzymes to be discussed can be summarized as Eq. (1),

$$S_{ox} + NAD(P)H + H^+ \rightleftharpoons S_{red} + NAD(P)^+ \qquad (1)$$

where S_{ox} and S_{red} represent the oxidized and reduced form of the substrate, respectively.

Dehydrogenase reactions are usually studied in the laboratory in buffers at constant pH, and this eliminates hydrogen ion concentration as a variable in a particular experiment. It is, moreover, generally agreed that the acquisition or loss of the hydrogen ion (Eq. 1) need not be represented as a distinct step for steady-state studies. This allows a valuable simplification of the kinetic equations.

2. SIMPLE SYSTEMS

Omission of the hydrogen ion allows Eq. (1) to be written as Eq. (2).

$$A + B \rightleftharpoons P + Q \qquad (2)$$

A represents the substance that gives rise to Q, and B the substance that gives rise to P. Referring to Eq. (1), B could be S_{ox}, in which case P would be S_{red}, A would be NAD(P)H and Q would be $NAD(P)^+$. Alternatively, A could be S_{ox}, in which case Q would be S_{red}, etc. If we do not at this stage specify whether A or B is coenzyme, an economy of diagrams is achieved.

Consideration of how the overall change represented by Eq. (2) might come about suggests two somewhat different types of process. In one type, A would interact with E (the enzyme molecule) to give a binary complex, EA, which would undergo internal hydrogen transfer to give another binary complex, E'Q, from which Q would dissociate, the modified enzyme E' then forming a binary complex, E'B, undergoing hydrogen transfer to EP, from which P would dissociate. This is represented schematically in Fig.1, and is commonly called a ping-pong mechanism (a term probably introduced by W.W. Cleland).

The other types of process involve a ternary complex, EAB, which could conceivably be formed (i) by interaction of

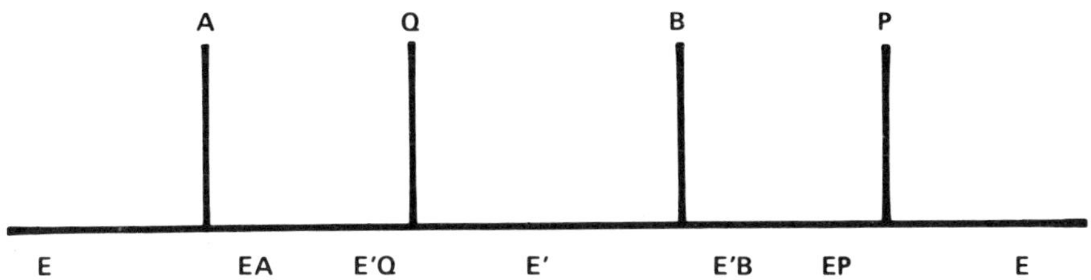

Fig. 1 A ping-pong mechanism for the overall change shown in Eq.2.

Fig. 2 An improbable mechanism in which a binary complex of reactants interacts with the enzyme to form a ternary complex

A and B to form a binary complex, AB, which then interacted with E to give EAB, (ii) by interaction of A and E to give EA, only EA reacting with B to give the ternary complex EAB, or (iii) by interaction of either A or B to give a binary complex (EA or EB), which then reacted with B or A to give in both cases the same ternary complex, EAB. Hydrogen transfer would occur in the ternary complex, giving EPQ, dissociation of EPQ occurring in a manner corresponding to the reverse of processes (i)-(iii). Some of these possibilities are represented schematically in Figs.2-5.

Mechanisms of the ping-pong type (Fig.1) may be found among hydrogen-transferring enzymes containing bound flavine moieties. Covalent modification of the enzyme also occurs during the functioning of glyceraldehyde-3-phosphate dehydrogenase (when a thiohemiacetal and acylenzyme are formed, see section 11.6, and Figs. 15 and 16) and glutamate dehydrogenase (when a Schiff base is formed, Hochreiter et al., 1972). It was once suggested that rat skeletal muscle glycerol dehydrogenase might have a mechanism of the ping-pong type (Toews, 1967), but confirmation would be desirable.

A mechanism of the kind represented in Fig.2 is improbable because it requires the formation of a significant concentration of the complex AB, and would forgo what is generally regarded as a principal advantage of an enzymic reaction, namely that the enzyme can bring together the reacting species. It is not thought to apply to any dehydrogenase.

The compulsory-order type of mechanism (Fig.3) is generally regarded as of great importance for dehydrogenases. A would usually be the coenzyme.

The alternative pathway mechanism (Fig.4) is also important. It is often called a random order mechanism, and we shall use this name so as not to introduce new terms, though the word random is perhaps not entirely satisfactory because the mechanism is not particularly haphazard, chance, uncontrolled, or irregular, and it includes both reactions in which the alternatives are of similar quantitative importance, and those in which certain of the pathways predominate. The compulsory-order mechanism (Fig.3) is a limiting case of the random order mechanism in which the alternatives below the central line in Fig.4 are quantitatively insignificant. Many dehydrogenase reactions may be of the random order type with marked predominance of one of the alternatives.

The mechanism represented in Fig.5 (which is also a limiting case of Fig.4) illustrates the possibility that the routes to the ternary complex from one side may not be the equivalent of the route from the other side. Evidence has been put forward that the major isozyme of horse liver alcohol dehydrogenase follows a mechanism similar to Fig.5, with A being NADH, B acetaldehyde, P ethanol, and Q NAD+ (Hanes et al., 1972).

The mechanism proposed for horse liver alcohol dehydrogenase many years earlier by Theorell & Chance (1951) (Fig.6) is a somewhat different limiting case (cf. Fig.3 and Fig.4). The acetaldehyde (B) or ethanol (P) react so quickly with the appropriate holoenzyme (EA or EQ) that a ternary

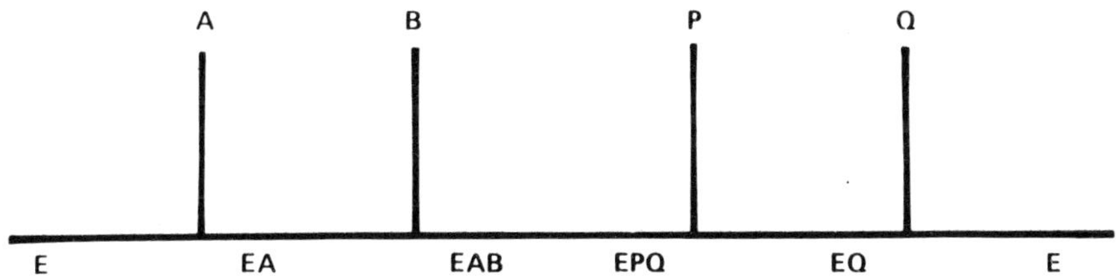

Fig. 3 A compulsory-order ternary complex mechanism

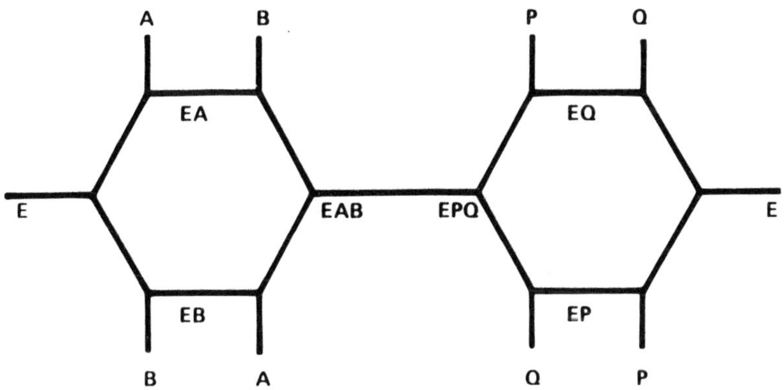

Fig. 4 An alternative pathway ternary complex mechanism (random order mechanism)

complex can be omitted from the kinetic equations.

3. KINETIC EQUATIONS

The derivation of kinetic equations from first principles, by the elimination of unknowns from simultaneous equations, is extremely tedious even for simpler mechanisms than those discussed above. A very much easier method (though one which still involves cumbersome algebra for all but the very simplest mechanisms) was devised by King & Altman (1956). Further short-cuts have since been devised (e.g. Cleland, 1975). Some abbreviated methods may, however, have dangers (see, for example, Cornish-Bowden, 1976). Computers can be used to generate steady-state rate equations (e.g. Hurst, 1969; Fromm, 1975).

4. RANDOM ORDER MECHANISM

Consideration of the random order mechanism (Fig.4) allows certain general points to be made. Other important mechanisms (Figs. 3 and 6) are limiting cases of it.

For application of the King & Altman (1956) procedure, it is convenient to re-draw Fig.4 as Fig.7. This enables the make-up of the rate equation to be assessed. There are 7 enzyme-containing species (including free enzyme) interconnected by 9 reversible steps, and there are 2 cycles with 4 steps, and 2 cycles with 5 steps (Fig.7). The number of terms for each enzyme-containing species will be 56. The denominator of the rate equation is the sum of the terms for each enzyme-containing species, and will consist of 392 terms. Each of these will consist of six rate constants (i.e. one less than the number of enzyme-containing species) and the associated concentration terms. The general equation is therefore too cumbersome to write out and manipulate, and, indeed, too complicated for the human mind to comprehend without simplification of some kind.

One simplification is to consider only the initial rate in the absence of products (Dalziel, 1958; Pettersson, 1969). The equation obtained relating 1/v and $[S]^{-1}$ is non-linear, but it can be arranged in the form of linear terms plus a non-linear "residue" (Pettersson, 1970). This "residue" always becomes zero as $[S]$ tends to zero (that is as $[S]^{-1}$ tends to infinity). The linear terms therefore represent the linear asymptote. The actual curves may be convex or concave as they steadily approach the linear asymptote; or, indeed they may be linear if the "residue" is zero for some other reason. Thus, the relation between 1/v and $[S]^{-1}$ becomes linear under various circumstances, and, taking into consideration the fact that there will always be limitations on the range of $[S]$ over which measurements of v can be made, and that there will always be some experimental error in even the best determinations of v, the plot of 1/v against $[S]^{-1}$ may in practice also appear to

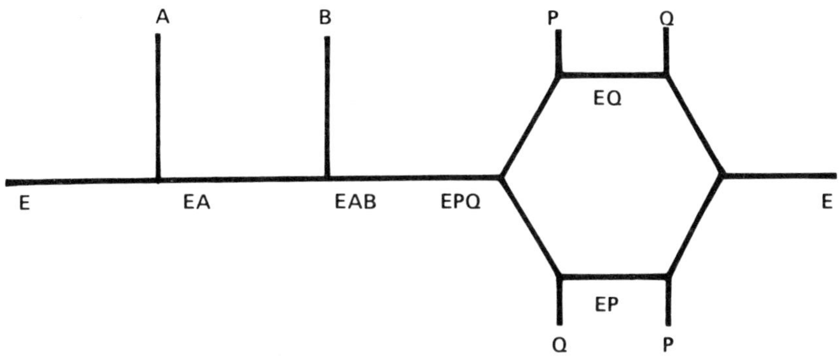

Fig. 5 A ternary complex mechanism, half with compulsory-order and half with alternative pathways

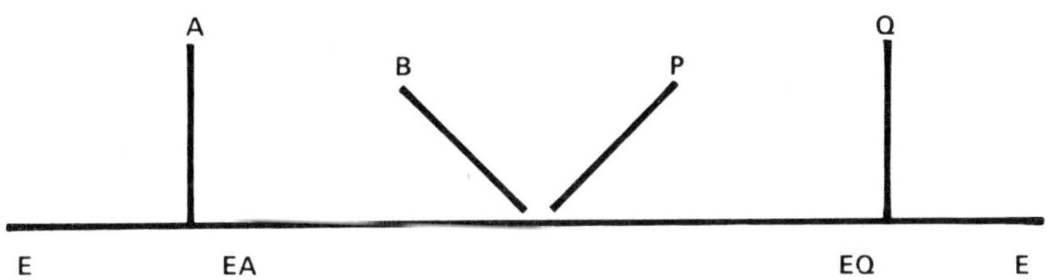

Fig. 6 The Theorell - Chance mechanism

be linear when the relationship is not truly linear. The apparent linearity of such a plot is therefore relatively uninformative; proper interpretations have been discussed by Pettersson (1972, 1974, 1977). The special cases of rapid equilibrium (Dalziel, 1958) and compulsory order (including the Theorell-Chance mechanism) (Dalziel & Dickinson, 1966a) give linear plots.

5. GENERAL RATE EQUATIONS AND SPECIAL CASES

In the case of two-substrate reactions, the variety of possible mechanisms and the complexity of many of the rate equations mean that when findings are shown to conform to a particular mechanism, it is a matter of concern to know what number of alternative mechanisms might predict rates similar in the respects tested. Wong & Hanes (1962) considered what they regarded as being the simplest general mechanism for two-substrate reactions, namely, a scheme equivalent to Fig.8. The mechanisms represented in Figs. 1,3,4,5 and 6 are all special cases of this. Means of distinguishing various special cases were tabulated and discussed (Wong & Hanes, 1962). Some of the cases were known from earlier work (e.g. Alberty, 1953, 1958; Dalziel, 1957, 1958; Ingraham & Makower, 1954; Koshland, 1953, 1955, 1956; Laidler, 1955a,b, 1956).

Soon afterwards, Cleland (1963a,b,c,d) published the schemes that are now very widely used for characterizing reactions of this type. General reviews appeared a little later (Cleland, 1967a, 1970). Further considerations, particularly of random-order mechanisms, were given by Pettersson (1974, 1977). Dalziel (1975) reviewed the kinetics of dehydrogenase reactions. For unbranched reactions, a procedure described by Ainsworth (1975) is also available to relate the mechanism to the kinetic constants determined experimentally. All these methods make systematic use of initial rate and product inhibition relationships.

The method by which the experimental data are processed is important if the best information is to be extracted from the data. When a hyperbolic relationship between v and [S] is expected (or when deviation from it is sought) a procedure still popular is to seek (either by plotting or by simple regression analysis) a linear relationship between $1/v$ and $[S]^{-1}$. The injudiciousness of this has often been pointed out (Johansen & Lumry, 1961; Wilkinson, 1961; Dowd & Riggs, 1965; Colquhoun, 1971). A comparison of seven methods (Atkins & Nimmo, 1975) indicated that the best procedures were those of Wilkinson (1961) and Eisenthal & Cornish-Bowden (1974). Developments of the latter procedure were subsequently described (Porter & Trager, 1977). Aspects of the statistical "error structure" of enzyme kinetic experiments have been analysed to improve the interpretation of results and to allow better experimental design (Cleland, 1967b; Storer et al., 1975; Siano et al., 1975; Askelöf et al., 1976). A method

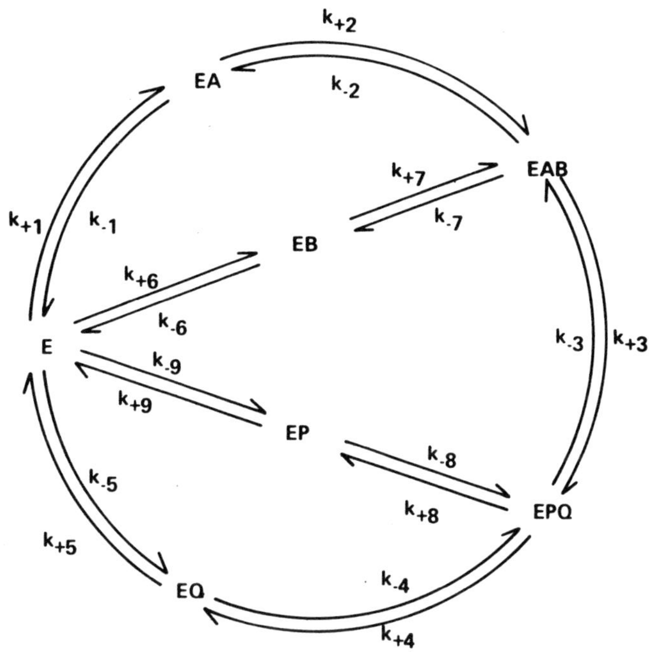

Fig. 7 Alternative pathway (random order) mechanism

of correcting initial estimates of V and K_m, which may be helpful in some circumstances, has been given (Glick et al., 1979).

Britton & Dann (1978) pointed out that because application of the method of Cleland (1963a,b) involves extrapolation, experimental errors or failure of the enzyme to conform to hyperbolic kinetics may have a large effect on the point of intersection of the reciprocal plots, and consequently on the conclusions that are drawn; also the method gives no indication of the relative contributions of alternative pathways. A method was developed that uses flux ratios (Britton, 1966, 1973, 1977; Britton & Clarke, 1968; Britton & Dann, 1978) and applied to rabbit muscle pyruvate kinase (Dann & Britton, 1978). This flux ratio method could prove useful for the study of dehydrogenases.

6. KINETIC ISOTOPE EFFECTS

The breaking of a bond to deuterium (or tritium) can be predicted to occur more slowly than the breaking of the corresponding bond to ^1H. Put very simply then, the position is as follows. If, when the rate of a dehydrogenase reaction is determined for the transfer of deuterium (or tritium) and for the transfer of ^1H, the ratio of these rates has about the "predicted" value, then the actual breaking of the bond to hydrogen may be the rate-determining step. If the ratio lies between 1 and the "predicted" value, then there may be no single rate-determining step; and if the ratio is unity, then some step other than the breaking of the bond to hydrogen may be rate determining (e.g. a binding or isomerization step).

The crux of the matter is: what is the "predicted" value? A well-known answer is "about 7 for deuterium, (and about $7^{1.44}$ for tritium). This derives from the assumption that the rate difference arises in a very simple way from contributions to the activation energy from changes in zero-point energy that occur when the reactants are converted into an activated complex. However, calculations based on a less simplified theory can lead to a "predicted" value for deuterium of much less than 7 (Westheimer, 1961). In chemical (non-enzymic) hydride transfer reactions, ratios ranging up to 5 have been observed (see Wiberg & Motell, 1963), while for chemical reactions involving proton transfer, ratios much less than 1 (e.g. Ford et al., 1967) and much more than 7 (e.g. Lewis & Robinson, 1968) have been found.

A method of estimating the isotope effect on an enzymic reaction was described by Northrop (1975). An equilibrium perturbation method was developed by Cleland's group (Schimerlik et al., 1975). A method for comparing $(V/K_m)_{^1H}$ and $(V/K_m)_{^3H}$ is available (Simon & Palm, 1966). Some discussion of dehydrogenases (mainly yeast and liver alcohol dehydrogenase, and glutamate dehydrogenase) was given by Klinman (1978). A few examples of isotope effects are shown in Tables 1 and 2.

A practical point that should be borne in mind is that nicotinamide coenzymes on standing in aqueous buffers or on

Table 1. Deuterium Isotope Effects on Steady—State Maximum Velocities[a]

Enzyme and Source	Substrate and Coenzyme	pH	$V_{^1H}:V_{^2H}$	Ref.
Lactate dehydrogenase Bovine heart	L-[2-^{2}H] lactate APAD+ b	8.6	1.0	c
Phosphoglycerate dehydrogenase *E. coli*	3-phospho-D-[2-^{2}H] glycerate APAD+	7.0	1.2	d
Octopine dehydrogenase *Pecten maximus L.*	Labelled D-octopine[e] NAD+	7.0	1.4	f
Cytosol malate dehydrogenase Pig heart	Labelled L-malate[g] NAD+	8.0	1.4[h]	i
Cytosol malate dehydrogenase Pig heart	[2,3-^{2}H]meso-tartrate NAD+	8.0	1.9[h]	i
Glucose-6-phosphate dehydrogenase Yeast	D-[1-^{2}H] glucose-6-phosphate NADP+	7.0	2.0	j
Glucose-6-phosphate dehydrogenase Red-cells	D-[1-^{2}H] glucose-6-phosphate NADP+	7.0	2.4	j
Oestradiol 17β-dehydrogenase Human placenta	[17α-^{2}H] oestradiol NAD+	9.4	2.6	k
Oestradiol 17β-dehydrogenase Human placenta	[17α-^{2}H] oestradiol NADP+	9.4	2.7	k

Footnotes to this Table are at the top of the next page.

Footnotes to Table 1.

a. For experimental conditions see references cited. For interpretation of isotope effects see also Northrop (1975)

b. $APAD^+$ is the 3-acetylpyridine analogue of NAD^+

c. Cantwell & Dennis (1970)

d. Dubrow & Pizer (1977)

e. N^2-(R-1-carboxy-[1-^{2}H]ethyl)-S-arginine

f. Doublet et al., (1975)

g. Presumably [2-^{2}H]-L-malate. See Kimball (1977)

h. Not maximum velocity (low coenzyme concentration)

i. Kimball et al., (1979)

j. Rose (1961)

k. Adams et al., (1965)

Table 2. Deuterium Isotope Effects on Steady-State Maximum Velocities for Alcohol Dehydrogenase from Horse Liver, EE Isozyme (Native and Modified by 4-Hydroxybutyrimidylation) and for Yeast Alcohol Dehydrogenase with Various Substrates and 4R-^{2}H NADH as Coenzyme.

	$V_{^1H}:V_{^2H}$		
	Liver		Yeast
	native[a]	modified[a]	native[b]
Acetaldehyde	1.1	1.5	4.6
Benzaldehyde	1.0	2.3	3.0
4-Chlorobenzaldehyde	1.0	1.8	3.3
4-Methylbenzaldehyde	1.1	2.4	5.4
4-Methoxybenzaldehyde	1.1	2.6	3.4

a. figures for pH 8.0 from Dworschack & Plapp (1977)

b. figures for pH 8.5 from Klinman (1972)

freeze drying, can give rise to potent inhibitors of some dehydrogenases (Dalziel, 1961a,b; 1962a,c; 1963a; Fawcett et al., 1961; Silverstein & Boyer, 1964a, 1966; Biellmann et al., 1979). So if hydrogen (^{1}H, ^{2}H, or ^{3}H) transfer from reduced coenzyme is studied, small differences in the content of such inhibitors, which might arise from different recent "histories" of the ^{1}H-, ^{2}H-, and ^{3}H-preparations, could be confused with isotope effects.

7. ISOTOPE EXCHANGE AT EQUILIBRIUM

At chemical equilibrium there is no net chemical reaction because the reaction is proceeding in both directions at equal rates. These rates can be measured by isotope exchange, and isotope exchange studies can be used to elucidate the mechanism of enzymic reactions (Boyer, 1959). The method is useful for determining (or confirming) what the order of addition of substrates is, whether the interconversion of ternary complexes is rate determining, for setting minimum values for some of the dissociation constants, and for detecting alternative reaction pathways ("partial randomness").

The use of systems at equilibrium achieves valuable simplifications, but the method is applicable to steady-state conditions generally (Cleland, 1967a).

The cases of two-substrate reactions under equilibrium conditions were analysed by Boyer & Silverstein (1963). Application of the method to dehydrogenase reactions is well illustrated by early studies on liver and yeast alcohol dehydrogenase (Silverstein & Boyer, 1964b), (which gave evidence of incompletely ordered reaction for both these enzymes, in disagreement with evidence of other kinetic methods at that time, e.g. Dalziel, 1962b, 1963a,b), on bovine heart and rabbit muscle lactate dehydrogenase (Silverstein & Boyer, 1964a) (which supported an ordered mechanism, as discerned by other kinetic methods, e.g. Hakala et al., 1956; Takenaka & Schwert, 1956; Zewe & Fromm, 1962; and Fromm, 1963); and on pig heart mitochondrial malate dehydrogenase (Silverstein & Sulebele, 1969) which supported an ordered mechanism, as proposed by Raval and Wolfe, 1962a,b, on other kinetic evidence).

A helpful explanation of the method of isotope exchange, which includes the derivation of equations, the interpretation of results, and some practical hints, was given by Plowman (1972), and a somewhat more mathematical presentation, including a new treatment (Ainsworth, 1977a), by Ainsworth (1977b).

8. SUBUNITS OF DEHYDROGENASES

Some dehydrogenases consist of a single polypeptide chain; octopine dehydrogenase from *Pecten maximus L.* (Doublet & Olomucki, 1975), aldehyde reductase from human liver

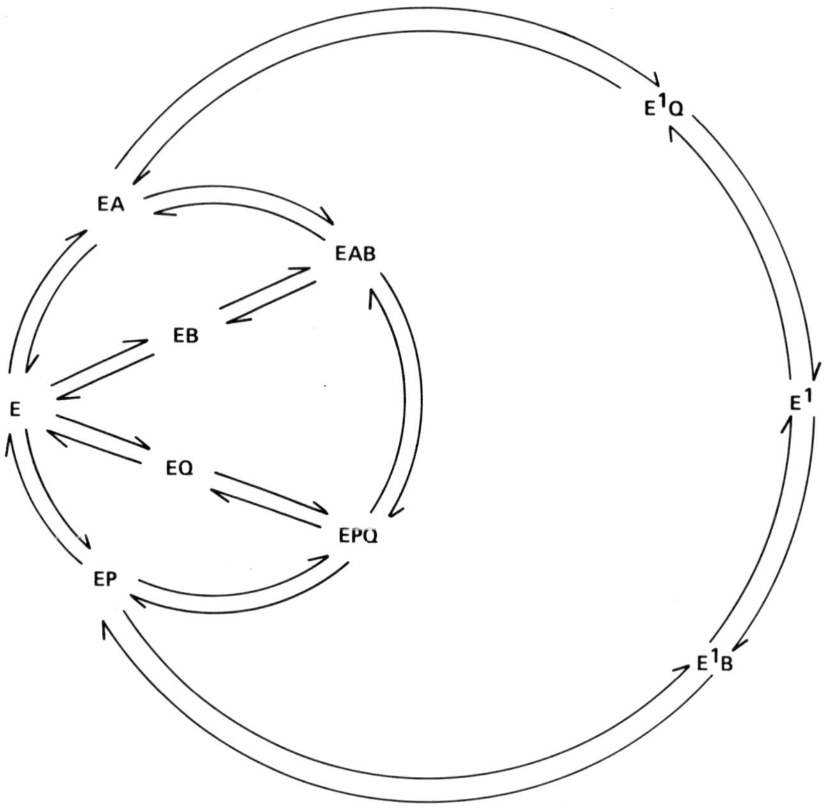

Fig 8. A general reaction scheme representing alternative pathways (as in Fig. 4) and enzyme modification (as in Fig. 1).

(Wermuth _et al._, 1977), and 20α-hydroxysteroid dehydrogenase from pig testis (Shikita & Tsuneoka, 1976) would be examples. However, most of the well-known dehydrogenases are oligomeric. Alcohol dehydrogenase from horse liver, and soluble malate dehydrogenase from pig and bovine heart are dimers; yeast alcohol dehydrogenase, pig and dogfish muscle lactate dehydrogenase, and lobster muscle glyceraldehyde-3-phosphate dehydrogenase are tetramers; bovine liver glutamate dehydrogenase is a hexamer. Polyol dehydrogenase from _Cephalosporium chrysogenum_ may be decameric (Birken & Pisano, 1976). 4-Aminobutanal dehydrogenase (Callewaert _et al._, 1964) and succinic semialdehyde dehydrogenase from Pseudomonas species (Rosemblatt _et al._, 1973) were described as trimers. Succinic semialdehyde dehydrogenase from pig brain (Blaner & Churchich, 1979) had about the same molecular weight (160,000) as the _Pseudomonas_ enzyme (164,000), so may have the same number of subunits. Trimers are unusual. Many soluble dehydrogenases are dimers or tetramers. Thus, of 11 malate dehydrogenases recently purified from mesophilic and thermophilic bacteria, 3 were dimers, and 8 were tetramers (Wright & Sundaram, 1979); aldehyde dehydrogenases from human erythrocytes (Inoue _et al._, 1979) and sheep liver (MacGibbon _et al._, 1979) are tetrameric; human liver pi-alcohol dehydrogenase is dimeric (Bosron _et al._, 1979); secondary-alcohol dehydrogenase from a _Pseudomonad_ is dimeric (Hou _et al._, 1979); soya bean cinnamyl alcohol dehydrogenase is dimeric (Wyrambic & Grisebach, 1979); glucose-6-phosphate dehydrogenase from various sources exists in multiple forms, dimers and tetramers being important (Chang _et al._, 1979; Gezici & Domagk, 1979, and references therein).

The existence of isozymes among dehydrogenases is well known, though their significance remains a matter of controversy (Huijing, 1979; Flores, 1979; Heffron, 1979; Svasti, 1979). Isozymes may differ in the values of their kinetic constants, substrate and coenzyme specify, and in other properties; a helpful compilation was given by Ureta (1978). Oligomers containing like isozymic subunits, and mixed isozymic subunits are well known in some cases, such as the H_4 and M_4 forms of rat lactate dehydrogenase, and the hybrid forms M_3H, M_2H_2, and MH_3 (Fine _et al._, 1963). When the subunits of a dehydrogenase are not identical, they are, nevertheless, usually of closely similar size. Bovine erythrocyte glucose-6-phosphate dehydrogenase is somewhat unusual in this respect, being a tetramer composed of subunits of molecular weight 58,000 and 62,000 (Gezici & Domagk, 1979). L-β-Hydroxyacid dehydrogenase from _Drosophila melanogaster_ is extremely unusual, being a dimer composed of subunits of greatly dissimilar molecular weight (23,500 and 40,000) (Cannistraro _et al._, 1979).

Regarding the formation of the active oligomers in vivo, many questions arise. The nascent polypeptide presumably folds first to form a protomer, which presumably does not have the exact conformation characteristic later of the protomer in the oligomer. Interaction of the protomers leads to formation of the stable oligomer, and conformational adjustments presumably accompany this process. At what stage does enzymic activity arise? Does the presence of coenzyme or substrate (or other ligand) guide the processes of folding and association? The answers to these and many related questions remain largely unknown, though some progress is being made (see, for example, Jaenicke et al., 1979).

For an oligomeric dehydrogenase, the question arises whether or not the subunits (after assembly into the native oligomer) function independently.

9. CO-OPERATIVE EFFECTS

The models for subunit interaction developed by Monod, Wyman & Changeux (1965) and by Koshland and co-workers (Koshland et al., 1966; Kirtley & Koshland, 1967; Koshland, 1970) are well known. Methods of plotting experimental data in such cases have been discussed (Buc, 1967; Horn & Börnig, 1969; Endrenyi et al., 1971; Cornish-Bowden & Koshland, 1975; Kegels, 1979). As indicated in Section 4, non-linear plots of 1/v against $[S]^{-1}$ are not necessarily evidence of subunit interactions. Detailed treatments are given by Bardsley & Childs (1975), Monneuse-Doublet et al., (1978), and Whitehead & Egmond (1979).

Positive co-operative effects may be explained by the model of Monod, Wyman & Changeux (1963) (Paulus & DeRiel,1975). A well-known dehydrogenase showing positive co-operative effects is yeast isocitrate dehydrogenase (Hathaway & Atkinson, 1963; Kuehn et al., 1971; Barnes et al., 1972).

The models of Koshland and co-workers can give rise to positive or negative co-operative effects. Several dehydrogenases have been reported to show negative co-operative effects. When there are several binding steps, such as the four substrate binding steps for a tetrameric enzyme, it is in principle possible to have a mixture of positive and negative co-operative effects.

An extreme form of negative co-operative effect is "half-of-the-sites" reactivity (see Levitzki & Koshland, 1976). Dehydrogenases which at one time or another have been regarded as showing "half-of-the-sites" reactivity are liver alcohol dehydrogenase, yeast alcohol dehydrogenase, glyceraldehyde-3-phosphate dehydrogenase (from rabbit muscle, sturgeon muscle and yeast), malate dehydrogenase, homoserine dehydrogenase (from *Escherichia coli*), and aldehyde dehydrogenase (yeast and horse liver) (see Seydoux et al., 1974) and dogfish lactate dehydrogenase (Lazdunski, 1974).

10. RAPID REACTION TECHNIQUES

Elucidation of the elementary steps in the reaction mechanism of an enzyme generally requires investigation of the events occurring during the period immediately following mixing, before the steady-state and so-called "initial rate" have been established. The period over which such observations

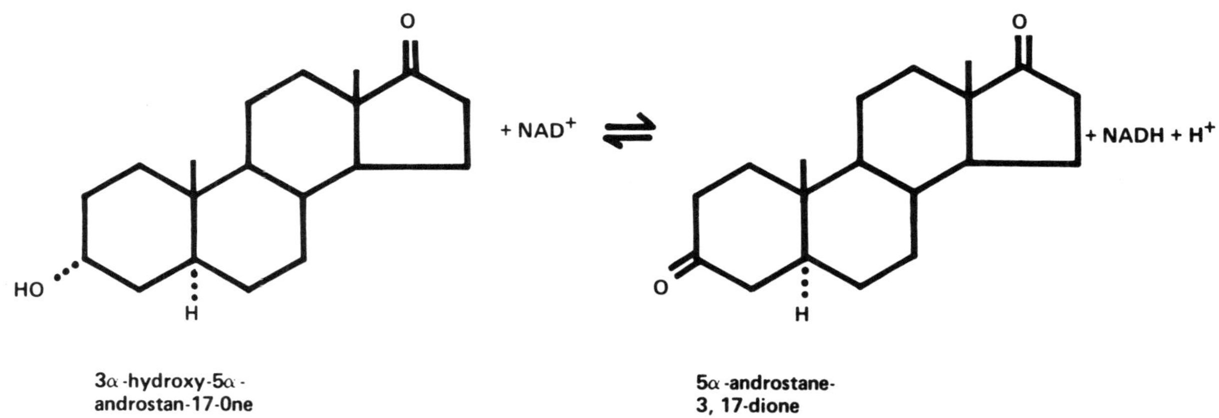

Fig. 9 3α-Hydroxysteroid dehydrogenase from Pseudomonas testosteroni is a monomeric enzyme reported to form dimers with altered activity (see text)

must be made ranges from something above 10^{-13}s (which is the period of molecular vibrations) up to something like 1s (by which time the steady-state will have been reached.) Stopped-flow methods (Gibson & Milnes, 1964; Seydoux & Bernhard, 1974) allow observations to be made down to about 10^{-3}s. Relaxation techniques (Eigen & de Maeyer, 1963; Eigen & Hammes, 1963; Gutfreund, 1971) allow observations down to about 10^{-10}s. An early application of relaxation methods to a dehydrogenase (Kirschner *et al.*, 1966) well illustrated the usefulness of the techniques. The transient-phase kinetics for enzyme systems involving two substrates have been worked out (Hijazi & Laidler, 1973; Pettersson, 1976, 1978).

11. KINETICS AND MECHANISM : SELECTED EXAMPLES

11.1 Aldehyde reductase from pig kidney (EC 1.1.1.2)

$$\begin{array}{c} CHO \\ | \\ HCOH \\ | \\ CH_2OH \end{array} + NADPH + H^+ \rightleftharpoons \begin{array}{c} CH_2OH \\ | \\ CHOH \\ | \\ CH_2OH \end{array} + NADP^+$$

This monomeric enzyme (molecular weight 36,000) requires NADP (Davidson & Flynn, 1979). The mechanism was studied at pH 7.0 and 25°C, using D-glyceraldehyde, glycerol, NADPH, and NADP$^+$. Analysis of initial rate and product inhibition findings by the method of Cleland (1963a,b) showed them to be consistent with an ordered bi bi mechanism (Fig.3) with coenzyme binding first and leaving last, and to exclude a random mechanism (Fig.4) and a Theorell-Chance mechanism (Fig.6). Using fluorescence techniques, it was shown that NADPH does bind to the enzyme in the absence of D-glyceraldehyde (Flynn *et al.*, 1975).

11.2 3α-Hydroxysteroid dehydrogenase from Pseudomonas testosteroni (EC 1.1.1.50)

This monomeric enzyme (molecular weight 47,100) uses NAD (Squire *et al.*, 1964) (Fig.9). The mechanism was studied at pH 7.0 and 25°C using 3α-hydroxy-5α-androstan-17-one, 5α-androstane-3,17-dione, NAD$^+$, and NADH (Skålhegg, 1975). Initial rate and product inhibition findings were analysed by the method of Cleland (1963a,b) and found to be consistent with an ordered bi bi mechanism (Fig.3) in which the coenzyme bound first and left last.

This monomeric enzyme formed dimers, especially at high protein concentrations. With 3α-hydroxy-5α-androstan-17-one as substrate, the dimer had about 10% of the activity of the monomer (Skålhegg, 1974). It was presumably because the kinetic experiments were done at low protein concentrations that formation of this dimer did not lead to more complex kinetic findings.

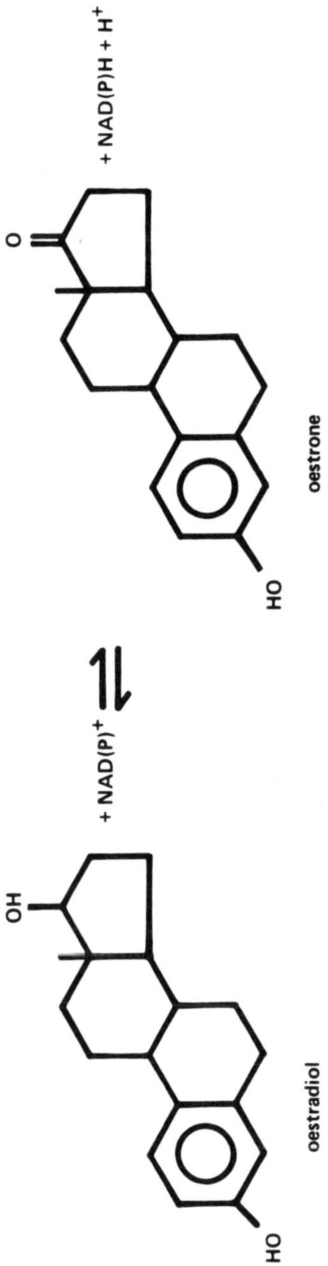

Fig. 10 Oestradiol 17β-dehydrogenase from human placenta shows somewhat different kinetics with NAD and with NADP (see text)

The monomer was active with an NAD^+ analogue containing a thioamide group instead of the amide, whereas the dimer was not active with this analogue (Skålhegg, 1975). These findings seem to require that the rates of dissociation and formation of the dimer are very slow.

11.3 Oestradiol 17β-dehydrogenase from human placenta (EC 1.1.1.62)

This enzyme consists of two identical subunits (molecular weight 2 x 34,000) (Burns et al., 1972). It utilizes NAD or NADP (Fig.10) and which coenzyme gives the higher reaction rate depends to some extent on the concentration and the substrate (e.g. Shaw & Jeffery, 1978).

Initial rate and product inhibition studies using oestradiol, oestrone, NAD^+, and NADH, or oestradiol-3-sulphate, oestrone-3-sulphate, NAD^+, and NADH were done at 25°C and at pH 7.0 (or pH 9.2; for technical reasons, NADPH was used as inhibitor instead of NADH in the experiment at pH 9.2) (Warren & Crist, 1967). These studies were interpreted as indicating an ordered reaction (Fig.3) in which the steroid bound first and left last (the reverse of the binding order of the previous two examples). Subsequent studies using isotope exchange at equilibrium were done at 25°C and pH 7.0 (or 9.0) (Betz, 1971), with oestradiol, oestrone, NAD^+, and NADH, or oestradiol, oestrone, $NADP^+$, and NADPH. These studies indicated an alternative pathway mechanism (Fig.4) in which, with NAD^+ and NADH, there was slight predomenance of the pathway with steroid binding first and leaving last. $NADP^+$ and NADPH bound more strongly to the enzyme than NAD^+ and NADH (Jarabak & Sack, 1969). The "K_m values" reported by various groups did not agree closely (Jarabak & Sack, 1969; Karavolas *et al.*, 1970; Warren & Crist, 1967), perhaps because of differences in buffer and pH. However, for a random mechanism (Fig.4) a linear plot of 1/v against $[S]^{-1}$ would not be expected (see Section 4) unless the reaction were of the rapid equilibrium type, in which hydride transfer in the ternary complex would be rate limiting. The deuterium isotope effect (Adams *et al.*, 1965) could be consistent with a rapid equilibrium random mechanism, but the rates of isotope exchange at equilibrium (Betz, 1971) appeared to exclude this.

The findings of Jarabak & Sack (1969) indicated that one NADPH bound per enzyme dimer. Groman *et al.*, (1975) found that only about 50% of oestrone covalently attached to the enzyme was converted into covalently-bound oestradiol upon addition of reduced coenzyme. Do these findings suggest the possibility of "half-of-the-sites" reactivity?

11.4 Alcohol dehydrogenase from horse liver (EC 1.1.1.1)

$$CH_3CH_2OH + NAD^+ \rightleftharpoons CH_3CHO + NADH + H^+$$

This is a dimeric enzyme with subunits of molecular

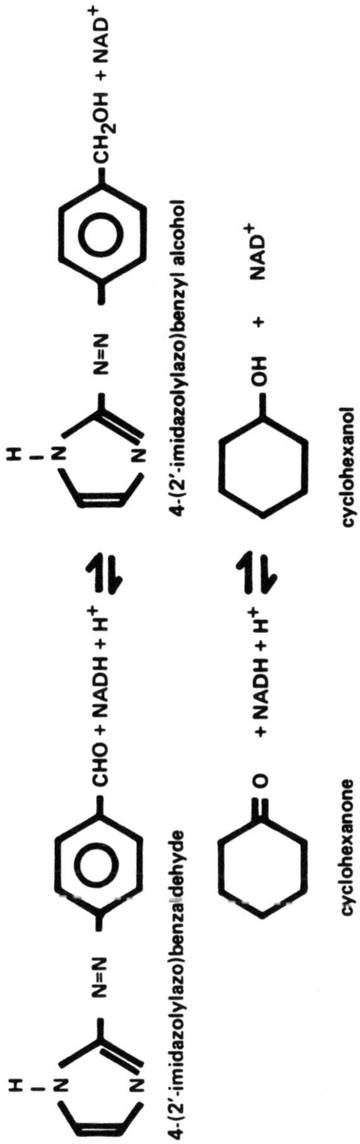

Fig. 11 Horse liver alcohol dehydrogenese shows slightly different kinetics with different substrates (see text)

weight about 40,000. It was for this enzyme that the Theorell-Chance mechanism (Fig.6) was first formulated (Theorell & Chance, 1951). There are several isozymes and modified forms (Lutstorf et al., 1970), and most of the numerous kinetic experiments that have been carried out over the years were probably done with mixtures in which the EE type predominated (Pietruszko et al., 1966).

Horse liver alcohol dehydrogenase has quite a broad substrate specificity (Fig.11). The primary alcohols, ethanol, propan-1-ol, butan-1-ol, and 2-methylpropan-1-ol (Dalziel, 1962b; Dalziel & Dickinson, 1966a) appeared from steady-state studies to follow a Theorell-Chance mechanism. Pre-steady-state studies (Brooks & Shore, 1971) indicated that hydride transfer was five times faster with propan-1-ol than with ethanol, and the maximum steady-state rate showed a significant isotope effect, indicating that hydride transfer was to some extent rate-determining. There was, however, also evidence for rate limitation by an isomerization step either of the ternary product complex or the enzyme-NADH binary complex (Shore & Gutfreund, 1970). Modification of the enzyme with methyl picolinimidate increased enzymic activity, evidently because the coenzyme dissociated faster from the modified enzyme, and for ethanol oxidation a deuterium isotope effect of 4.8 on V_{max} was found for the modified enzyme under steady-state conditions (Plapp et al., 1973). From steady-state (Dalziel & Dickinson, 1966a) and isotope exchange (Silverstein & Boyer, 1964b) studies, the formation of abortive complexes E.NADH. alcohol and E.alcohol has been suggested, especially for the secondary alcohol cyclohexanol, for which a partly random mechanism was indicated by isotope exchange experiments (Dalziel & Dickinson, 1966b; Ainslee et al., 1972). Quite a stable abortive complex was formed with the aromatic substrate 4-(2'-imidazolylazo)benzaldehyde (McFarland et al., 1974).

Various secondary alcohols gave steady-state kinetics that were not consistent with an ordered mechanism (Dalziel & Dickinson, 1966a; Dickinson & Dalziel, 1967) and stopped-flow and deuterium transfer experiments with propan-2-ol showed that hydride transfer in the ternary complex was slow and rate-limiting at high substrate concentrations (Brooks & Shore, 1971). Detailed studies by Dutler's group using cyclohexanol/cyclohexanone, and benzaldehyde/benzyl alcohol established the importance of considering in these cases a general model that included abortive complexes (Fig.12) (Hadorn, 1975; Meier, 1975).

Acetaldehyde undergoes hydration (Fig.13) and in aqueous solution the hydrate accounts for about 55% of the acetaldehyde. Other aldehydes are hydrated to varying extents. Ordinary aqueous solutions of ketones, on the other hand, do not contain substantial concentrations of hydrate unless the ketone contains electron-withdrawing substituents. (Triketohydrindene, for example, forms a crystalline hydrate and pyruvate is about 6% hydrated, but acetone is not hydrated, though it does undergo rapid transient hydration in acidic and basic solutions). The question therefore arises; if the non-hydrated form of the oxo compound is the true substrate or product of the enzymic reaction, can the hydrated form also

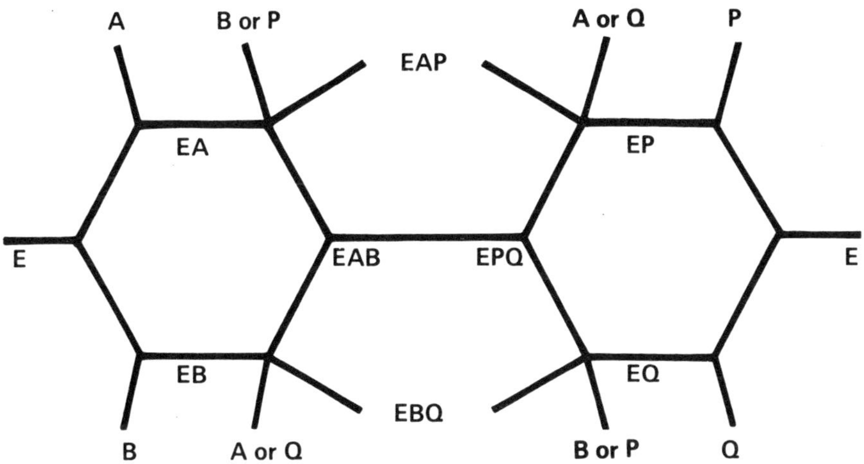

Fig. 12 An alternative pathway ternary complex mechanism with abortive ternary complexes

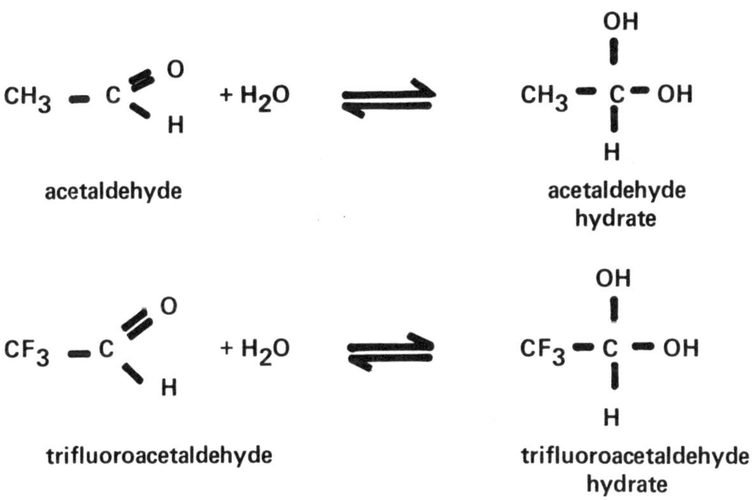

Fig. 13 Aldehydes are extensively hydrated in aqueous solution (see text)

bind, and if so, to which species. In the case of acetaldehyde, the non-hydrated form was found to be the substrate, the enzyme did not catalyse the dehydration of the hydrated form, and inhibition by the hydrated form was not observed even at the moderately high concentration of 0.4mM (Abdallah *et al.*, 1979). The hydrated form of trifluoroacetaldehyde, on the other hand, was an inhibitor, evidently combining with the E.NAD$^+$ complex (Abdallah *et al.*, 1979).

The question of whether or not the subunits of the dimer react independently has attracted much attention. Stopped-flow and temperature-jump relaxation studies on the reduction of acetaldehyde (Shore & Gutfreund, 1970) indicated that the two active sites of the dimeric enzyme were equivalent. However, studies with aromatic aldehydes led to suggestions that the subunits did not react independently, but in a reciprocating manner ("half-of-the-sites" reactivity) (Bernard *et al.*, 1970; McFarland & Bernard, 1972; Luisi & Favilla, 1972; Luisi & Bignetti, 1974; Dunn , 1974; McFarland & Chu, 1975). Hijazi & Laidler (1973) drew attention to the form of the transient-phase kinetic equations, and suggested that findings interpreted as showing "half-of-the-sites" reactivity were in fact consistent with a Theorell-Chance mechanism, though the work of Hardman & Blackwell (1974) cast doubts on this. Assumptions in the evidence for "half-of-the-sites" activity regarding the absorption coefficients of complexes involving NAD$^+$ and NADH were also criticised (Hadorn *et al.*, 1975). Detailed studies of kinetic transients using stopped-flow techniques (Hadorn *et al.*, 1975; Weidig et al., 1977; Kvassman & Pettersson, 1976, 1978) and a quite different technique using rapid sampling of the reaction product (Kordal & Parsons, 1979) showed no evidence for "half-of-the-sites" reactivity, or anything other than identical and independent catalytic sites. However, single-turnover experiments in which the effect of small *para* substituents in the benzaldehyde substrate, and the deuterium isotope effect were investigated, led Dunn *et al.*, (1979) to conclude that the compulsory-ordered mechanism proposed by Theorell & Chance (1951) and Kvassman & Pettersson (1976) is correct but insufficient to account for the biphasic kinetics of the pre-steady-state, and that the simplest development of the mechanism to explain the findings is the involvement of ligand-dependent protomer-protomer interactions which critically influence protomer catalytic activity.

11.5 Sorbitol dehydrogenase from sheep liver (EC 1.1.1.14)

Estimates of the molecular weight of this enzyme were 115,000 (Smith, 1962) and 147,000 (Walsall *et al.*, 1978), and of the subunit molecular weight 36,000 and 41,000 (Walsall *et al.*, 1978). It seems likely that it is a tetramer, but on the basis of these estimates alone a trimer would not be excluded.

Initial rate and product inhibition studies (Christensen *et al.*, 1975) at 23.5°C and pH 7.0 were interpre-

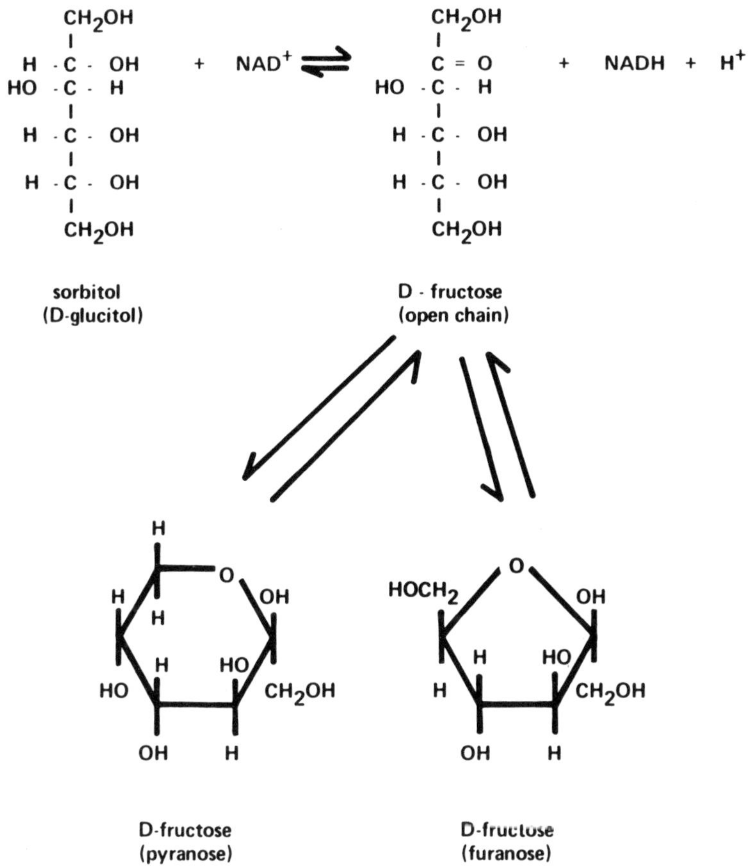

Fig. 14 Sorbitol dehydrogenase from sheep liver is said to react by a rapid-equilibrium random mechanism with one abortive complex (see text)

ted as showing a rapid equilibrium random mechanism with an abortive complex, E.sorbitol.NADH (cf. Fig.4 and Fig.8; formulae Fig.14).

Several dehydrogenases show some abortive complex formation, and incompletely ordered mechanisms, but a rapid equilibrium random mechanism is unusual.

11.6 Glyceraldehyde-3-phosphate dehydrogenase from rabbit muscle (EC 1.2.1.12)

$$\begin{matrix} CHO \\ | \\ HCOH \\ | \\ CH_2OPO_3^{2-} \end{matrix} + HPO_4^{2-} + NAD^+ \rightleftharpoons \begin{matrix} CO.OPO_3^{2-} \\ | \\ HCOH \\ | \\ CH_2OPO_3^{2-} \end{matrix} + H^+ + NADH$$

Glyceraldehyde-3-phosphate dehydrogenases from many sources are tetramers (molecular weight 4 x 36,000). In cases in which the complete sequence was determined, such as, for example, lobster muscle (Davidson et al., 1967), pig muscle (Harris & Perham, 1968), yeast (Jones & Harris, 1972), and Thermus aquaticus (Hocking & Harris, 1976), the subunits were identical.

The enzymes from yeast and rabbit muscle attracted much interest during the 1960's ("the age of allostery") because they showed marked and different co-operative effects for the binding of NAD^+ (Kirchner et al., 1966; Conway & Koshland, 1968; De Vijlder & Slater, 1968). Work on enzymes from many sources has been reviewed (Harris & Waters, 1976). The lobster muscle enzyme (Buehner et al., 1974), and the enzyme from Bacillus stearothermophilus (Biesecker et al., 1977) gave crystals favourable for X-ray crystallography. The rabbit liver enzyme was investigated to seek from the kinetics an explanation of how, in the conditions of the aerobic hepatocyte, the enzyme functions in the direction of gluconeogenesis (Smith & Velick, 1972). The thermal stability of the muscle enzyme from various lower vertebrates was greater for those enzymes that were less reactive towards Ellman's reagent, 5,5'-dithiobis-(2-nitrobenzoic acid) (Leibman & Nesterov, 1979). The yeast enzyme used most tetrahedral oxy anions tested in place of phosphate as acyl acceptor (but not SO_4^{2-} or SeO_4^{2-}, ReO_4^- or $GeO(OH)_3^-$) (Byers et al., 1979), and this may prove useful for further kinetic studies because of the different reactivities of the intermediates. Studies using mutant strains of Escherichia coli (Hillman, 1979) offer other interesting possibilities for future work.

In the case of sturgeon muscle enzyme, the kinetics of the transient phase of the reductive dephosphorylation of 1,3-diphosphoglycerate, led to the proposal that a first steady-state, involving tetrameric enzyme with all four sites acylated, was followed by establishment of a second steady-state in which only two sites were acylated, and that this "half-of-the-sites" species predominated in the final steady-state (Kellershohn & Seydoux, 1979).

Early kinetic studies with the rabbit muscle enzyme

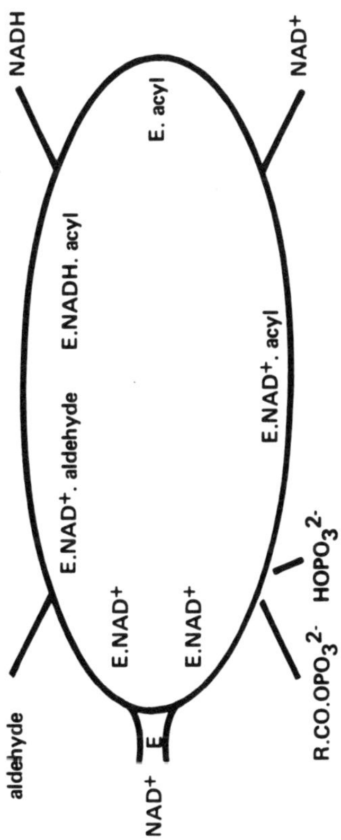

Fig. 15 Simplified scheme of a kinetic mechanism for rabbit muscle glyceraldehyde-3-phosphate dehydrogenase (see text). Subunit interactions are not shown

(Furfine & Velick, 1965) were thought to indicate a rapid equilibrium random mechanism involving a quaternary complex. Subsequent evidence was regarded as indicating an ordered mechanism (Fahien, 1966; Orsi & Cleland, 1972), and, when glyceraldehyde-3-phosphate was substrate, the order of binding was thought to be NAD^+, glyceraldehyde-3-phosphate, and phosphate (or arsenate), with release of NADH as the final and rate-determining step (Orsi & Cleland, 1972). Later, initial rate and product inhibition patterns were done at 25°C and pH 7.6 using either glyceraldehyde, glyceraldehyde-3-phosphate, or 3-hydroxypropionaldehyde phosphate as substrate, and arsenate or phosphate as acyl acceptor; also, ADPR was used as a dead-end inhibitor (Meunier & Dalziel, 1978). The findings were analysed according to equations derived for three-substrate reactions (Dalziel, 1969). It was concluded that the mechanism could not be ordered, but could involve enzyme-substitution. Inhibition with high concentrations of phosphate or arsenate indicated competition for a binding site for glyceraldehyde-3-phosphate, but was not found with glyceraldehyde as substrate (Meunier & Dalziel, 1978). A scheme is shown (Fig.15), but the actual mechanism must be rather more complicated.

The acyl enzyme is believed to be a thiol ester formed by nucleophilic attack by cys-149 on the carbonyl carbon of the aldehyde (Fig.16) (Harris et al., 1963). Reaction of cys-149 is probably facilitated by hydrogen bonding to his-176, which becomes equivalent to transfer of the thiol hydrogen to the histidine ("proton abstraction") as the sulphur-carbon bond develops to form the thiohemiacetal.

Regarding subunit interactions, Meunier & Dalziel (1978) noted deviations from "Michaelis-Menten behaviour" consistent with negative co-operative effects in binding NAD^+, and instability of the species $E(NAD^+)_3$ and $E(NAD^+)_4$.

Until 1979, it seemed quite clear that the binding of NAD^+ or NADH to the rabbit muscle enzyme showed negative co-operativity (Velick et al., 1953; Conway & Koshland, 1968) and that the greatest difference in the free energy of binding was between the second and third coenzyme molecules (Bell & Dalziel, 1975a), which corresponded to a change in the conformation of the protein when two coenzyme molecules were bound (Boers & Verhoeven, 1973; Bell & Dalziel, 1975b; Henis et al., 1979). However, preparation of the enzyme by a new method (Scheek & Slater, 1978) gave enzyme that behaved somewhat differently, notably in showing only a slight negative co-operative effect for coenzyme binding, and allowed the interpretation that NAD^+ and NADH compete not for the non-acylated enzyme, but for acylated enzyme (Scheek et al., 1979). Further work is needed before the significance of these findings can be assessed.

11.7 Glutamate dehydrogenase from bovine liver (EC 1.4.1.3)

This enzyme (molecular weight 6 x 56,000) has six identical subunits (Moon et al., 1972). It utilizes NAD or NADP. Each molecule (hexamer) was shown to bind six reduced coenzyme molecules at "active" sites, and six at "non-active"

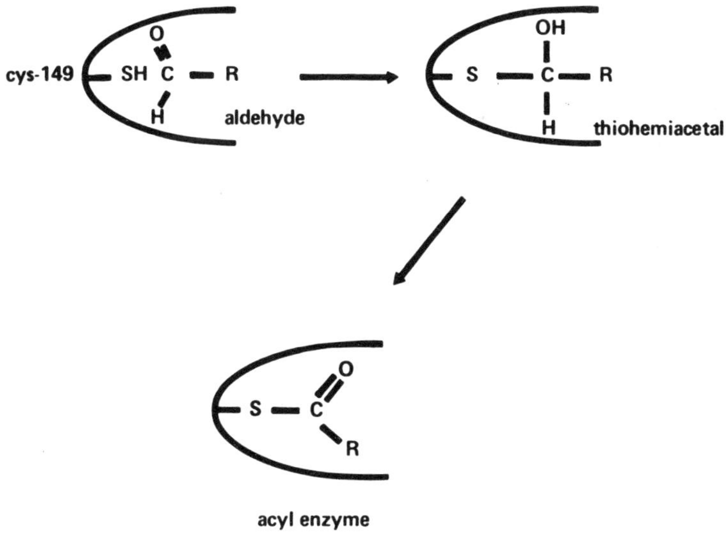

Fig. 16 Formation of substituted enzyme during the glyceraldehyde-3-phosphate reaction (after findings by Segal & Boyer (1953), Harris et al (1963), Meunier & Dalziel (1978), and others). Interactions involving his - 176 (not shown) probably contribute to the high reactivity of cys-149

sites, but the "non-active" sites had about 10 times higher affinity for NADH than for NADPH (Krause *et al.*, 1974). ADP is an activator, and GTP is an inhibitor. The enzyme also acts as an alanine dehydrogenase, and many activators of the glutamate dehydrogenase activity inhibit the alanine dehydrogenase activity, and *vice versa*. Models were proposed by Tomkins *et al.*, (1965) and Cross & Fisher (1970). The large body of work on this and other glutamate dehydrogenases was reviewed (Goldin & Frieden, 1972; Fisher, 1973; Sund *et al.*, 1975; Smith *et* al., 1975; Eisenberg *et al.*, 1976). More recently, self-association of the hexamers was studied by pressure relaxation kinetics (Halvorson, 1979), conformation changes in spin-labelled hexamer were studied by ESR (Zantema *et al.*, 1979) and information on the allosteric sites was obtained by using 5'-*p*-fluorosulphonylbenzoylguanosine as an affinity labelling reagent for the GTP-binding site (Pal & Colman, 1979), and by affinity chromatography using suitable bound ligands (Brodelius & Kaplan, 1979).

The need to combine kinetic and structural studies was particularly clear in the case of this enzyme because of the obvious complexity of the early kinetic findings.

12. CONCLUSION

Techniques are available that give detailed information on the kinetic and related structural properties of dehydrogenases. Molecular events not evident from steady-state studies can be delineated by studying the transient phase. Studies on natural structures (native enzymes) can be supplemented by studies on forms in which lysines or histidines have been modified, zinc has been replaced by cobalt, etc. Special substrates quite different from the natural substrates can be devised to test particular aspects of a mechanism.

Related structural studies can give increasingly accurate location of atoms. Light atoms not sufficiently clear from X-ray crystallography can, in principle, be located by neutron diffraction. Conformation changes and the environment of bound ligands are often revealed by resonance techniques.

The question, then, is not (as perhaps it was some years ago) what is it possible to do, but rather what do we want to find out.

Diverse aims for kinetic studies on dehydrogenases can be distinguished. One is to define the chemical mechanism of reaction. Cleland (1977) reviewed methods for achieving this, and gave the general mechanism as follows:

$$\mathrm{NADH} + \overset{\displaystyle R_1}{\underset{\displaystyle R_2}{\mathrm{C}}}{=}\mathrm{O}\cdots\mathrm{H}^{+}{-}\mathrm{X} \rightleftharpoons \mathrm{NAD}^{+} + \mathrm{H}{-}\overset{\displaystyle R_1}{\underset{\displaystyle R_2}{\mathrm{C}}}{-}\mathrm{O}{-}\mathrm{H}\cdots\mathrm{X}$$

where X is an assisting catalytic group, and is histidine in

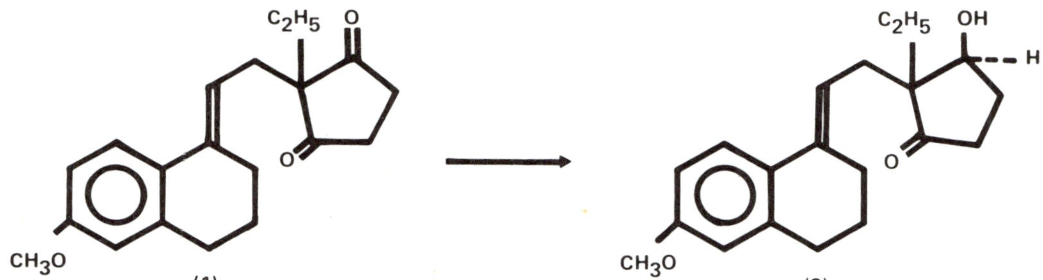

(1) (2)

Fig. 17 Compound (1) has no enantiomorphs but the oxo groups are prochiral centres. A microbial enzyme converts compound (1) stereospecifically into compound (2), the enantiomorph needed for drug synthesis

lactate and glyceraldehyde-3-phosphate dehydrogenases, but in horse liver alcohol dehydrogenase is the active site zinc, to which one histidine, two cysteine residues and one molecule of water are ligands. Cleland (1977) also stressed the value of studying pH effects on steady-state kinetics. Knowles (1976) emphasized the need for caution in interpreting pH effects.

Stopped-flow studies require much larger amounts of enzyme than steady-state studies, so the practical choice may be between one stopped-flow experiment and numerous steady-state measurements. Which will be more helpful will depend upon the point at issue.

Another aim of kinetic studies is to elucidate how metabolic pathways function. What is needed for this purpose is information on the behaviour of the enzyme with natural substrates under conditions close to those in vivo.

A third aim is to enable the structure and function of enzymes to be compared from an evolutionary point of view. With what degree of detail the mechanism should be established for this purpose will depend upon the circumstances. Differences in structure associated with differences in regulatory properties are an obvious area of interest.

A fourth aim is to permit the design and assessment of substrate analogues and inhibitors for possible use as drugs (or herbicides, or insecticides, as the case might be). Modifications that make prostaglandins poor substrates for the inactivating enzyme 15-hydroxyprostaglandin dehydrogenase have been of interest, for example.

There are also various practical objectives for which more limited kinetic studies are required. In this category would be the design of an assay for an enzyme, for example, the H isozyme of lactate dehydrogenase in blood for clinical purposes, or the design of an assay for a substrate, such as ethanol in blood for forensic purposes.

The use of separated enzymes for industrial synthesis is increasing. Dehydrogenases have much to offer. The conversion of compound 1 (Fig.17) into compound 2, for example, is of interest to pharmaceutical industry. Compound 1 is symmetrical, whereas compound 2 is a particular enantiomorph. But the oxo groups of compound 1 are prochiral centres and a microbial enzyme carries out the required conversion stereospecifically (Smith, 1968). For the design of industrial processes, some knowledge of the kinetics is needed.

Solid-supported enzymes are desirable for many industrial applications. This raises further kinetic issues (see, for example, Paul et al., 1978), which are not dealt with here.

13. REFERENCES

Abdallah, M.A., Biellmann, J.-F. & Lagrange, P. (1979) Biochemistry 18, 836-838.

Adams, J.A., Jacobson, H.I., Levy, H.R. & Talalay, P. (1965) Steroids, Suppl. 1, 75-84.

Ainslie, G.R. & Cleland, W.W. (1972) J. Biol. Chem. 247, 946-951.

Ainsworth, S. (1975) J. theor. Biol. 50, 129-151.

Ainsworth, S. (1977a) J. theor. Biol. 64, 381-389.

Ainsworth, S. (1977b) In "Steady-state enzyme kinetics", pp.182-201, Macmillan, London.

Alberty, R.A. (1953) J. Am. Chem. Soc. 75, 1928-1932.

Alberty, R.A. (1958) J. Am. Chem. Soc. 80, 1777-1782.

Askelöf, P., Korsfeldt, M. & Mannervik, B. (1976) Eur. J. Biochem. 69, 61-67.

Atkins, G.L. & Nimmo, I.A. (1975) Biochem. J. 149, 775-777.

Bardsley, W.G. & Childs, R.E. (1975) Biochem. J. 149, 313-328.

Barnes, L.D., McGuire, A.J. & Atkinson, D.E. (1972) Biochemistry, 11, 4322-4328.

Bell, J.E. & Dalziel, K. (1975a) Biochim. Biophys. Acta, 391, 249-258.

Bell, J.E. & Dalziel, K. (1975b) Biochim. Biophys. Acta, 410. 243-251.

Bernhard, S.A., Dunn, M.F., Luisi, P.L. & Schack, P. (1970) Biochemistry, 9, 185-192.

Betz, G. (1971) J. Biol. Chem. 246, 2063-2068.

Biellmann, J.-F., Lapinte, C., Haid, E. & Weimann, G. (1979) Biochemistry, 18, 1212-1217.

Biesecker, G., Harris, J.I., Thierry, J.C., Walker, J.E. & Wonacott, A.J. (1977) Nature (London) 266, 328-333.

Birken, S. & Pisano, M.A. (1976) J. Bacteriol. 125, 225-232.

Blaner, W.S. & Churchich, J. (1979) J. Biol. Chem. 254, 1794-1798.

Boers, W. & Verhoeven, J.W. (1973) Biochim. Biophys. Acta, 328, 1-9.

Bosron, W.F., Li, T.-K., Dafeldecker, W.P. & Vallee, B.L. (1979) Biochemistry, 18, 1101-1105.

Boyer, P.D. (1959) Arch. Biochem. Biophys. 82, 387-410.

Boyer, P.D. & Silverstein, E. (1963) Acta Chem. Scand., 17, S195-S202.

Britton, H.G. (1966) Arch. Biochem. Biophys. 117, 167-183.

Britton, H.G. (1973) Biochem. J. 133, 255-261.

Britton, H.G. (1977) Biochem. J. 161, 517-526.

Britton, H.G. & Clarke, J.B. (1968) Biochem. J. 110, 161-179.

Britton, H.G. & Dann, L.G. (1978) Biochem. J. 169, 29-37.

Brodelius, P.E. & Kaplan, N.O. (1979) Arch. Biochem. Biophys., 194, 449-456.

Brooks, R.L. & Shore, J.D. (1971) Biochemistry, 10, 3855-3858.

Buc, H. (1967) Biochem. Biophys. Res. Comm., 28. 59-64.

Buehner, M., Ford, G.C., Moras, D., Olsen, K.W. & Rossmann, M.G. (1974) J. Mol. Biol., 90, 25-49.

Burns, D.J.W., Engel, L.L. & Bethune, J.L. (1972) Biochemistry, 11, 2699-2703.

Byers, L.D., She, H.S. & Alayoff, A. (1979) Biochemistry, 18, 2471-2480.

Callewaert, D.M., Rosemblatt, M.S. & Tchen, T.T. (1974) J. Biol.

Chem., 249, 1737-1741.
Cannistraro, V.J., Borack, L.I. & Chase, T. (1979) Biochim. Biophys. Acta, 569, 1-5.
Cantwell, A.M. & Dennis, D. (1970) Biochem. Biophys. Res. Commun., 41, 1166-1170.
Chang, H.-L., Holten, D. & Karin, R. (1979) Can. J. Biochem., 57, 396-401.
Christensen, U., Tüchsen, E. & Andersen, B. (1975) Acta Chem. Scand., B29, 81-87.
Cleland, W.W. (1963a) Biochim. Biophys. Acta, 67, 104-137.
Cleland, W.W. (1963b) Biochim. Biophys. Acta, 67, 173-187.
Cleland, W.W. (1963c) Biochim. Biophys. Acta, 67, 188-196.
Cleland, W.W. (1963d) Nature (London), 198, 463-465.
Cleland, W.W. (1967a) Ann. Rev. Biochem., 36, 77-112.
Cleland, W.W. (1967b) Adv. Enzymol., 29, 1-32.
Cleland, W.W. (1970) In "The enzymes", (Boyer, P.D. ed.) 3rd edn. Vol. 1, pp.1-65 Academic Press, New York.
Cleland, W.W. (1975) Biochemistry, 14, 3220-3224.
Cleland, W.W. (1977) Adv. Enzymol., 45, 273-387.
Colquhounm D. (1971) Lectures on biostatistics, pp.257-272, Clarendon Press, Oxford.
Conway, A. & Kosland, D.E. Jr. (1968) Biochemistry 7, 4011-4022.
Cornish-Bowden, A. (1976) Biochem. J. 159, 167.
Cornish-Bowden, A. & Koshalnd, D.E. Jr. (1975) J. Mol. Biol. 95, 201-212.
Cross. D.G. & Fisher, H.F. (1970) J. Biol.Chem. 245, 2612-2621.
Dalziel, K. (1957) Acta Chem. Scand. 11, 1706-1723.
Dalziel, K. (1958) Trans. Faraday Soc. 54, 1247-1253.
Dalziel, K. (1961a) Biochem. J. 80, 440-445.
Dalziel, K. (1961b) Nature (London), 191, 1098-1099.
Dalziel, K. (1962a) Biochem. J. 84, 240-244.
Dalziel, K. (1962b) Biochem. J. 84, 244-254.
Dalziel, K. (1962c) Nature (London) 195, 384-385.
Dalziel, K. (1963a) J. Biol. Chem. 238, 1538-1543.
Dalziel, K. (1963b) J. Biol. Chem. 238, 2850-2858.
Dalziel, K. (1963c) Nature (London), 197, 462-464.
Dalziel, K. (1969) Biochem. J. 114, 547-556.
Dalziel, K. (1975) In "The enzymes", 3rd edn. (Boyer, P.D. ed.), Vol. 11, pp.1-60, Academic Press, New York.
Dalziel, K. & Dickinson, F.M. (1966a) Biochem. J. 100, 34-46.
Dalziel, K. & Dickinson, F.M. (1966b) Biochem. J. 100, 491-500.
Dann, L.G. & Britton, H.G. (1978) Biochem. J. 169, 39-54.
Davidson, W.S. & Flynn, G.T. (1979) Biochem. J. 171, 595-601.
Davidson, B.E., Sajgo, M., Noller, H.F. & Harris, J.I. (1967) Nature (London) 216, 1181-1185.
DeVjilder, J.J.M. & Stater, E.C. (1968) Biochim. Biophys. Acta 167, 23-34.
Dickinson, F.M. & Dalziel, K. (1967) Biochem. J. 104, 165-172.
Doublet, M.-O. & Olomucki, A. (1975) Eur. J. Biochem. 59, 175-183.
Doublet, M.-O., Olomucki, A., Baici, A. & Luisi, P.L. (1975) Eur. J. Biochem. 59, 185-191.
Dowd, J.E. & Riggs, D.S. (1965) J. Biol. Chem., 240, 863-869.
Dubrow, R. & Pizer, L.I. (1977) J. Biol. Chem., 252, 1539-1551.
Dunn, M.F. (1974) Biochemistry, 13, 1146-1151.
Dunn, M.F., Bernhard, S.A., Anderson, D., Copeland, A., Morris,

R.G. & Roque, J.-P. (1979) Biochemistry 18, 2346-2354.
Dworschack, R.T. & Plapp, B.V. (1977) Biochemistry, 16, 2716-2725.
Eigen, M. & de Maeyer, L. (1963) In "Techniques of Organic Chemistry" (Friess, S.L., Lewis, E.S. & Weissberger, A. eds.), Vol. 8 pp.895-1054 New York, Interscience.
Eigen, M. & Hammes, G.G. (1963) Adv. Enzymol. 25, 1-38.
Eisenberg, H., Josephs, R. & Reisler, E. (1976) Adv. Protein. Chem., 30, 101-181.
Eisenthal, R. & Cornsih-Bowden, A. (1974) Biochem. J. 139, 715-720.
Endrenyi, L., Chan, M.-S. & Wong, J.T.-F. (1971) Can. J. Biochem. 49, 581-598.
Fahien, L.A. (1966) J. Biol. Chem., 241, 4115-4123.
Fawcett, C.P., Ciotti, M.M. & Kaplan, N.O. (1961) Biochim. Acta 54, 210-212.
Fine, I.H., Kaplan, N.O. & Kuftinec, P. (1963) Biochemistry 2, 116-121.
Fisher, H.F. (1973) Adv. Enzymol., 39, 369-417.
Flores, R. (1979) TIBS, 4, N32-N33.
Flynn, T.G., Shires, J. & Walton, D.J. (1975) J. Biol. Chem., 250, 2933-2940.
Ford, W.T., Graham, E.W. & Cram, D.J. (1967) J. Am. Chem. Soc., 89, 4661-4669.
Fromm, H.J. (1963) J. Biol. Chem., 238, 2938-2944.
Fromm, H.J. (1975) Initial rate enzyme kinetics Springer-Verlag, Berlin.
Furfine, C.A. & Velick, S.F. (1965) J. Biol. Chem., 240, 844-855.
Gezici, S. & Dinagk, G.F. (1979) Int. Biochem., 10, 427-432.
Gibson, Q.H. & Milnes, L. (1964) Biochem. J., 91, 161-171.
Glick, N., Landman, A.D. & Roufogalis, B.D. (1979) TIBS, 4, N82-N83.
Goldin, B.R. & Frieden, C. (1972) Curr. Top. Cell. Regul., 4, 77-117.
Groman, E.V., Schultz, R.M. & Engel, L.L. (1975) J. Biol. Chem., 250, 5450-5454.
Gutfreund, H. (1971) Ann. Rev. Biochem., 40, 315-341.
Hadorn, M. (1975) Doctoral Diss. ETH 5558, Zürich.
Hadorn, M., John, V.A., Meier F.K. & Dutler, H. (1975) Eur. J. Biochem., 54, 65-73.
Hakala, M.T., Glaid, A.J. & Schwert, G.W. (1956) J. Biol. Chem. 221, 191-209.
Halvorson, H.R. (1979) Biochemistry, 18, 2480-2487.
Hanes, C.S., Bronskill, P.M., Gurr, P.A. & Wong, J.T.-F. (1972) Can. J. Biochem., 50, 1385-1413.
Hardman, M.J. & Blackwell, L.F. (1974) Biochim. Biophys. Acta, 350, 247-252.
Harris, J.I. & Perham, R.N. (1968) Nature (London) 219, 1025-1028.
Harris, J.I. & Waters, M. (1976) In "The enzymes" (Boyer, P.D. ed.), 3rd edn., Vol. 13, pp.1-49, Academic Press, New York.
Harris, J.I., Meriwether, B.P. & Park, J.H. (1963) Nature

(London), 198, 154-157.
Hathaway, J.A. & Atkinson, D.E. (1963) J. Biol. Chem., 238, 2875-2881.
Heffron, J.J.A. (1979) TIBS, 4, N133.
Henis, Y.I., Levitzki, A. & Gafni, A. (1979) Eur. J. Biochem., 97, 519-528.
Hijazi, N.H. & Laidler, K.J. (1973) Can. J. Biochem., 51, 832-840.
Hillman, J.D. (1979) Biochem. J., 179, 99-107.
Hochroitcr, M.C., Patek, D.R. & Schellenberg, K.A. (1972) J. Biol. Chem., 247, 6271-6276.
Hocking, J.D. & Harris, J.I. (1976) In "Enzymes and proteins from thermophilic microorganisms" (Zuber, H., ed.), Birkhauser Verlag, Basel.
Horn, A. & Börnig, H. (1969) FEBS Letters, 3, 325-329.
Hou, C.T., Patel, R.N., Laskin, A.I., Barnabe, N. & Marczak, I. (1979) FEBS Letters, 101, 179-183.
Huijing, F. (1979) TIBS, 4, N132.
Hurst, R.O. (1969) Can. J. Biochem., 47, 941-944.
Ingraham, L.I. & Makower, B. (1954) J. Phys. Chem., 58, 266-270.
Inoue, K., Nishimukai, H. & Yamasawa, K. (1979) Biochim. Biophys. Acta, 569, 117-123.
Jaenicke, R., Rudolph, R.& Heider, I. (1979) Biochemistry, 18, 1217-1223.
Jarabak, J. & Sack, G.H. (1969) Biochemistry, 8, 2203-2212.
Johansen, G. & Lumry, R. (1961) Compt. Rend. Trav. Lab. Carlsberg, 32, 185-214.
Jones, G.M.T. & Harris, J.I. (1972) FEBS Lett., 22, 185-189.
Karavolas, H.J., Baedecker, M.L. & Engel, L.L. (1970) J. Biol. Chem., 245, 4948-4952.
Kegeles, G. (1979) FEBS Letters, 103, 5-6.
Kellershohn, N. & Seydoux, F.J. (1979) Biochemistry, 18, 2465-2470.
Kimball, D.F. (1977) Ph.D. dissertation, University of Oregon.
Kimball, D.F., Peterson, L., McLoughlin, D.J. & Wolfe, R.G. (1979) Arch. Biochem. Biophys., 195, 66-73.
King, E.L. & Altman, C. (1956) J. Phys. Chem., 60, 1375-1378.
Kirschner, K., Eigen, M., Bittman, R. & Voight, B. (1966) Proc. Nat. Acad. Sci., U.S.A., 56, 1661-1667.
Kirtley, M.E. & Koshland, D.E. (1967) J. Biol. Chem., 242, 4192-4205.
Klinman, J.P. (1972) J. Biol. Chem., 247, 7977-7987.
Klinman, J.P. (1978) Adv. Enzymol., 46, 415-494.
Knowles, J.R. (1976) Critical Reviews of Biochemistry, 4, 165-173.
Kordal, R.J. & Parsons, S.M. (1979) Arch. Biochem. Biophys., 194, 439-448.
Koshland, D.E. Jr. (1953) Biol. Revs., 28, 416-436.
Koshland, D.E. Jr. (1955) Discuss. Faraday Soc., 20, 142-148.
Koshland, D.E. Jr. (1956) J. Cell. comp. Physiol., 47, Suppl.1 217-234.
Koshland, D.E. Jr. (1970) In "The enzymes", (Boyer, P.D., ed.) 3rd edn., pp.342-396 Academic Press, New York.
Koshland, D.E. Jr., Nemethy, G. & Filmer, D. (1966) Biochemistry 5, 365-385.

Krause, J., Bühner, M. & Sund, H. (1974) Eur. J. Biochem., 41, 593-602.
Kuehn, G.D., Barnes, L.D. & Atkinson, D.E. (1971) Biochemistry, 10, 3945-3951.
Kvassman, J. & Pettersson, G. (1976) Eur. J. Biochem., 69, 279-287.
Kvassman, J. & Pettersson, G. (1978) Eur. J. Biochem., 87, 417-427.
Laidler, K.J. (1955a) Trans. Faraday Soc., 51, 528-539.
Laidler, K.J. (1955b) Discuss Faraday Soc., 20, 83-96.
Laidler, K.J. (1956) Trans. Faraday Soc., 52, 1374-1382.
Lazdunski, M. (1974) Progress in Bioorganic Chemistry, 3, 81-140.
Leibman, D.Y. & Nesterov, V.P. (1979) Int. J. Biochem., 10, 645-649.
Levitzki, A. & Koshland, D.E. (1976) Current Topics in Cellular Regulation, 10, 1-40.
Lewis, E.S. & Robinson, J.K. (1968) J. Am. Chem. Soc., 90, 4337-4344.
Luisi, P.L. & Bignetti, E. (1974) J. Mol. Biol., 88, 653-670.
Luisi, P.L. & Favilla, R. (1972) Biochemistry, 11, 2303-2310.
Lutstorf, U.M., Schürch, P.M. & von Wartburg, J.P. (1970) Eur. J. Biochem., 17, 497-508.
McFarland, J. & Bernhard, S.A. (1972) Biochemistry, 11, 1486-1493.
McFarland, J.T. & Chu, Y.-H. (1975) Biochemistry, 14, 1140-1146.
McFarland, J.T., Chu, Y.-H. & Jacobs, J.W. (1974) Biochemistry, 13, 65-69.
MacGibbon, A.K.H., Motion, R.L., Crow, K.E., Buckley, P.D. & Blackwell, L.F. (1979) Eur. J. Biochem., 96, 585-595.
Meier, F.K. (1975) Doctoral Diss. ETH 5559, Zürich.
Meunier, J.-C. & Dalziel, K. (1978) Eur. J. Biochem., 82, 483-492.
Monneuse-Doublet, M.-O., Olomucki, H. & Buc, J. (1978) Eur. J. Biochem., 84, 441-448.
Monod, J., Wyman, J. & Changeux, J. (1965) J. Mol. Biol., 12, 88-118.
Moon, K., Piszkiewicz, D. & Smith, E.L. (1972) Proc. Natl. Acad. Sci., U.S.A., 69, 1380-1383.
Northrop, D.B. (1975) Biochemistry, 14, 2644-2651.
Orsi, B.A. & Cleland, W.W. (1972) Biochemistry, 11, 102-109.
Pal, K.P. & Colman, R.F. (1979) Biochemistry, 18, 838-845.
Paul, F., Coulet, P.R., Gautheron, D.C. & Engasser, J.-M. (1978) Biotechnology & Bioengineering, 22, 1785-1796.
Paulus, H. & DeRiel, J.K. (1975) J. Mol. Biol., 97, 667-671.
Pettersson, G. (1969) Acta Chem. Scand., 23, 2717-2726.
Pettersson, G. (1970) Acta Chem. Scand., 24, 1271-1274.
Pettersson, G. (1972) In "Structure and Function of Oxidation-Reduction Enzymes" (Åkeson, Å. & Ehrenberg, A., eds.) pp.739-743, Pergamon Press, Oxford.
Pettersson, G. (1974) Eur. J. Biochem., 46, 1-4.
Pettersson, G. (1976) Eur. J. Biochem., 69, 273-278.
Pettersson, G. (1977) Biochem. Biophys. Acta, 484, 199-207.
Pettersson, G. (1978) Acta Chem. Scand., B32, 437-446.
Pietruszko, R., Clark, A., Graves, J.M.H. & Ringold, H.J. (1966) Biochem. Biophys. Res. Comm., 23, 526-534.

Plapp, B.V., Brooks, R.L. & Shore, J.D. (1973) J. Biol. Chem., 248, 3470-3475.
Plowman, K.M. (1972) In "Enzyme kinetics" pp.76-90, McGraw Hill Book Co., New York.
Porter, W.R. & Trager, W.F. (1977) Biochem. J., 161, 293-302.
Raval, D.N. & Wolfe, R.G. (1962a) Biochemistry, 1, 263-269.
Raval, D.N. & Wolfe, R.G. (1962b) Biochemistry, 1, 1112-1117.
Rose, I.A. (1961) J. Biol. Chem., 236, 603-609.
Rosemblatt, M.S., Callewaert, D.M. & Tchen, T.T. (1973) J. Biol. Chem., 248, 6014-6018.
Scheek, R.M. & Slater, E.C. (1978) Biochim. Biophys. Acta, 526, 13-24.
Scheek, R.M., Berden, J.A., Hooghiemstra, R., & Slater, E.C. (1979) Biochim. Biophys. Acta, 569, 124-134.
Schimerlik, M.I., Rife, J.E. & Cleland, W.W. (1975) Biochemistry, 14, 5347-5354.
Segal, H.L. & Boyer, P.D. (1953) J. Biol. Chem., 204, 262-272.
Seydoux, F. & Bernhard, S.A. (1974) Biophys. Chem. 1, 161-174.
Seydoux, F., Malhorta, O.P. & Bernhard, S.A. (1974) Critical Reviews in Biochemistry, 2, 227-257.
Shaw, M.A. & Jeffery, J. (1978) Biochem. Soc. Trans., 6, 205-207.
Shikita, M. & Tsuneoka, K. (1976) FEBS Letters, 66, 4-7.
Shore, J.D. & Gutfreund, H. (1970) Biochemistry, 9, 4655-4659.
Siano, D.B., Zyskind, J.W. & Fromm, H.J. (1975) Arch. Biochem. Biophys., 170, 587-600.
Silverstein, E. & Boyer, P.D. (1964a) J. Biol. Chem., 239, 3901-3907.
Silverstein, E. & Boyer, P.D. (1964b) J. Biol. Chem., 239, 3908-3914.
Silverstein, E. & Boyer, P.D. (1966) In "Biochemical Preparations" (Maehly, A.C., ed.), Vol. 11, pp.89-95, John Wiley, New York.
Silverstein, E. & Sulebele, G. (1969) Biochemistry, 8, 2543-2550.
Simon, H. & Palm, D. (1966) Angew. Chem. Int. Ed. Engl., 5, 920-933.
Skålhegg, B.A. (1974) Eur. J. Biochem., 46, 117-125.
Skålhegg, B.A. (1975) Eur. J. Biochem., 50, 603-609.
Smith, H. (1968) J. Reprod. Fert. Suppl., 5, 3-12.
Smith, M.G. (1962) Biochem. J., 83, 135-144.
Smith, C.M. & Velick, S.F. (1972) J. Biol. Chem., 247, 273-284.
Smith, E.L., Austen, B.M., Blumenthal, K.M. & Nyc. J.F. (1975) In "The Enzymes" (Boyer, P.D. ed.) 3rd edn., Vol. 11, pp.294-367, Academic Press, New York.
Squire, P.G., Delin, S. & Porath, J. (1964) Biochim. Biophys. Acta, 89, 409-421.
Storer, A.C., Darlison, M.G. & Cornish-Bowden, A. (1975) Biochem. J., 151, 361-367.
Sund, H., Markau, K. & Koberstein, R. (1975) In "Biological macromolecules, subunits in biological systems, part C" (Timasheff, S.N. & Fasman, G.D., eds.), Vol. 7, pp.225-287 Marcel Dekker, New York.
Svasti, J. (1979) TIBS, 4, N133-N134.
Takenaka, Y. & Schwert, G.W. (1956) J. Biol. Chem., 223, 157-

170.
Theorell, H. & Chance, B. (1951) Acta Chem. Scand., 5, 1127-1144.
Toews, C.J. (1967) Biochem. J., 105, 1067-1073.
Tomkins, G.M., Yielding, K.L., Curran, J.F., Summers, M.R. & Bitensky, M.W. (1965) J. Biol. Chem., 240, 3793-3798.
Ureta, T. (1978) Current Topics in Cellular Regulation, 13, 233-258.
Velick, S.F., Hayes, J.E. & Harting, J. (1953) J. Biol. Chem., 203, 527-544.
Walsall, E.P., Lyons, S.A. & Metzger, R.P. (1978) Comp. Biochem. Physiol., 59B, 213-218.
Warren, J.C. & Crist, R.D. (1967) Arch. Biochem. Biophys., 118, 577-584.
Weidig, C.F., Halvorson, H.R. & Shore, J.D. (1977) Biochemistry, 16, 2916-2922.
Wermuth, B., Münch, J.D.B., & von Wartburg, J.-P. (1977) J. Biol. Chem., 252, 3821-3828.
Westheimer, F.H. (1961) Chem. Rev., 61, 265-273.
Whitehead, E.P. & Egmond, M.R. (1979) Biochem. J., 177, 631-639.
Wiberg, K.B. & Motell, E.L. (1963) Tetrahedron, 19, 2009-2023.
Wilkinson, G.N. (1961) Biochem. J., 80, 324-332.
Wong, J.T.-F. & Hanes, C.S. (1962) Canad. J. Biochem. & Physiol., 40, 763-804.
Wright, I.P. & Sundaram, T.K. (1979) Biochem. J., 177, 441-448.
Wyrambic, D. & Grisebach, H. (1979) Eur. J. Biochem., 97, 503-509.
Zantema, A., Vogel, H.J. & Robillard, G.T. (1979) Eur. J. Biochem., 96, 453-463.
Zewe, V. & Fromm, H.J. (1962) J. Biol. Chem., 237, 1668-1675.

STRUCTURE AND MECHANISM OF LIVER ALCOHOL DEHYDROGENASE, LACTATE DEHYDROGENASE AND GLYCERALDEHYDE-3-PHOSPHATE DEHYDROGENASE

Carl-Ivar Brändén and Hans Eklund

STRUCTURE AND MECHANISM OF LIVER ALCOHOL DEHYDROGENASE, LACTATE DEHYDROGENASE AND GLYCERALDEHYDE-3-PHOSPHATE DEHYDROGENASE

Carl-Ivar Brändén and Hans Eklund

HISTORICAL BACKGROUND

In 1962 Michael Rossmann, Herman Watson and Carl Brändén in the Cambridge laboratory of Max Perutz and John Kendrew occasionally discussed future crystallographic projects which might illuminate structure - function relationships within families of proteins. At that time the structure of myoglobin had been determined to high resolution and that of haemoglobin to low resolution. A small number of additional studies of monomeric protein molecules were at the low resolution stage but very little work had been started on larger oligomeric enzyme molecules.

Since the late Ieuan Harris, who pioneered the sequence studies of dehydrogenases, worked in the same laboratory it was quite natural that we became interested in the X-ray crystallographic studies of this family of enzymes. We thought that by studying these we might discover not only some structure - function relationships but also gain insight into coenzyme and substrate specificity and the role of subunit interactions in oligomeric enzymes. Thus Watson initiated a study of lobster glyceraldehyde-3-phosphate dehydrogenase, GAPDH, in Cambridge, Rossmann set up a crystallographic laboratory to study dogfish lactate dehydrogenase, LDH, at Purdue University and Brändén started to study liver alcohol dehydrogenase, LADH, at the Agricultural University in Uppsala. We chose this particular dehydrogenase mainly because we could establish a collaboration with Theorell and his group in Stockholm who had studied the kinetics of LADH since 1948. Some years later Len Banaszak in St. Louis initiated a study of soluble malate dehydrogenase, MDH, from pig heart.

The first attempt to deduce any structural relationships between these enzymes was made at a meeting on dehydrogenases in Konstanz in 1969 where Rossmann and Brändén compared low resolution models of LDH and LADH (Brändén & Rossmann, 1970). It was exciting to find that already at that stage we could observe that one half of the subunits showed structural similarities. It was also realized that the subunits were joined together in the molecules differently with respect to the similar parts. In retrospect this has turned out to be an astonishingly accurate deduction from low resolution models.

The first structure that was solved at high resolution was LDH (Adams et al., 1970). When the structure of MDH was obtained a few years later (Hill et al., 1972) the surprising and completely unexpected result was that almost the complete subunit structures were very similar. The next structure that was solved was LADH (Brändén et al., 1973) which showed a clear separation of the subunit structures into two domains. One domain bound the coenzyme and had a structure very similar to a corresponding part of the LDH subunit structure, the so called Rossmann fold. An idealized diagram of this domain is shown in Figure 1. The other domain was completely different from the rest of the LDH structure. Shortly afterwards Rossmann and his coworkers obtained the structure of lobster GAPDH (Buehner et al., 1973) and more recently Wonacott and coworkers that of GAPDH from Bacillus stearothermophilus (Biesecker et al., 1977). Again the same theme was obtained. Thus one could see a general structural pattern emerge among the NAD-dependent dehydrogenases (Rossmann et al., 1974; Ohlsson et al., 1974; Rossmann et al., 1975): similar function, coenzyme binding, is achived by similar domain structures whereas different functions, substrate binding and catalysis require quite different domain structures, except when the substrates are as similar as malate and lactate.

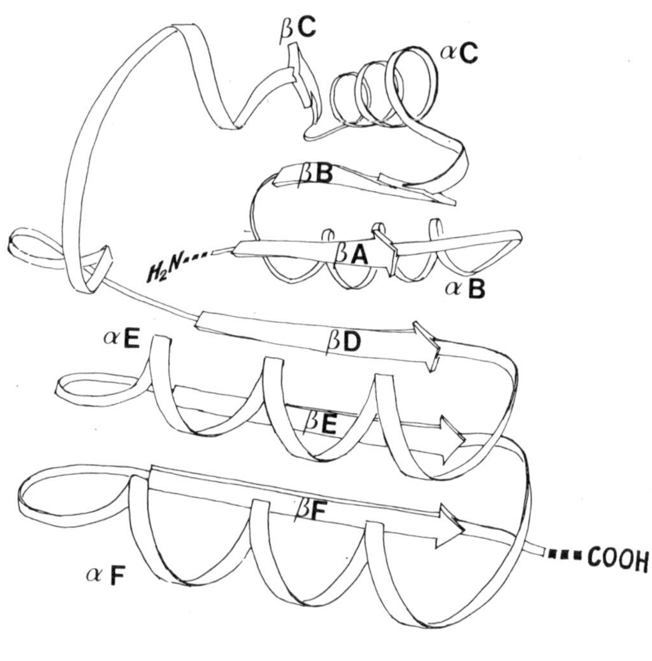

Fig. 1. Schematic diagram of an idealized NAD-binding domain. (Ohlsson et al., 1974).

One surprising aspect was that no statistically significant sequence similarities were present in these structurally very similar domains.

Subsequently it was found that one domain of phosphoglycerate kinase (Blake & Evans, 1974; Bryant et al., 1974) exhibited the Rossmann fold and that various variations of this fold were present in adenylate kinase (Schulz et al., 1974), the second domain of phosphoglycerate kinase, phosphoglycerate mutase (Campbell et al., 1974) and triose-phosphate isomerase (Banner et al., 1975). These results initiated a still continuing discussion of the evolutionary relationships between these domains (Rossmann et al., 1974, 1975). The difficulties are, on the one hand, how to assess evolutionary relationships which are so distant in time that sequence similarities are absent and, on the other hand, that we have very few criteria for assessing the role of structural stability of particular folds.

In the following we will concentrate mainly on those aspects of the dehydrogenase structures which are correlated to their enzymatic functions. We will restrict the discussion to those three dehydrogenases, LDH, LADH and GAPDH for which both sequence information and high resolution structures are available. Previous recent reviews on the individual enzymes have been given by Holbrook et al. (1975), Brändén et al. (1975) and by Harris and Waters (1976). The various complexes that have been studied by X-ray diffraction are summarized in Table 1. The amino acid sequence of MDH has not been determined so structural mechanistic deductions of that enzyme are premature in spite of the high resolution structure determination (Hill et al., 1972; Banaszak & Bradshaw, 1975) and are thus not included here. A low resolution structure of NADP dependent 6-phosphogluconate dehydrogenase (Adams et al., 1977) is also available and the extension to high resolution is eagerly awaited.

Table 1. X-ray investigations of LADH, LDH and GAPDH.

Enzyme complex	Res.(Å)	Reference
LADH		
Apo horse	2.4	Eklund et al. (1976)
+ADP-ribose	2.9	Abdallah et al. (1975)
+8-bromo-ADP-ribose	4.5	-"-
+ADP	3.7	Zeppezauer et al. (1975)
+3-iodo-PyAD + Imidazole	4.5	Samama et al. (1977)
+PyAD + Imidazole	4.5	-"-
+PyAD	4.5	-"-
+5-iodo-salicylate	4.5	Einarsson et al. (1974)
+1,10-phenanthroline	4.5	Boiwe & Brändén (1977)
+Imidazole	2.9	-"-
+Iodoacetate (modified)	4.5	Zeppezauer et al. (1975)
Holo horse		
+NADH + DMSO	2.9	Eklund & Brändén (1979)
+NAD(H) + bromobenzylalcohol	4.5	Plapp et al. (1978a)
+NAD + trifluoroethanol	4.5	-"-

Table 1. cont.

+NAD(H) + solvent alcohol	4.5	Plapp *et al.* (1978a)
LDH		
Apo dogfish M_4	2.0	Adams *et al.* (1970) Eventoff *et al.* (1977)
+Tetraiodofluorescein	4.0	Wassarman & Lentz (1971)
+Adenosine	2.8	Chandrasekhar *et al.* (1973)
+AMP	2.8	-"-
+ADP	2.8	-"-
+Citrate	2.8	Adams *et al.* (1973)
+NAD	5.0	Adams *et al.* (1970)
Holo dogfish M_4		
+NAD + pyruvate	3.0	White *et al.* (1976)
+NAD + oxalate	3.0	-"-
+NADH + oxamate	3.0	-"-
Holo pig M_4		
+NADH + oxamate	6.0	Hackert *et al.* (1973)
Holo pig H_4		
+NADH+ oxamate	6.0	Eventoff *et al.* (1975)
Apo mouse C_4	2.9	Musick & Rossmann (1979)
GAPDH		
Holo lobster	2.9	Buehner *et al.* (1974a) Moras *et al.* (1975)
+8-bromo-NAD	2.9	Olsen *et al.* (1976b)
-NAD ("meso")	2.9	-"-
+Citrate	2.9	Olsen *et al.* (1976a)
+Trifluoroacetone	2.9	Garavito *et al.* (1977b)
Holo B. stearothermophilus	2.7	Biesecker *et al.* (1977)
Apo B. stearothermophilus	6.0	Wonacott & Biesecker (1977)
Holo human	5.0	Watson *et al.* (1972)

STRUCTURAL REQUIREMENTS FOR DEHYDROGENASES

The reactions catalyzed by the dehydrogenases under discussion are in principle very simple reactions. In the simplest form an alcohol group of a metabolite is oxidized by the removal of a proton from the hydroxyl group and the transfer of a hydride ion from the reacting carbon, the carbon atom that binds the hydroxyl group, to the oxidized form of the coenzyme. A second reaction is sometimes coupled to the dehydrogenation such as phosphorylation in the case of GAPDH and decarboxylation in the case of 6-phosphogluconate dehydrogenase. An essential point for a discussion of the structural requirements for these reactions is that a hydride ion is directly transferred from the reacting carbon of the substrate to the C-4 atom in the nicotinamide moiety of NAD^+ (Popják, 1970).

These enzymes must thus recognize and bind NAD in such a way that the nicotinamide is correctly positioned in the active site. The binding energy of NAD to an enzyme is potentially very high. Fersht (1977) estimates that a dissociation constant of 10^{-20} M could occur if these potential binding energies were realized. Since reasonably high K_m values are an important component of efficient catalysis it is thus essential that some of the possible interactions between enzyme and coenzyme do not occur. Thus there is a conflict between the requirements of recognizing NAD and positioning the nicotinamide on the one hand and sufficiently weak binding for efficient catalysis on the other. This conflict may be one of the important reasons why the dehydrogenases are relatively slow enzymes, frequently with the dissociation of coenzyme as the rate limiting step.

Furthermore, these enzymes must recognize and bind the substrates with various degrees of specificity in such a way that the reacting carbon is correctly positioned and oriented in the active site. However, the substrate molecule and the nicotinamide moiety may not be rigidly fixed relative to each other. The electronic rearrangement of the reacting carbon from sp^2-hybridization in the oxo-compounds to sp^3 in the alcohol requires different preferred pathways for hydride transfer and thus different relative orientations. Thus it is generally found in these structures that the nicotinamide moiety is not rigidly fixed but positioned in a hydrophobic cavity which is sufficiently large to allow different orientations of the ring. Dutler (1977) has described such possible movement for alcohol dehydrogenase in terms of a wind-screen wiper arrangement.

Direct transfer of a hydride ion is facilitated in a hydrophobic surrounding where water is excluded. Thus there is also a conflict between this requirement and the fact that the active site must be open and available for positioning of both nicotinamide and substrate. We will see that this apparent conflict is resolved by induced conformational changes of the protein structures such that the active site is shielded from the solution after binding of coenzyme and substrate.

SUBUNIT STRUCTURES

The amino acid residues of each subunit of these dehydrogenases are organized into two units, a coenzyme binding domain which has a similar structure in all dehydrogenases examined and a catalytic domain which differs in different dehydrogenases. This is illustrated in Figures 2 and 3 which show schematic diagrams of the subunit structures of LADH and GAPDH. The first domain contains most of the residues responsible for coenzyme binding while the second contains most of the residues involved in substrate binding, substrate specificity and catalytic activity. The location of the different domains within the amino acid sequences is shown in Table 2. The catalytic centres are in the interface between these two domains. The spatial organization of the domains with respect to each other as well as the quaternary structure of the subunit arrangement differ.

Table 2. Location of domains within the aminoacid sequences.

Enzyme	Coenzyme binding	Catalytic	Extra helix
LADH	176-318	1 -175, 319-374	168-183
LDH	22 -164,(1-20)	165-331	249-265
GAPDH	1 -147	148-334	313-334

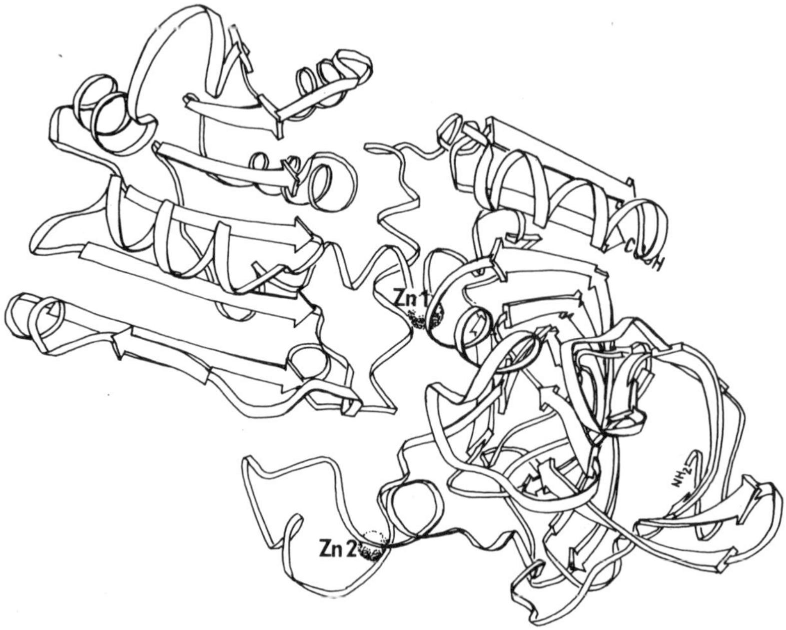

Fig.2. Schematic diagram of one subunit of LADH. (Designed by B. Furugren).

1. Coenzyme binding domain

The structures and similarities of the coenzyme binding domains in the dehydrogenases have been described in detail (Ohlsson et al., 1974; Rossmann et al., 1974, 1975) and the evolutionary and structural implications have since been extensively discussed. This domain is formed by two β-α-β-α-β-units forming a six stranded parallel pleated sheet surrounded by helices. An idealized coenzyme binding domain is shown in Figure 1. The two units of three strands are joined together by an approximate two-fold axis between and parallel to the two middle strands βA and βD. This has important consequences for the arrangement of the residues at the carboxyl end of the central region of the pleated sheet. The helices that are linked to these two central strands are on opposite sides of the sheet. Thus the loops that join these adjacent strands with the corresponding helices leave the strands in opposite directions. By this arrangement a pseudosymmetrical crevice is obtained, starting from one side of the sheet, passing over the edge of the sheet between the two central strands and continuing on the other side. The coenzyme molecules bind in this crevice in all three dehydrogenases with the adenine and nicotinamide ends on opposite sides of the sheet. The pyrophosphate group makes a bend over the edge of the sheet between the two central strands.

A very similar arrangement for coenzyme binding is also found in the NADP dependent enzyme dihydrofolate reductase (Matthews et al., 1978). Here the parallel strands are arranged somewhat differently, but the two adjacent central strands βB and βE and the corresponding helices form a similar crevice where the coenzyme molecule binds.

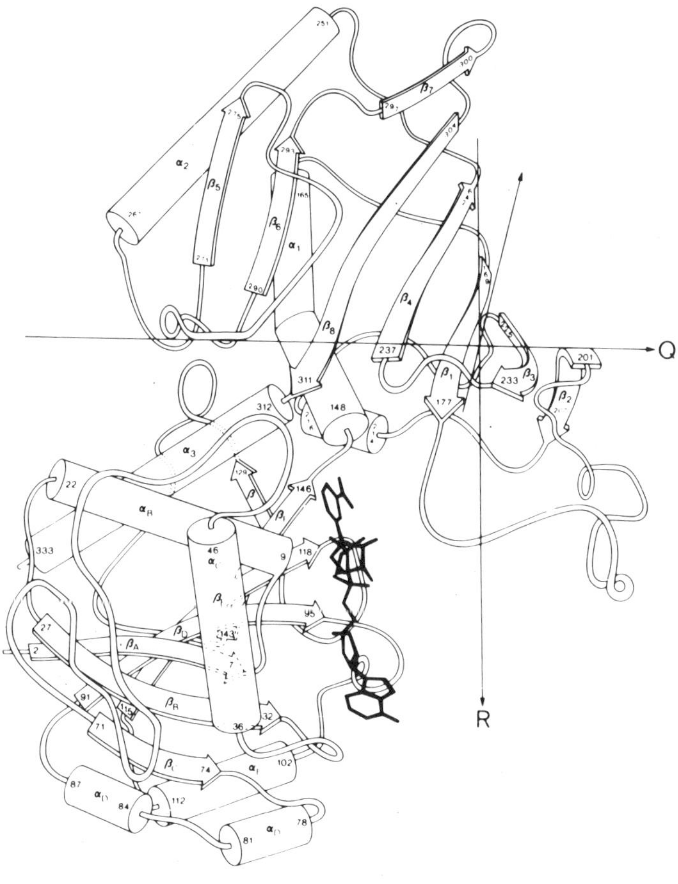

Fig. 3. Schematic diagram of one subunit of GAPDH from B. stearothermophilus. (Reprinted from Biesecker et al., 1977).

There are some structural differences between the coenzyme binding domains of the dehydrogenases outside this central NAD-binding region (Ohlsson et al., 1974). The functionally most interesting difference is found for LDH where there is a long flexible loop that joins strand βD and helix αE. This loop is involved in functionally important conformational changes.

One additional structurally homologous α-helix is associated with all three coenzyme binding domains. The positions in the amino acid sequences of this helix are different, however, in the different dehydrogenases (see Table 2). In LADH this helix (called αA) sequentially preceeds βA while it is sequentially the last part of GAPDH after the catalytic domain. In LDH the corresponding helix (called α3G) is sequentially in the middle of the parts that build up the catalytic domain.

2. Catalytic domains

While the coenzyme binding domain consists of sheets of parallel strands and connecting helices, the catalytic domains in all these dehydrogenases are to a large extent built up by antiparallel strands of pleated sheet going back and forth through the domains.

The catalytic domain of LDH is quite symmetrical, being built up by two similar folding units, $\alpha\beta_3\alpha$, with two α-helices separated by three antiparallel strands of pleated sheet. These units are connected by helix α2G (Figure 4).

The catalytic domain of LADH has a complicated structure of pleated sheet with four helices (Figure 2). The domain is built up from two sequentially different parts (Table 2). The first of these units consists of long antiparallel stretches, while the last part is a β-α-β-unit similar to what is found in the coenzyme binding domains. The whole domain can be described as consisting of three pleated sheet regions β-I, β-II and β-III. β-I is a six stranded sheet where two strands at one end are parallel while the others are antiparallel. β-II is an extension of the strands in β-I after a sharp bend and contains one additional strand. These are antiparallel. β-III is a cylindrical formation involving six strands which are mainly antiparallel. This domain binds two zinc atoms one of which has been related to catalytic activity. This zinc atom is bound with protein ligands Cys 46, His 67 and Cys 174. The second zinc atom is bound by cysteines 97, 100, 103 and 111.

The catalytic domain in lobster GAPDH consists of nine strands of pleated sheet, most of which are antiparallel, and with three roughly parallel α-helices on one side of the sheet. The structure of GAPDH from *Bacillus stearothermophilus* (Figure 3) is very similar but the interaction of the β-sheet regions is different from that of the lobster enzyme and gives rise to a modified hydrogen bonding pattern. The pleated sheet is here described in terms of eight strands. An important feature of the catalytic domain in GAPDH is an irregular S-shaped loop of polypeptide chain (residues 178-201) between the first two strands of pleated sheet.

3. The active site

In all three dehydrogenases the active site is located in a cleft between the two domains with the nicotinamide anchored to one side of the cleft by interactions with the coenzyme binding domain. The substrate interacts with residues from the catalytic domain on the other side of the cleft. One reason for the complicated structural arrangement of the catalytic domains must be to provide a framework for proper positioning of these residues so that the substrate can be correctly oriented with respect to the nicotinamide. Figure 5 illustrates the three different subunits viewed in a direction where the coenzyme binding domains and the bound coenzyme molecules are in the same orientation. It is apparent from this

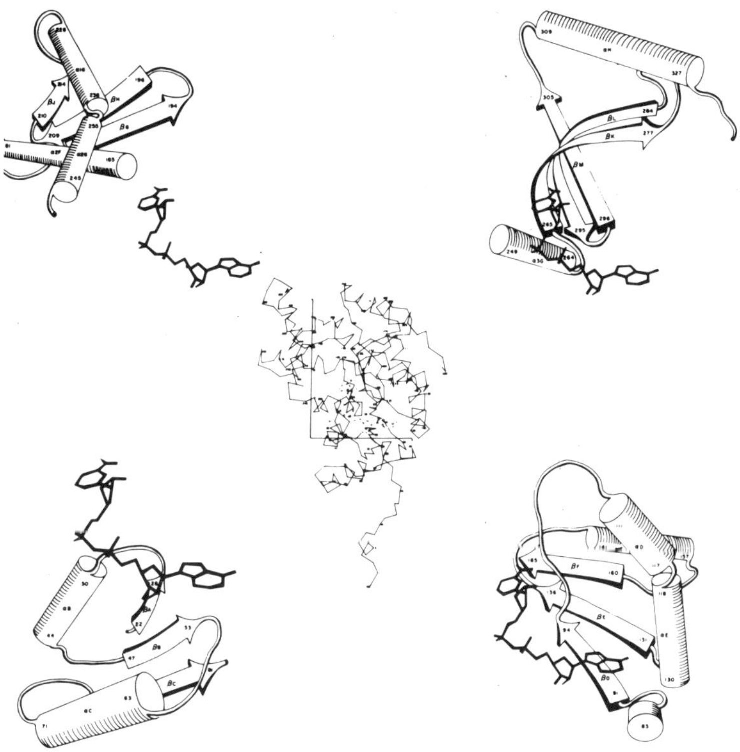

Fig. 4. Schematic diagram of one subunit of LDH with "blow-up" of each of the four principal domain structures. (Holbrook *et al.*, 1975)

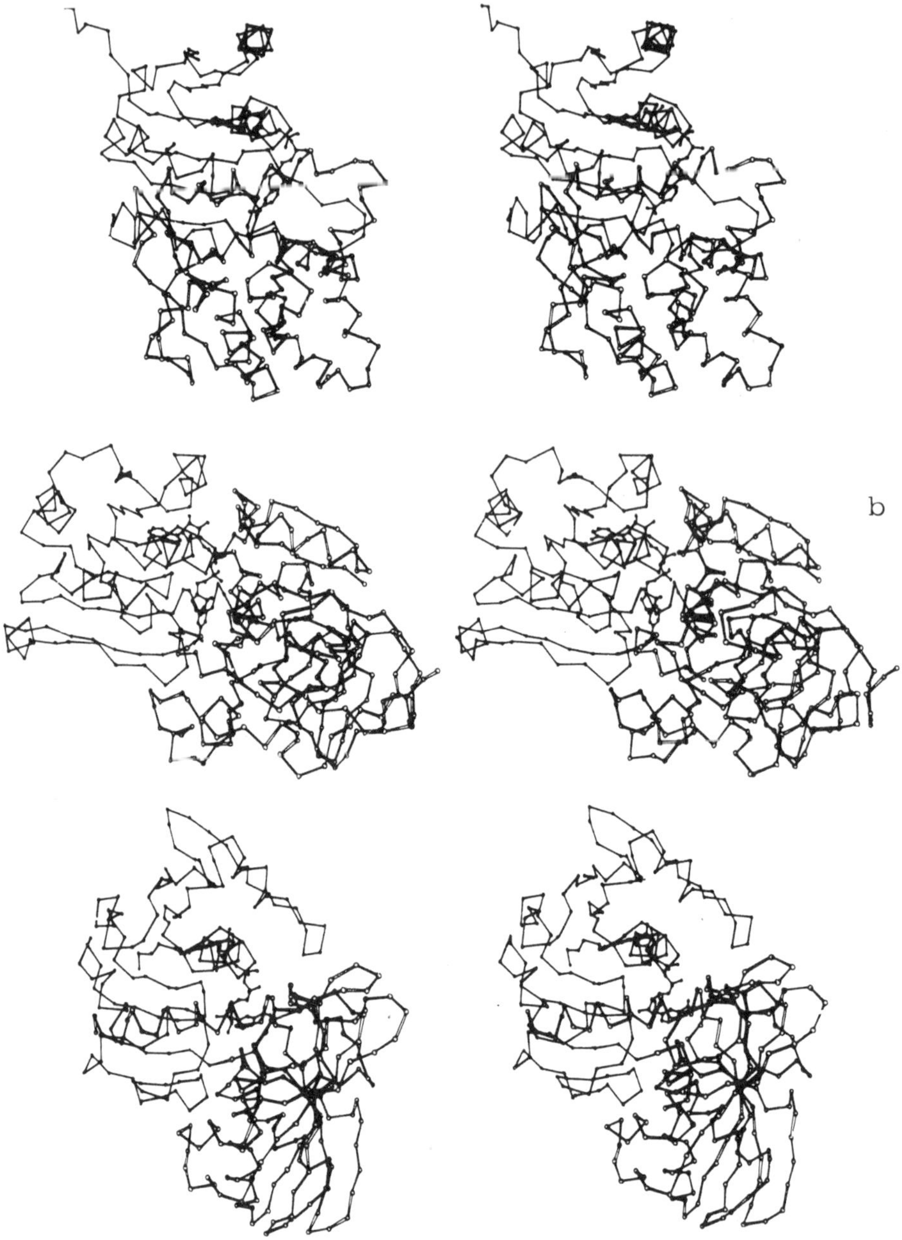

Fig. 5. Stereodiagrams of the subunits of a) LDH, b) LADH, c) GAPDH viewed with the NAD-binding domain in a similar orientation.

figure that the catalytic domains are then differently oriented. Thus the proper positioning of the catalytic groups requires a different structural orientation of the framework of the catalytic domains in all three dehydrogenases.

QUATERNARY STRUCTURE

The dehydrogenases, for which the three dimensional structures are known, are dimers (LADH and MDH) and tetramers (LDH and GAPDH). The subunits in LDH and GAPDH are related by three two-fold axes called P, Q and R to tetramers of 222 symmetry. These are crystallographic in both the apo and ternary complexes of dogfish LDH, while in all other cases none or only one of these two-fold axes are present as crystal symmetries. The deviations from strict 222 symmetry are however small in these cases.

The Q-axes are similarly located in these tetramers with regard to the coenzyme binding domains and relate αB, αC and α3G of one subunit to the corresponding parts of another subunit. This axis is present in all known dehydrogenases except LADH and is exactly two-fold in crystals of human holo-GAPDH. The interactions across the Q-axis are frequent in LDH and MDH but few in GAPDH. This has led to the concept that GAPDH is a dimer of dimers where strongly associated dimers are related by weak interactions across the Q-axis.

By far the largest number of interactions in GAPDH are across the P-axis between the antiparallel pleated sheets of the catalytic domains. The interactions across the third molecular two-fold axis, R, are almost exclusively related to the S-loop of residues 178-201. The catalytic sites are quite close to the R-axis. The S-loop thus directly connects the coenzyme binding site of one subunit to the catalytic site of the R-axis related subunit. A dimer related by the two-fold P-axis has been suggested to be the most stable association permitting the fine manipulations related to the catalytic activity across the R-axis.

In contrast the most extensive contacts in LDH are across the Q-axis. These contacts also represent dimer contacts in MDH. Contacts along the P-axis are primarily between side chains of β-structures in the catalytic domain. The contacts along the R-axis are almost entirely related to the N-terminal arm. The absence of the N-terminal residues, and the existence of the dominant Q-axis attractions will cause a break down of the tetramer into Q-axis related dimers, as have been observed in MDH.

Although the directions of the P- and R-axes with respect to the coenzyme binding domains are rather similar in LDH and GAPDH the interactions across these axes are different. The centres of the tetramers are thus quite differently located. In LDH the centre is at the amino end of the pleated sheet in the

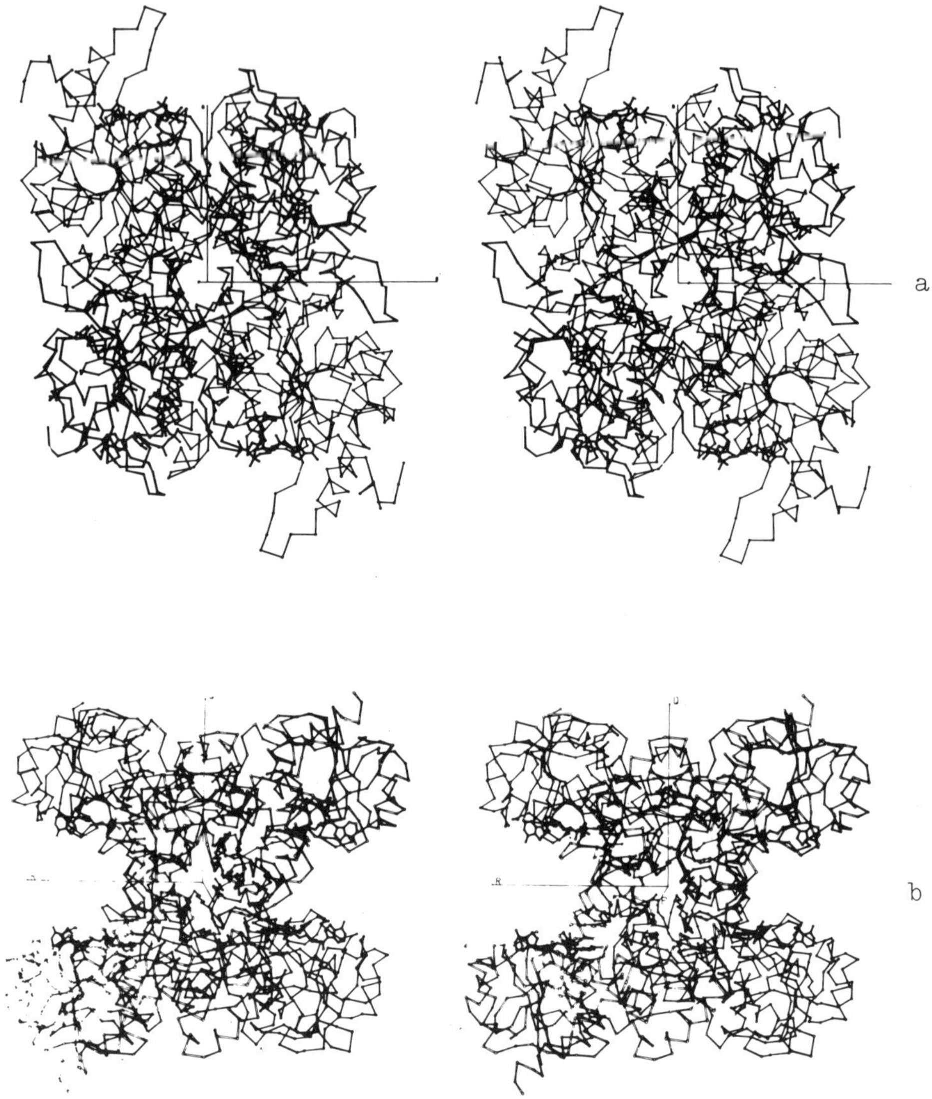

Fig. 6. Stereodiagrams of the molecules of a) LDH (Holbrook et al. 1975) and b) GAPDH (Harris & Waters, 1976).

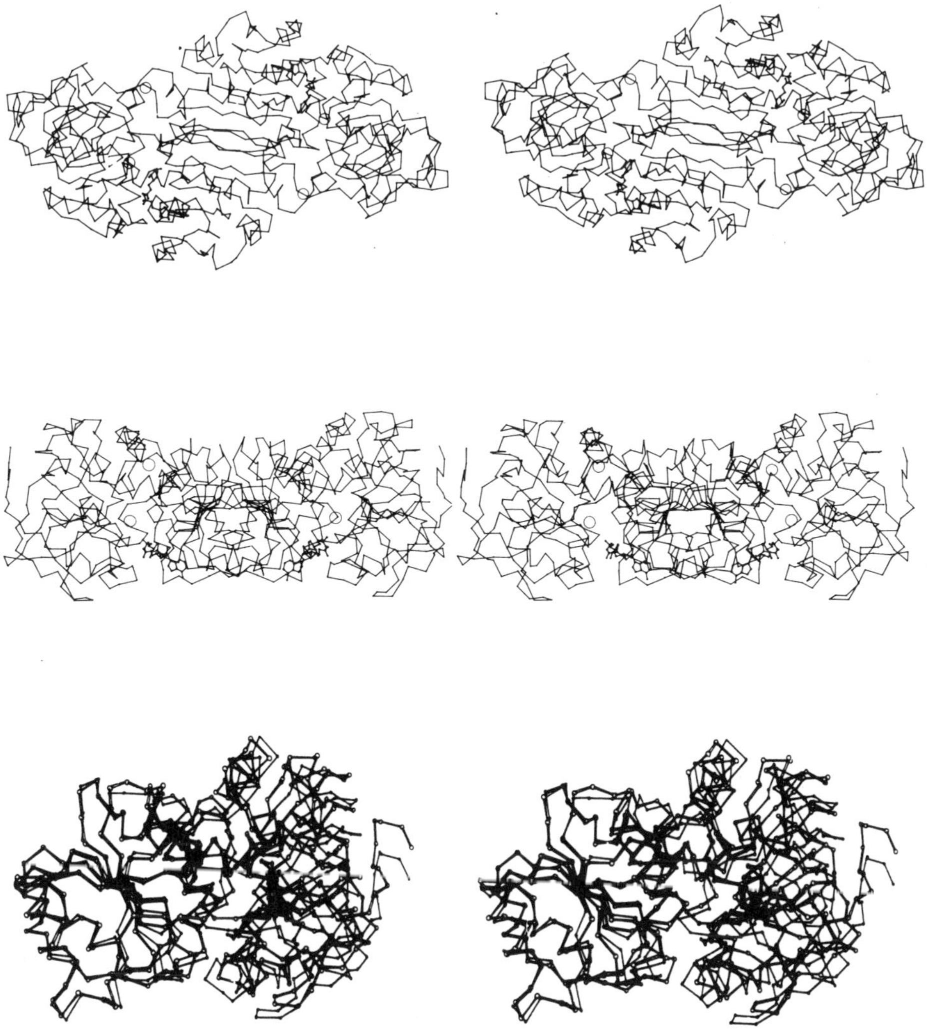

a

b

c

Fig. 7. Stereodiagrams of LADH. The molecule viewed a) along the two-fold axis, b) perpendicular to the two-fold axis. c) The subunits of apoenzyme (thick lines) and holoenzyme (thin lines) superimposed.

coenzyme binding domain, while in GAPDH the centre is close to the catalytic domain at the carboxyl end of the same sheet. As a consequence the tetramer in LDH is arranged so that the coenzyme molecules are bound far away from each other, while in GAPDH these are bound closely together (Figure 6). This arrangement in GAPDH is belived to be related to the observed cooperative properties.

The subunit association is quite different in LADH. The direction of the two-fold axis is roughly in the direction of the Q-axis found in other dehydrogenases with respect to the coenzyme binding domain but the axis relates quite different regions of the enzyme. The parallel pleated sheet strands of the two coenzyme binding domains of the molecule are joined together across the two-fold axis by antiparallel β-binding between the βF strands. The pleated sheet structure thus extends through the whole dimer and comprises 12 strands. Furthermore, the strands of each subunit, corresponding in position to helix αF in LDH, are connected antiparallel to each other. These two strands and a connecting helix of each subunit enclose a large area 20 x 30 Å of entirely hydrophobic residues. The molecule is thus built up by a central core of two coenzyme binding domains with the two catalytic domains on each side, separated from the core by the active sites (Figure 7).

Yeast alcohol dehydrogenase, YADH, is a tetramer. The subunit structure as deduced from sequence comparisons (Jörnvall et al., 1978), is similar to that of the liver enzyme except one deleted region of 21 residues. The arrangement of the subunits into a tetrameric molecule could not however be deduced from sequence comparisons alone, although some interaction regions could be proposed.

CONFORMATIONAL CHANGES

The active site of dehydrogenases must be easily accessible from the solution to allow binding of both coenzyme and substrate. Furthermore, the site must be open for substrate binding after the coenzyme has bound when substrate binding occurs after coenzyme binding. This is nicely illustrated in the LADH molecule where there are two entrances to the active site, one for the coenzyme and another for the substrate.

However, when the catalytic activity takes place, there is a requirement for a hydrophobic environment in the active site. This is achieved in these enzymes by a combination of coenzyme and substrate binding and a conformational change of the enzyme which closes off the active site. In this way the active site becomes completely shielded from the solution.

Enzymes are, on the one hand, tightly packed rigid bodies with a large number of internal interactions but at the same time they have a certain flexibility in some regions which allow them to adapt themselves to different requirements for each step of the catalysis. The details of this flexibility may however be different and we shall see that there are completely different ways for LDH and LADH to achieve similar effects.

In LDH this conformational change involves a large movement of a loop, including helix αD which in the holoenzyme structure covers the active site. The largest movement of a main chain atom in this loop is as large as 10.2 Å (White et al., 1976). In addition other structural elements close to the active site move to a smaller extent (Figure 8).

In LADH there is no structural feature in the coenzyme binding domain that corresponds to the flexible loop of LDH. Thus conformational changes must be different and are indeed comprised of a movement of the whole catalytic domains relative to the coenzyme binding domains (Figures 7c and 9). This has the effect that the cleft between the two domains becomes narrower. The loop around residues 55 and part of the preceding helix, residues 47 - 53, move closer to the bound coenzyme, about 4 Å for residue 55. In order to allow this movement, a loop of residues 294 - 297 on the other side of the cleft changes its conformation with the largest change of 8-9 Å (Figure 9b). Apart from this change the domains essentially keep their internal structures, whereas differences in the conformation mainly concern some of the sidechains at the surface of the molecule.

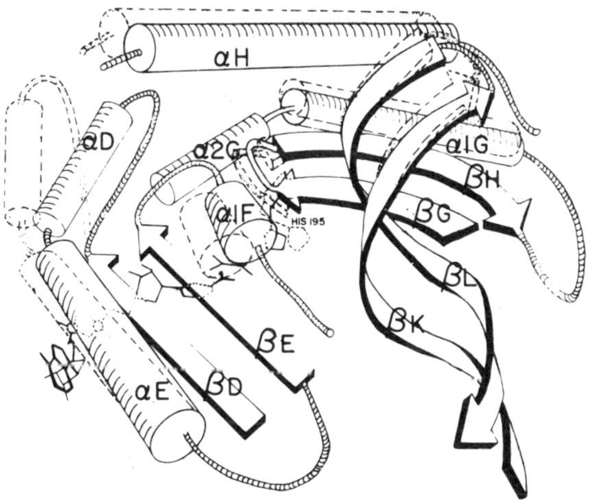

Fig. 8. Diagrammatic representation of conformational changes in secondary structural elements of an LDH subunit. Conformation of ternary complex is drawn in full line, while the apostructure is stippled.

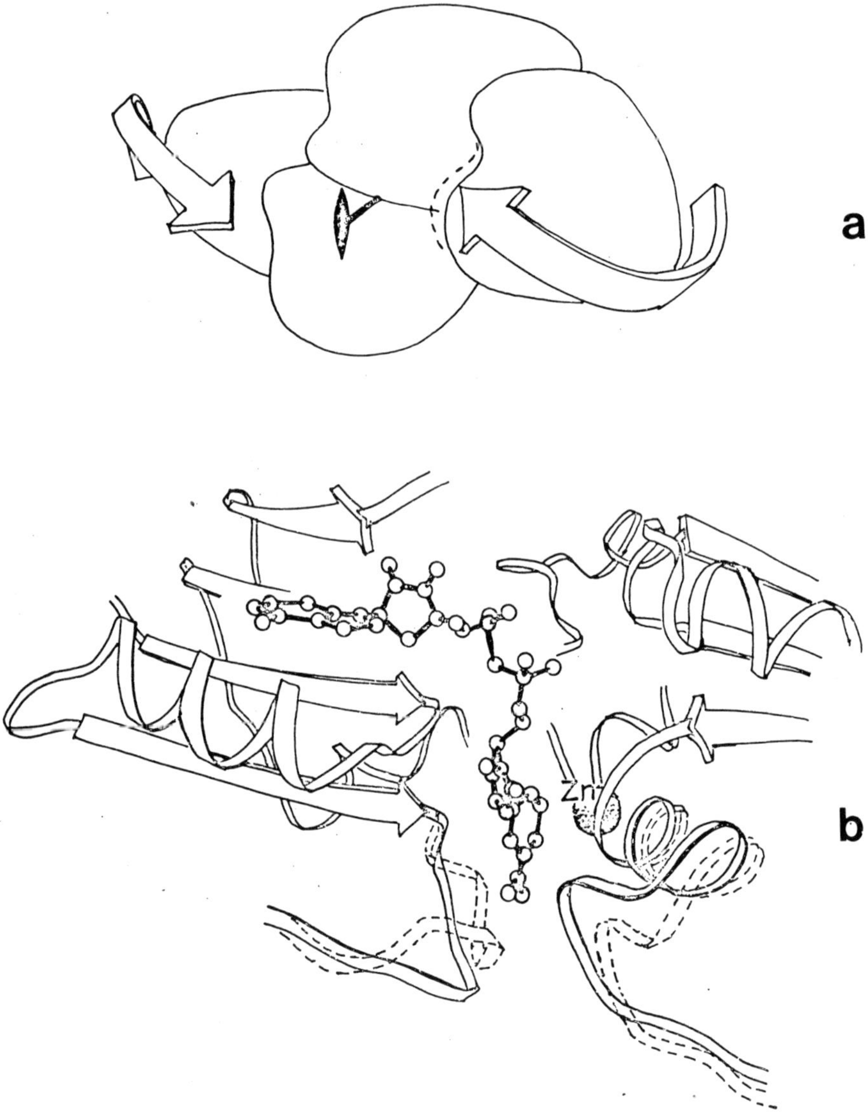

Fig. 9. a) A schematic drawing of the main conformational changes of the alcohol dehydrogenase molecule in the transition from apoenzyme to ternary complex.
b) The domain-domain interaction in the ternary complex with the active site and the coenzyme. Apoenzyme structure in broken lines.

While the nature of the conformational changes are very different in LADH and LDH they achieve a similar effect. The loop around residue 55 is in approximately the same region in space relative to the coenzyme binding domain as loop 103 - 115 in LDH. The combined effect of the conformational changes and the coenzyme and substrate binding is to shield off the active site from the solution. As we shall see this effect has been suggested to play an important role in lowering the activation energy for the hydride transfer step. Thus, similar functional effects have been produced by conformational changes of completely different structural regions in these two enzymes.

The apo- and holoenzyme structures for GAPDH from Bacillus stearothermophilus have been compared at 6 Å resolution (Wonacott & Biesecker, 1977). The comparison shows, similarly to LADH, that the conformational differences involve changes in the structural relation between the domains. However, there is a significant difference between these two cases. In LADH the central core of the molecule is formed by the two coenzyme binding domains and the catalytic domains move towards this core upon ternary complex formation. In GAPDH the catalytic domains form a central core of the molecule, and as a consequence, the NAD binding domains move relative to this core. The positions of helices αD and αE differ by about 5 Å. In addition the positions of the S-loops appear to be different which results in the S-loop being more accessible to solvent in the apoenzyme. The movements of the coenzyme binding domains produce a contraction of the tetramer on binding NAD to apo GAPDH. It should be noted that the S-loop in GAPDH is located in the same area relative to the active site as the loop regions in LDH and LADH.

1. Hinge regions

For LADH and GAPDH, where there are movements of whole domains, there must exist hinge regions around which these movements take place. The hinge region in LADH is the hydrophobic core between the two crossed helices α2 and α3, which form the two covalent connections between the domains. By a small sliding movement of these two helices the rotation of the catalytic domain can be accomplished without any large structural changes of the domains except in the interaction area between the two domains in the active site cleft on the opposite side of the molecule.

In GAPDH the COOH-terminal helix α3 is associated with the coenzyme binding domain, where it fits into a groove. This area appears to function as a hinge allowing relative rotation of the two domains.

2. Requirements for conformational changes

From what is known so far about the conformational changes in the dehydrogenases there is no general rule regarding the factors that induce these conformational changes. On the contrary, different parts of the coenzyme are responsible in each case. Thus groups of the coenzyme necessary for the conformational changes seem to be the pyrophosphate as well as substrate binding in LDH, the nicotinamide ribose in GAPDH and the nicotinamide in LADH.

Arginine 101 in LDH forms an ion pair with the pyrophosphate group of the coenzyme in the holoenzyme structure of ternary complexes. Its formation may be one necessary condition for the conformational change of this loop and its subsequent rearrangement upon substrate binding in the subunit. The enzyme conformation in binary complexes has been more difficult to study. Soaking apoenzyme crystals with coenzyme causes a loss of crystal symmetry, presumably by some initial conformational changes in the loop.

The factors responsible for the large conformational changes in LADH are being actively studied. Three types of experiments indicate that proper positioning of the nicotinamide ring in its binding site is one important factor. Binding studies have shown that coenzyme fragments do not change the conformation of the protein (Abdallah et al., 1975). The ADP-ribose part of the coenzyme binds to the apoenzyme structure and induces no conformational change. The second class of experiments is binding studies of coenzyme analogues which are modified in the nicotinamide ring. Binary complex formation with pyridine adenine dinucleotide (PyAD), 3-iodo-PyAD and 5-methyl-NAD do not cause conformational changes of the enzyme. In these complexes the nicotinamide ring was found to be bound on the surface of the molecule in a pocket about 15 Å from the active site (Figure 11b) (Samama et al., 1977). Tetrahydro-NAD also binds to the apoenzyme crystals but in a conformation similar to the coenzyme binding found in ternary complexes.

The third class of experiments is the binding of coenzyme to enzyme molecules which have a modified environment of the active site zinc atom. Enzyme with imidazole bound to the active site zinc atom was found to bind NADH to the "apo"-form without any large conformational changes. The imidazole apparently produces an unfavourable environment for the nicotinamide ring. The conformation of bound NADH resembles closely the conformation of bound coenzyme in GAPDH, with its B-side directed in an unproductive way against the substrate site (Zeppezauer, E., unpublished observations). The coenzyme molecule also binds without any gross conformational changes of the protein (Zeppezauer, E., unpublished observations) to enzyme molecules that have been modified by carboxymethylation of the zinc ligand Cys 46. Here the coenzyme seems to have a conformation similar to that found in the ternary complexes.

Thus in order to achieve the conformational transition from apo- to holostructure there is a requirement both for proper positioning of the nicotinamide part of the coenzyme and a suitable environment around the active site zinc atom.

From studies of the binding of NAD-analogues to crystals of apo-GAPDH it seems that the ribose of the nicotinamide plays an essential role in effecting the structural changes associated with NAD-binding. It has been suggested that the anion binding site which bridges the nicotinamide ribose and residues of the S-loop is the crucial interaction which triggers the conformational changes (Wonacott & Biesecker, 1977).

COENZYME BINDING

NAD is a flexible molecule which can adopt many different conformations. When bound in a productive manner to dehydrogenases it has essentially the same conformation in all the cases investigated so far. A schematic diagram of this conformation is given in Figure 10. This is a reflection of the fact that the interaction areas between these enzymes and the various moieties of the coenzyme are similar due to the close structural similarities of the coenzyme binding domains. This conformation of bound coenzyme, which is the same for NAD^+ and NADH is different from that found (Saenger et al., 1977) in crystals of the Li-salt of NAD (see Figure 11), and also from

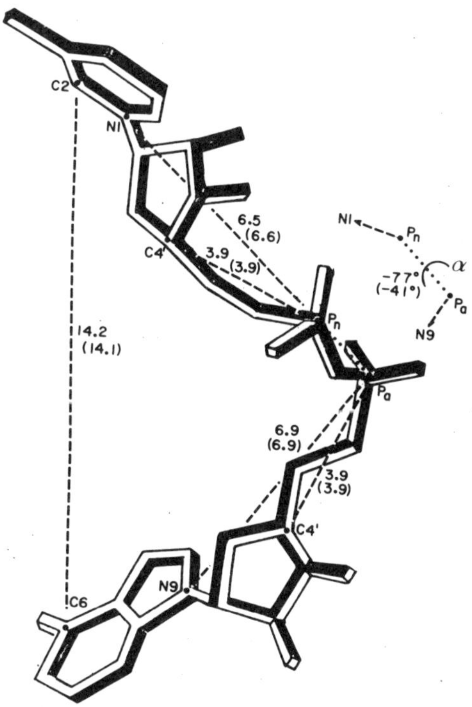

Fig 10. Schematic diagram of the conformation of NAD bound to A-specific dehydrogenases. (Banaszak & Bradshaw, 1975).

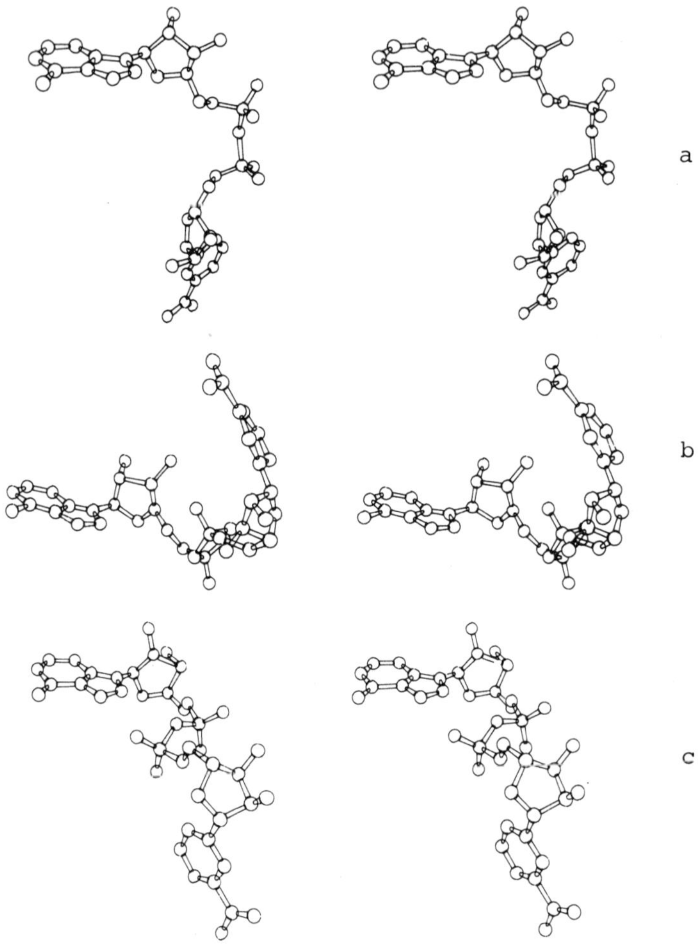

Fig. 11 Stereodiagrams of coenzyme conformations in
a) NADH bound to LADH
b) 5-methyl-NAD bound to LADH
c) The Li-salt of NAD.

those believed to be present in solutions of free NAD. Moreover this conformation is energetically quite unfavourable (Saenger *et al.*, 1977). It is thus apparent that the interaction with these enzymes forces the coenzyme into a conformation which would never be obtained for NAD free in solution. We will now examine these interactions in more detail. A list of the reported contacts for LDH, LADH and GAPDH is given in Table 3 and a schematic diagram of some important sidechain contacts are given in Fig 12.

1. Adenine

The adenine ring is in a hydrophobic cavity about 20 Å from the active centre. In both GAPDH and LADH there are bulky hydrophobic sidechains on both sides of the ring which form a slot into which the ring is positioned. A few of the possible hydrogen bonds from the N-atoms of the ring are realized but in no case is there any firm structural evidence of specificity for adenine in these interactions. The ring is oriented in such a way that the NH_2-group points away from the molecule into the solution. This orientation is the molecular basis for the successful application of affinity chromatography (Lindberg *et*

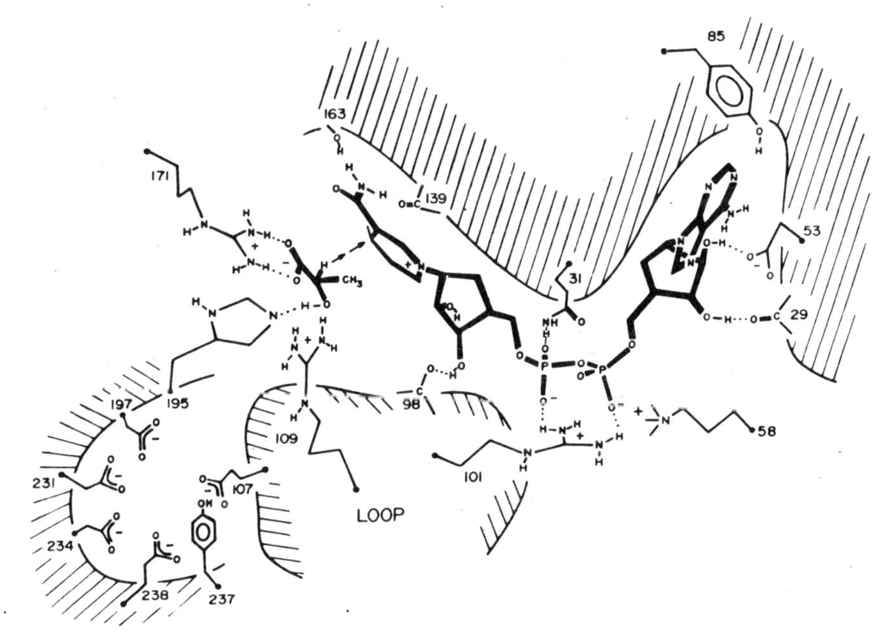

Fig. 12 Diagrammatic representation of the principle groups in the active centre of LDH while binding lactate and NAD (Holbrook *et al.*, 1975)

al., 1973) with NAD attached to the gel through a spacer to this NH_2-group. The cavity is generally larger than required for adenine binding and can accomodate a variety of different inhibitor molecules. Thus it has been shown that the dye tetra-iodo fluorescein binds to LDH (Wassarman & Lentz, 1971) and that salicylate (Einarsson et al., 1974) and the naphthalene group of 8-anilino-1-naphthalene sulphonate bind to LADH in this cavity (Eklund, H., unpublished observations). The fact that a large variety of compounds with aliphatic or aromatic hydrophobic groups are inhibitors of these dehydrogenases is in all probability due to the binding properties of this cavity.

2. Adenosine ribose

A comparison of the binding of the adenosine ribose part to these enzymes has provided an explanation in molecular terms for the discrimination between the binding of NAD and NADP. The 2´- hydroxyl group of the ribose moiety of NAD is hydrogen bonded to a structurally equivalent aspartate residue in all these dehydrogenases. It has been suggested (Rossmann et al., 1975) that the extra phosphate group of NADP will cause both steric and charge repulsion with this aspartate residue and thus weaken the binding considerably. Solution studies of the coenzyme analogue 2´-dNAD (Suhadolnik et al., 1977) have shown that removal of this OH-group changes the kinetic constants appreciably. These effects strongly indicate a non-productive positioning of the nicotinamide presumably transmitted through a different interaction with the adenosine deoxyribose to the enzyme. In the NADP-dependent enzyme dihydrofolate reductase (Matthews et al., 1978) there is no aspartate residue at the equivalent position. Instead there is an arginine residue from a different strand, corresponding to βA in the dehydrogenases, which interacts with the phosphate group attached to the 2´-position in NADP.

The 3´-hydroxyl group hydrogen bonds to lysine residues in both LDH and LADH. The structurally equivalent residue in GAPDH is Pro 188 in the flexible S-loop of the R-axis related subunit. This lysine in LADH, Lys 228, is particularly interesting since it has been shown (Plapp et al., 1973; Dworschack et al., 1975) that chemical modifications of this residue increase the turnover number for alcohol oxidation tenfold due to increase of the dissociation rate of the enzyme-coenzyme complex. These modified enzyme molecules have been crystallized (Plapp et al., 1978b) and are now being studied in order to obtain a molecular explanation for this rate increase.

The adenosine ribose moiety binds so close to the carboxyl end of the parallel pleated sheet that the last residue of βA is invariably glycine in all these dehydrogenases from all observed species. A sidechain in this position would prevent the observed binding mode and possibly push the coenzyme into a different conformation.

Table 3. Enzyme - coenzyme interactions in LDH, LADH and GAPDH.

Underlined residues on the same line are structurally equivalent according to the alignement by Rossmann *et al.* (1975).

Coenzyme moiety	Residues LDH	LADH	GAPDH	Type of contact
Adenine	Val 27	Phe 198	Asn 6	Interior of pocket
	Val 52	Val 222	Asn 31	-"-
	Asp 53	Asp 223	Asp 32	H-bond to ring N
	Val 54	Ile 224	Phe 34	Hydrophobic
	Tyr 85			H-bond to ring N
	Ile 96			Hydrophobic
	Ala 98	Ile 269	Thr 96	-"-
	Ile 119		Phe 99	-"-
	Ile 123			-"-
		Arg 271		H-bond to N6
Adenosine ribose	Gly 28	Gly 199	Gly 7	
	Asp 53	Asp 223	Asp 32	H-bond to O-2´
	Lys 58	Lys 228	C=O 8	H-bond to O-3´
	Asp 30			H-bond to O-3´
	Met 55	Ile 200		Hydrophobic
			Leu 187	Hydrophobic from
			Pro 188	adjacent subunit
Pyrophosphate	N 31	N 202	N 10	H-bond to main chain
	N 32	N 203	N 11	-"-
	Arg 101	Arg 47	Asn 180	Charged or polar
	Lys 58	Arg 369		-"-
			Gly 99	
			Ala 31	
Nicotinamide ribose	C=O 98	C=O 269	C=O 96	H-bond to O-3´
	N 140	N 294	anion	H-bond to O-2´
		His 51	Ser 119	H-bond to O-3´
	Gly 99	Gly 270	Gly 97	
	Val 32	Val 203		Hydrophobic
	Thr 97			
	Pro 139	Gly 293	Ala 120	-"-
		Val 294		-"-
Nicotinamide	Val 32	Val 203	Ile 11	Hydrophobic
	Val 138	Val 292		-"-
	Asn 140	Val 294		
	Leu 167	Thr 178	Tyr 317	
	Thr 246			
	Ile 250			
			Cys 149	Charge transfer
Carboxamide	C=O 139	C=O 292		H-bond
	Ser 163	N 319	Asn 313	-"-

3. Pyrophosphate

The pyrophosphate part makes a bend around the edge of the pleated sheet. By this conformation the adenine end is positioned on one side of the sheet and the nicotinamide end on the other side. Important interactions in stabilizing this conformation seem to be hydrogen bonds from the oxygen atoms to corresponding main chain nitrogen atoms at the beginning of helix αB. The pyrophosphate binding site is thus close to the carboxyl end of the parallel pleated sheet and interacts with the main chain at the amino end of helix αB. In almost all enzymes that contain several parallel strands of pleated sheet an anion binding site important for activity is located close to the carboxyl end of the sheet. It has been pointed out (Hol *et al.*, 1978) that the dipole moment of the helices that surround these sheets create a favorable environment for an anion binding site at the amino end of these helices. It certainly seems more than a coincidence that in all these three dehydrogenases the pyrophosphate moiety of the coenzyme is attached by hydrogen bonds to main chain amino groups of structurally equivalent residues at the amino end of helix αB.

Additional interactions with the pyrophosphate are provided by positively charged side chains in LADH and LDH and by an asparagine residue in GAPDH. In LADH one of these side chains is Arg 47 which in the holoenzyme bridges the crevice between the domains and clamps down onto the pyrophosphate group. The importance of this particular interaction has recently been illustrated in yeast alcohol dehydrogenase. Comparisons of the primary structure of YADH (Jörnvall, 1977) with the tertiary structure of LADH have shown that the subunit structures must be very similar (Jörnvall *et al.*, 1978). In "wild-type" YADH the residue equivalent to Arg 47 is His 44. Recently a yeast mutant was obtained in which this histidine residue was replaced by an arginine (Wills & Jörnvall, 1979). The affinity for the coenzyme in this mutant enzyme is increased by a factor of five compared to the wild type enzyme.

4. Nicotinamide ribose

The precise conformation of the polypeptide chain in the coenzyme binding domain is important not only for positioning the pyrophosphate moiety but also the nicotinamide ribose part. Thus O-3′ of this ribose forms hydrogen bonds to the main chain carbonyl oxygen atoms of structurally equivalent residues in all three dehydrogenases. These bonds position the ribose so close to the main chain that there is no space available for a side chain in the next residue which is invariably glycine. O-2′ is hydrogen bonded to main chain amino groups of structurally equivalent residues of a different loop in LDH and LADH. In GAPDH O-2′ forms a hydrogen bond with an oxygen atom from the phosphate group of the substrate. Sidechains in the environment of this binding site are essentially hydrophobic. It is important to realize that hydrogen bonds between ribose hydroxyl groups and polar groups on the surface of the enzyme contribute very little to the overall binding energy. In the unliganded state both enzyme and coenzyme form hydrogen bonds by interactions of these atoms with water molecules in the solvent. The specific hydrogen bonds in the complex are however in all probability essential for the proper positioning of the coenzyme molecule.

5. Nicotinamide

The nicotinamide ring is positioned at the active site with one side of the ring plane facing the substrate and the other side close to a hydrophobic region of the coenzyme binding domain. LADH and LDH are A-specific with respect to hydride transfer to the nicotinamide ring whereas GAPDH is B-specific (Popják, 1970). In structural terms this is accomplished by similar binding modes to the enzymes, similar positions of carbon atom C-1 of the substrates with respect to nicotinamide and similar conformations for the coenzyme with the important exception of a rotation of 180° in GAPDH around the glycosidic bond between the nicotinamide ring and ribose. The positions of the pyridine rings are thus similar but the carboxamide parts are quite differently oriented. In all three cases the carboxamide is in the vicinity of polar groups to which hydrogen bonds could be formed. The similarities in the binding mode of the nicotinamide ring can be illustrated by the fact that in all three cases the rings are in hydrophobic contact with structurally equivalent hydrophobic sidechains (Val 203 in LADH). Eventoff et al. (1977) have suggested that the carboxamide forms different hydrogen bonds in the M_4 and H_4 isozymes of LDH as a result of an amino acid sequence difference which might effect the binding of the pyrophosphate groups.

Binding of NAD^+ to GAPDH apoenzyme produces a characteristic absorption band, the Racker band (Racker & Krimsky, 1952). In the *Bacillus* enzyme Cys 149 is ideally positioned to form a charge transfer interaction with the nicotinamide ring which could explain the absorption properties. There is no indication in this structure of a covalent bond between these groups, as has been suggested for two of the subunits in the lobster enzyme (Moras et al., 1975).

6. Binding of coenzyme analogues

It is apparent from these comparisons of coenzyme binding that the special fold of the coenzyme binding domain provides a framework for the proper alignment of the different moieties of the coenzyme molecule. There are substantial variations in the nature of the actual sidechains at the various binding sites for these moieties. In none of these binding sites is there a discriminating specificity for binding the particular chemical moiety present such as, for example, the specificity pocket in serine proteases. In addition there do not seem to be any particularly strong binding sites, with the possible exception of adenosine, but rather a large number of weak interactions which contribute to bind the coenzyme molecule in an energetically unfavourable conformation. These results have a number of important implications for work using coenzyme analogues, some of which will be discussed here.

One of the main difficulties in using the dehydrogenases for biotechnological purposes, e.g. in enzyme reactors, is the instability of NAD towards hydrolysis. Since these enzymes exhibit no particular steric specificity for the moieties present in NAD it should in principle be possible to use chemically different molecules as hydride acceptors. It should be a challenging task for an organic chemist to design a chemically stable coenzyme analogue based on the structural requirements for binding to these dehydrogenases and disregarding the actual chemical nature of NAD.

A large number of papers has been published describing the effect on the activity of different chemical modifications of various parts of the NAD-molecule. Most of the conclusions reached from these studies have been based on the implicit assumption that the modified coenzyme analogues bind in essentially the same way as the unmodified coenzyme. The first clear evidence, to our knowledge, that this assumption is not generally valid came from the modification studies in Woenckhaus' laboratory using a reactive coenzyme analogue in which adenine was replaced by 5-bromoacetyl-4-methyl-imidazole. This analogue is active as a hydrogen acceptor in the LADH enzymic reaction (Woenckhaus & Jeck, 1971). It also causes complete inactivation of LADH by specific modification of Cys 174 (Jörnvall et al., 1975). In the absence of knowledge of the tertiary structure of LADH a reasonable conclusion would have been that Cys 174 is close to the adenine binding site of NAD. However, since it was known that Cys 174 is a zinc ligand close to the nicotinamide binding site it was concluded that this analogue can bind in two different ways. In the "correct" orientation with nicotinamide close to zinc it is catalytically active. In the "wrong-way" orientation it should bind with the nicotinamide group in the adenine binding pocket and with the reactive adenine analogue close to zinc where it can alkylate Cys 174.

A surprising result was obtained when we investigated the binding of pyridine-adenine dinucleotide to LADH (Samama et al., 1977). We found that this analogue bound in a very different manner from NADH and with a different conformation. The pyridine ring is not in the vicinity of the zinc atom in the interior of the subunit, but assumes a completely different position on the surface of the molecule about 15 Å away from the nicotinamide position of bound NADH. Since the adenosine moiety binds in the same position as in NADH the conformational differences are mainly accomplished by a difference in the orientation and position of the pyrophosphate group so that the sugar phosphate chain takes a completely different direction with respect to the enzyme in this complex. This is illustrated in Figure 11.

This analogue does not induce the conformational transition but binds to the apoenzyme structure. However, the binding mode is not a simple function of the conformational state of the enzyme since NADH can also bind to the apoenzyme structure with the nicotinamide ring in a similar position as in holoenzyme (see below). Thus small modifications on the coenzyme molecule have resulted in drastic differences in the interaction with the enzyme. The fine balance between these different binding modes is illustrated by the behaviour of 3-iodo-pyridine-AD. This enzymatically active analogue binds like pyridine-AD in the binary complex with LADH or in the ternary complex with imidazole and does not induce a conformational transition. However in ternary complexes with oxygen ligands such as alcohol or DMSO bound to zinc, the conformational transition is induced (Samama, J.-P., unpublished observations) and the analogue binds like NADH to the holoenzyme structure.

Even the unmodified coenzyme can bind in different ways to LADH under certain conditions. Theorell & McKinley-McKee (1961) noted that the inhibitor imidazole could act as an activator when low substrate concentrations were used. They attributed this to the ability of imidazole to decrease the affinity of the enzyme for NADH. Eila Zeppezauer in our laboratory has studied this enzyme-NADH-imidazole complex to high resolution and finds that imidazole prevents the conformational transition so that NADH is bound to the apoenzyme structure. Since imidazole binds to the active zinc atom at the substrate binding site one can thus regard the substrate site as an allosteric site with respect to coenzyme binding. The mode of binding is very similar to NADH bound to the holoenzyme structure with one very interesting exception. The nicotinamide part has rotated 180° around the glycosidic bond so that it binds in a B-specific fashion. A comparison with NAD bound to GAPDH shows that positional differences are less than 2 Å between all corresponding atoms of the nicotinamide rings, using the same procedure of alignement as in previous studies (Ohlsson <u>et al</u>., 1974). The carboxamide group interacts with Arg 369. There are small conformational changes of the atoms around the active site including a movement of the zinc atom of about 1 Å presumably to make room for this different position of the carboxamide which otherwise would be very crowded.

These examples illustrate that the interactions between a biological macromolecular receptor molecule and a small effector molecule, in this case an enzyme - coenzyme interaction, can be drastically changed either by minor modifications on the effector molecule or by the presence of other molecules which bind at different sites.

SUBSTRATE BINDING AND CATALYTIC MECHANISM

The mechanisms of the reactions that are catalyzed by these dehydrogenases are defined when all the intermediates, complexes and conformational states of the enzyme are characterized and the rate constants for their interconversions determined. These crystal structures have provided information on the presence and position of possible catalytic groups in the active site and on the conformational differences of those states of the enzymes which have been isolated in crystalline form. Although it has not yet been possible to obtain the detailed structure of a productive enzyme-coenzyme-substrate complex, the use of substrate analogues and model building has served as adequate substitutes.

This structural information has provided a framework which has been used in conjunction with kinetic and other studies in solution to derive possible mechanisms. Since there are a number of recent reviews devoted to this subject (Holbrook *et al.*, 1975; Brändén *et al.*, 1975; Harris & Waters, 1976; Fersht, 1977) we will only summarize here the pertinent results and add a few recent details especially about LADH. In our presentation we are particularly indebted to the excellent synthesis of structural and kinetic data given by Alan Fersht (1977).

1. LDH

LDH catalyzes the oxidation of L-lactate to pyruvate. In order to determine the structure of the substrate binding site Rossmann and coworkers have studied the ternary complexes of LDH-NAD-oxalate, LDH-NADH-oxamate and LDH-NAD-pyruvate (White *et al.*, 1976). The last complex has a covalent link between the C-4 of the nicotinamide ring and the methyl carbon of pyruvate. All three complexes crystallize isomorphously in the holoenzyme conformation with the flexible loop in the "down" position. From these studies an approximate substrate binding site could be deduced. By model building of the true substrate into this position the following picture of the possible interactions between enzyme and substrate has emerged (see Fig. 13). The carboxylate group of the substrate forms a salt bridge with the side-chain of Arg 171. The imidazole ring of His 195 is conveniently located to interact with the oxygen atom bound to the reacting carbon of the substrate. The hydroxyl group of lactate can form a hydrogen bond with the unprotonated ring while the carbonyl group of pyruvate can form a hydrogen bond with the protonated form. These interactions form the major recognition sites for the substrate. Because the reacting carbon of lactate is a chiral centre, these two specific interactions are sufficient to discriminate between the D- and the L-form of the substrate. The hydrogen atom attached to the reacting carbon points towards C-4 of the nicotinamide ring in L-lactate whereas the methylgroup would be oriented in that direction for the D-form.

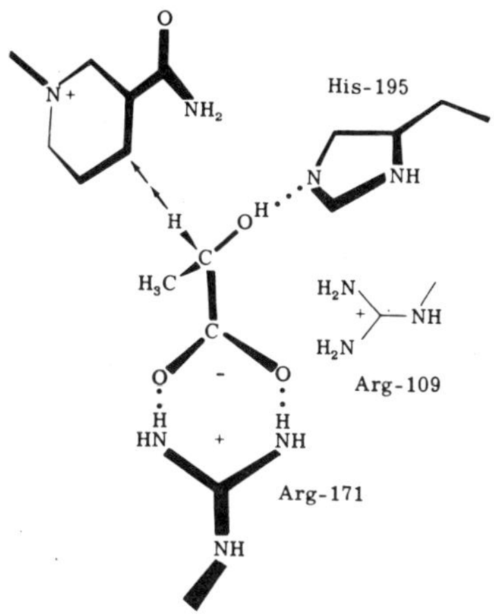

Fig. 13 Schematic representation of substrate binding in LDH. (Holbrook et al., 1975)

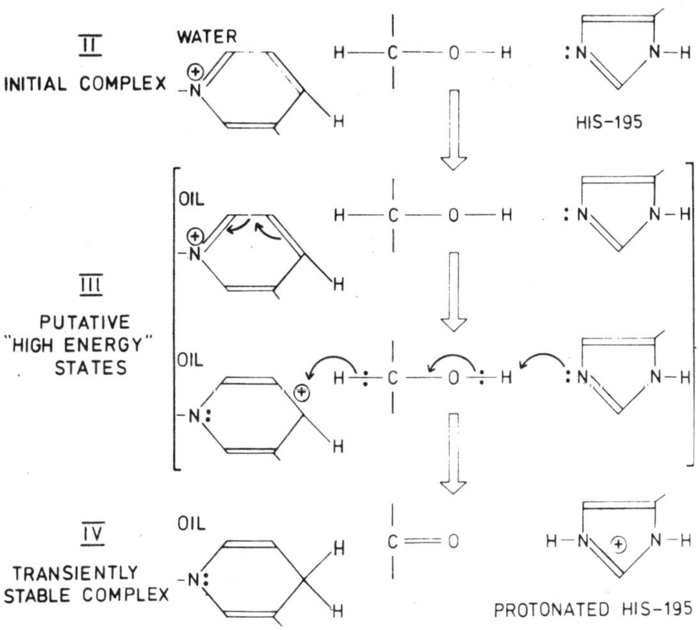

Fig. 14 Interconversion of the ternary complexes in LDH according to Parker and Holbrook (1977).

On the basis of evidence then available Holbrook et al. (1975) suggested a mechanism for the catalytic action which was later modified by Parker and Holbrook (1977) into an oil-water-histidine mechanism when the complete sequence information became available. The first step in the mechanism is the rapid binding of NAD^+ with the loop still in the "up" position so that the active centre is exposed and the pyridinium cation could be neutralized with a solvent anion. The next steps involve binding of lactate and the conformational transition of the flexible loop to the "down" position. By this change the coenzyme and substrate are enclosed in the active site and the nicotinamide ring obtains a hydrophobic environment with no negative counterions. This high energy state is reduced by the electronic rearrangement in the ring to leave a transient $C\text{-}4^+$species which extracts a hydride ion from the bound lactate and induces a flow of electrons from the unprotonated form of His 195. When this rearrangement is completed a transiently stable intermediate of NADH-pyruvate and the protonated form of His 195 has been formed. This interconversion is illustrated in figure 14. Thus hydride transfer and proton abstraction from the substrate occur simultaneously in LDH. The role of His 195 is dual; to orient the substrate and to act as an acid-base catalyst stabilizing the negative charge that develops on the oxygen of pyruvate during reduction and removing the proton from lactate during oxidation. The driving force for this reaction is suggested to be the very favorable interactions which can be made when the loop closes onto the protein surface. This conformational change is essential to the mechanism since it provides the hydrophobic environment which facilitates the electron flow.

This beautiful mechanism is consistent with a large number of experimental observations. There are, however, some loose ends for which satisfactory answers still remain to be found. The mechanism is ordered which implies that lactate or pyruvate cannot bind to the enzyme in the absence of coenzyme. Although His 195 moves slightly during the conformational change there are virtually no positional differences of Arg 171 which apparently forms a strong interaction with the substrate. Thus it is not obvious why lactate does not bind to the apoenzyme form. Furthermore, there is an additional charged residue Arg 109 which is brought into the vicinity of the active site by the conformational transition. No function for this residue is implied in the mechanism just described. Garavito et al. (1977a) have suggested, on the basis of positional similarities of this residue and Arg 231 in GAPDH, that these positive charges are involved in the destabilization of the Michaelis complex in favour of the transition state. The nature of this destabilization and its importance relative to other factors are however unknown. It might be appropriate to ask whether these two loose ends are connected and try to examine whether Arg 109 is important for substrate binding in some indirect way.

2. LADH

LADH catalyzes the oxidation of various primary and secondary alcohols to aldehydes and ketones, respectively. Although the enzyme has a broad substrate specificity it does not oxidize several important cell metabolites that contain alcohol groups, such as lactate and malate.

In order to determine the binding mode of substrates we have studied crystals grown from an equilibrium mixture of LADH, NAD^+ and p-bromobenzylalcohol first to 4.5 Å resolution and subsequently to 2.9 Å resolution (Eklund, H., unpublished results). These crystals have been shown to contain appreciable amounts of bound NADH (Bignetti et al., 1979). The structure of the stable nonreactive complex between tetrahydro-NAD and trans-4-N,N-dimethylaminocinnamaldehyde (DACA) has also been determined to high resolution (Samama, J.-P., unpublished results). This complex has similar spectral properties to the catalytically active transient LADH-NADH-DACA complex (Dunn et al., 1975). Finally we have studied complexes with NAD and the inhibitor trifluoroethanol to low resolution (Plapp et al., 1978a) and with NADH and DMSO to both low (Eklund & Brändén, 1979) and high resolution (Eklund, H., unpublished results). In all these complexes the substrate or inhibitor molecule bind in a similar fashion which is consistent with a productive binding mode of a substrate. It has thus been assumed that this binding mode reflects the binding of a substrate in a productive ternary complex, illustrated in Figure 15.

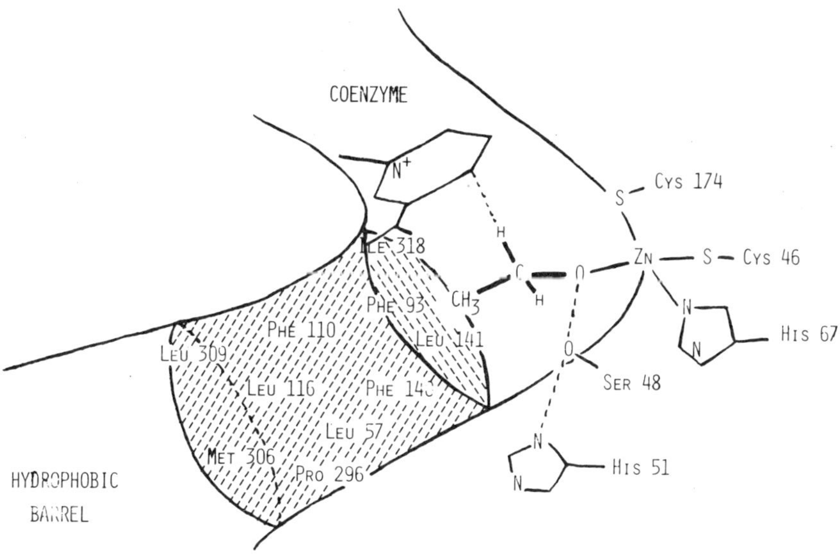

Fig. 15 Schematic representation of substrate binding in LADH.

The substrate oxygen atom binds directly to the active site zinc atom. In the apoenzyme structure this position is occupied by a water molecule. The oxygen atom of the substrate is furthermore hydrogen bonded to the side-chain of Ser 48 which in turn is hydrogen bonded to a side-chain nitrogen atom of His 51 (see Figure 15). Ser 48 is the only polar group in the vicinity of the reactive part of the substrate apart from the zinc atom and the zinc ligands Cys 46, Cys 174 and His 67. The side-chains of the substrates are in a hydrophobic barrel which is lined by about ten hydrophobic side-chains from the catalytic domain of one subunit and the coenzyme binding domains of both subunits. The hydrogen atom to be transferred according to the observed stereochemistry points towards C-4 of the nicotinamide. The hydrophobic barrel can accomodate a wide variety of non-polar side-chains by non-specific hydrophobic interactions. The only specific substrate recognition site is thus for the alcohol or the aldehyde oxygen atom. This is in contrast to LDH or GAPDH which in addition recognize adjacent charged groups. Such substrates are prevented from binding to LADH by the hydrophobicity of the barrel.

Using the apoenzyme structure as a framework for a number of solution studies Brändén and coworkers (Eklund *et al.*, 1974; Brändén *et al.*, 1975) suggested the mechanism illustrated in Figure 16. The first step is a rapid initial binding of NAD^+. By isomerization of this E-NAD^+ complex the pK_a of the zinc bound water molecule is shifted from about 9.6 to about 7.6. This isomerization to the holoenzyme structure is facilitated in the presence of alcohol in which case the pK drops to less than 4.5 (Shore *et al.*, 1974). As a result of this isomerization a hydroxyl group remains bound to zinc and a proton is released. It was suggested that this proton is released through the system of hydrogen bonds involving Ser 48 and His 51 according to the scheme shown in Figure 17. It has recently been shown by theoretical calculations (Tapia *et al.*, 1979) that the electric field created by the protein framework facilitates this proton tunneling along the hydrogen bonds. The presence of NAD^+ and zinc with a small positive charge further enhances this effect. Incidentally this system can be regarded as a proton pump which, as a result of NAD^+ binding and subsequent isomerization, pumps protons from the interior hydrophobic pocket to the surface of the molecule.

In the next step of the mechanism alcohol displaces the zinc bound hydroxyl group and binds directly to zinc as an alcoholate group. The hydroxyl ion thus extracts the proton from alcohol by the formation of water possibly in a transient penta-coordinate complex. Interconversion of the ternary complexes by hydride transfer is facilitated by the electrophilic nature of the zinc atom. In the last and rate determining step NADH dissociates away and the enzyme isomerizes back to the apoenzyme structure with water bound to the catalytic zinc.

The recently determined high resolution structure of the holoenzyme complexes with substrates is entirely consistent with this mechanism. Furthermore, the system involving Ser 48 and His 51 is preserved in this structure in spite of the fact that this region is subjected to extensive conformational changes. In addition, the second nitrogen atom of the imidazole group of His 51 hydrogen bonds to the 2´-hydroxyl group of the nicotinamide ribose. There is thus now a firm structural basis for incorporation of these features into mechanistic deductions

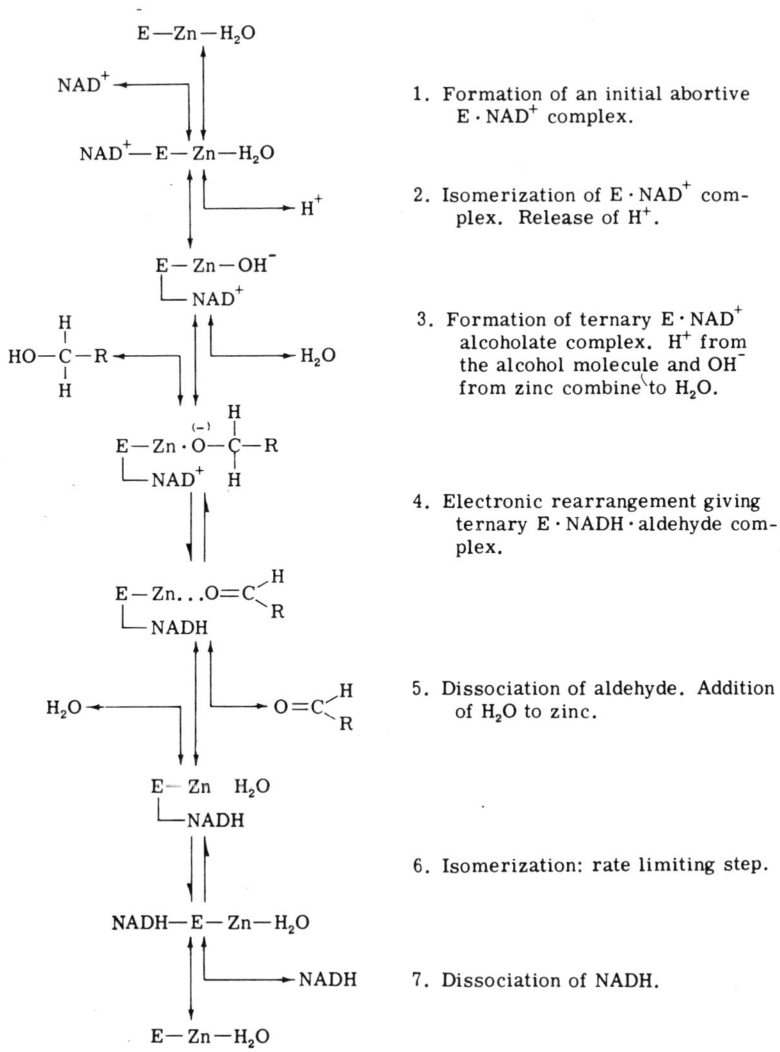

Fig. 16 Suggested mechanism for the catalytic action of LADH by Brändén *et al*. (1975).

about the catalytic events that take place within the holoenzyme conformation. On the basis of transient-state kinetic data Brooks et al. (1972) reported that the hydride transfer step in alcohol oxidation required a group of pK 6.4 to be in the unprotonated state, whereas McFarland & Chu (1975) claimed that there is no pH-dependence for the same step in aldehyde reduction from pH 6 to pH 10. Kvassman and Pettersson (1978) have confirmed these results and suggest that the pH dependence of the apparent rate constant for hydride transfer from alcohol substrates to NAD^+ derives from a proton dependent pre-equilibrium step occurring at the ternary complex level. They also suggest that this group would have the catalytic function of ensuring that the substrate is bound as an alcoholate ion at physiological pH.

These kinetic observations can now be correlated with the holoenzyme structure in the following way. The kinetically observed group with pK 6.4 is His 51. The hydrogen bond system His 51 - Ser 48 - substrate oxygen atom is essential for a productive binding of the alcohol substrate in the form of a negatively charged alcoholate ion. Protonation of His 51 would require breaking the hydrogen bond to the 2´-hydroxyl group of the nicotinamide ribose and thus give a different orientation of the imidazole ring which would disrupt this hydrogen bond system. The interconversion of the ternary complexes according to this scheme is depicted in Figure 18. In this mechanism hydride transfer and proton abstraction are separate steps.

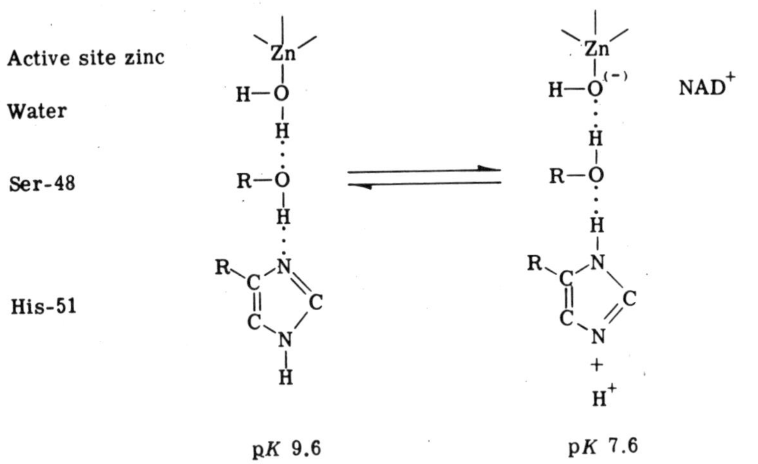

Fig. 17 Schematic diagram illustrating the mechanism for the proton release in LADH induced by NAD^+ binding.

This mechanism has been questioned on the basis of kinetic evidence (Blackwell & Hardman, 1975; Klinman, 1976) which indicates that no charge develops on the substrate as a result of hydride transfer. This is a characteristic feature of general acid-base catalysis. Thus it has been argued (Dworschack & Plapp, 1977) that the charge difference between an alcoholate ion bound to zinc and an aldehyde bound to zinc is not compatible with this kinetic evidence.

This argument is, however, not necessarily correct. It is well known that alcoholate ions bind strongly to zinc ions. In fact Meerwein showed more than fifty years ago that alcohols act as strong acids in the presence of zinc chloride due to the formation of zinc-alcoholate compounds and the liberation of protons. Thus an alcoholate ion bound to zinc is very different from a free alcoholate ion and is more similar to a neutral alcohol molecule. Furthermore the interaction between an aldehyde and zinc is quite weak. Thus the protein ligands-zinc-alcoholate system can work in a fashion similar to the general acid-base catalyzed reaction with respect to charge differences of the oxidized and reduced forms of the substrate. The zinc ion neutralizes the alcoholate ion by soaking up electrons which are redonated simultaneously with hydride transfer. For a detailed discussion on the properties of zinc ions as electrophilic catalysts the review by Dunn (1975) is recommended. One crucial question is whether the specific active site zinc atom in LADH can function as an effective electron sink in this way. Very little is known about the electronic properties of a system consisting of a zinc atom liganded to two cysteine-sulphur atoms and a histidine nitrogen

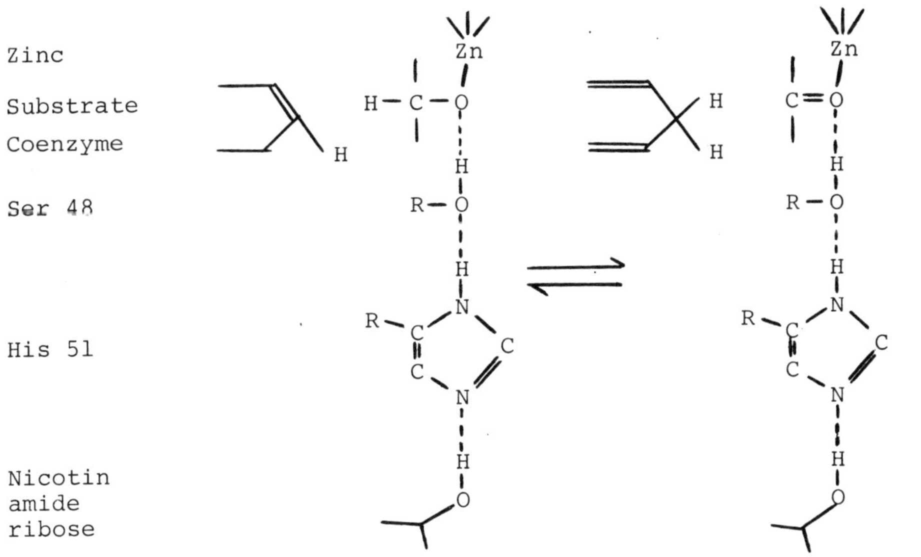

Fig. 18 Interconversion of ternary complexes in LADH.

atom inside a protein molecule.

The activation of NAD^+ for hydride transfer in this mechanism would be rather similar to that for LDH. The negative charge of the alcoholate ion would be at least partly neutralized by zinc and no other negatively charged groups are present to neutralize NAD^+. As in LDH the conformational change from apo- to holoenzyme structure closes off the active site and the nicotinamide ring obtains a very hydrophobic environment suitable for the formation of a transient C-4+ species.

Dworschack and Plapp (1977) have suggested a modification of this mechanism in order to incorporate an acid-base system. They propose that binding of the substrate to zinc does not displace the water molecule or hydroxyl ion but that they are both bound in a penta-coordinate complex. The water molecule could thus act as a general acid and the hydroxyl as the general base. This is an intellectually attractive mechanism but there is no evidence from the holoenzyme structures that a water molecule is bound to zinc in addition to the substrate oxygen atom. This point cannot, however, be settled definitely until a representative holoenzyme structure has been extended to higher resolution and refined.

It has also been suggested (Sloan *et al.*, 1975) on the basis of distances calculated from nmr-measurements of Co-substituted LADH that the oxygen atom of the substrate does not bind directly to zinc but binds via a zinc-bound water molecule. The high resolution holoenzyme structures, however, show quite clearly that both alcohol and aldehyde as well as DMSO bind directly to the zinc atom.

3. GAPDH

GAPDH catalyzes the reversible oxidative phosphorylation of glyceraldehyde-3-phosphate to 1,3-diphosphoglycerate. There are two main reaction sequences in the mechanism, first a dehydrogenation of the substrate followed by a phosphorylation. The structures of the enzyme-substrate complexes have been deduced from model building experiments based on the holoenzyme structures from both lobster (Buehner *et al.*, 1974b; Moras *et al.*, 1975) and *Bacillus stearothermophilus* (Biesecker *et al.*, 1977). The following structural picture of the previously proposed reaction mechanism (Orsi & Cleland, 1972; Harrigan & Trentham, 1973) has emerged from these two independent investigations (see Figure 19).

The 3-phosphate group of the substrate interacts with the side-chains of Arg 231 and Thr 179 and with the 2′-hydroxyl of the nicotinamide ribose. The C-2 hydroxyl group of the substrate hydrogen bonds to Ser 148 and the aldehyde group is close to the side-chain of the reactive cysteine 149. In this position the substrate is poised for the first step of the reaction sequence (Figure 19). A hemithioacetal is formed between C-1 of the substrate and Cys 149. By this reaction the carbonyl group is converted into an alcohol which can form a hydrogen bond to His 176. These interactions orient the substrate so that the C-1 hydrogen points towards C-4 of the nicotinamide ring. The dehydrogenation reaction can now occur in a way similar to the situation in LDH with His 176 as the general base catalyst.

The NADH molecule formed in this step must now be replaced by a NAD^+ molecule for the phosphorylation reaction to proceed. There is a specific binding site for the attacking orthophosphate close to Cys 149 created by hydrogen bonds to the amide nitrogen atoms of residue 149 and 150 as well as to hydroxyl groups from Ser 148, Thr 150 and C-2 of the substrate. From this position orthophosphate can react with the thioester, an acyltransfer occurs and the acyl-phosphate product is formed. It is believed that the requirement of NAD^+ for this reaction is due to the ability of its positive charge to polarize the sulphur atom of the hemithioacetal.

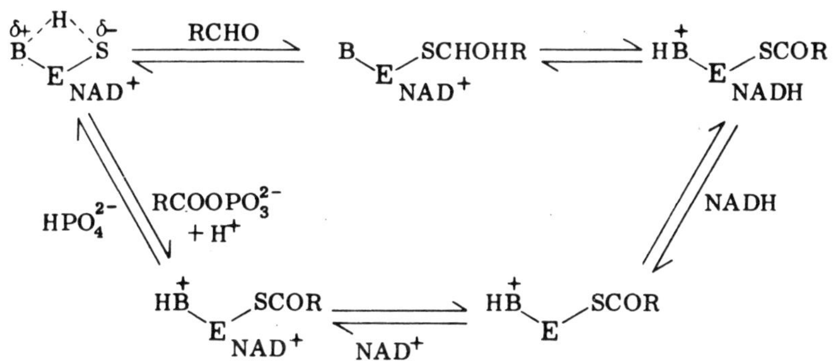

a

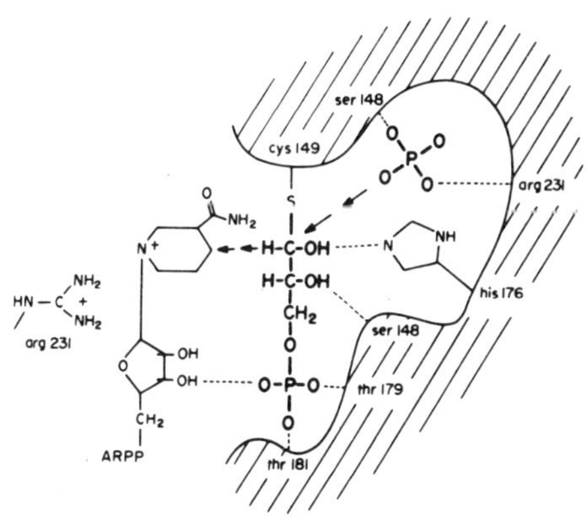

b

Fig 19 a Reaction mechanism of GAPDH according to Harrigan and Trentham (1973).
b Schematic representation of substrate binding in GAPDH. (Harris & Waters, 1976). Ser 148 is shown twice.

4. Cooperativity in GAPDH

Each molecule of NAD^+ binds with a decreasing affinity to muscle GAPDH, which thus exhibits negative cooperativity. Two different models have been proposed to explain this functional non-equivalence of chemically identical subunits, one based on ligand induced conformational transitions (Stallcup & Koshland, 1973) and the other based on preexisting molecular asymmetry in the form of a dimer of asymmetric dimers (Bernhard & MacQuarrie, 1973). Crystallographic studies of the human (Watson et al., 1972; Buehner et al., 1974a) holoenzyme have established that the Q-axis in these crystals is an exact two-fold symmetry axis. Both the lobster and the Bacillus enzyme crystallize with the complete tetramer in the asymmetric unit and exhibit aproximate 222 symmetry. The high resolution structural work on these two enzymes shows excellent agreement on the main features of the subunit conformations but they disagree on the extent of asymmetry between the subunits.

Biesecker et al. (1977) find for the Bacillus enzyme that, within their limits of errors, all four subunits as well as the bound coenzyme molecules are identical.

For the lobster enzyme, Rossmann and coworkers originally described (Moras et al., 1975) a number of structurally different features between the two subunits that are related by the R-axis. These features included differences in the conformation of the coenzyme molecules and in one of the anion binding sites. In subsequent papers it was shown that these anion binding sites were identical (Olsen et al., 1976a) and, furthermore, that a bromine-labeled coenzyme analogue was bound with the same conformation in all subunits (Olsen et al., 1976b). Until these structures have been refined it is very difficult to be certain that the observed differences or identities in the electron densities reflect preexisting structural asymmetry or symmetry in the holoenzymes. However, Garavito et al. (1977b) have recently demonstrated structural asymmetry in an abortive ternary complex of the lobster enzyme. This complex is produced by the covalent modification of Cys 149 with trifluoroacetone. Different conformations of this molecule as well as differences in the orientations of some of the catalytic groups were found in the subunits related by the R-axis.

The flexible S-loop has many of the properties expected of a structure which transmits a structural change from one subunit to another in a cooperative system. It connects directly the NAD-binding site of one subunit to the catalytic site of the R-axis related subunit. Furthermore there is evidence that its structure in the apoenzyme is quite different from that in the holoenzyme. It has been suggested by both groups that this S-loop plays a central role in transmitting information between subunits so that negative cooperativity occurs.

5. Comparison of active sites

Although the catalytic domains in these three dehydrogenases have very different structures and the catalytic groups are quite different there are some obvious similarities in their catalytic mechanisms. In LDH and GAPDH there is a histidine residue that acts as an acid-base catalyst by forming a hydrogen bond with the carbonyl or alcohol group of the substrate. LADH is essentially similar with the zinc ion acting as the electron sink. Furthermore, these groups orient the substrates for hydride transfer and also stabilize the transition states for the reactions.

In all three cases there are conformational transitions between apoenzyme and holoenzyme. In LDH and LADH these have been suggested to be important in activating NAD^+ for hydride transfer and, quite probably, they are also important in creating the productive substrate binding site. This would then be a molecular explanation for the ordered mechanism. These structural changes are an interesting case of convergent function since the residues involved in closing the active sites in LDH and LADH are present in different domains of these two structures. In GAPDH the flexible S-loop is situated in the same region in space relative to the active site.

In order to explore other possible convergent functions or unrecognized similarities the structures of these active sites have been compared in detail (Garavito et al., 1977a; Argos et al., 1978). These comparisons have revealed some new and interesting results. The first concerns the observation (Popják, 1970) that the A- and B-side stereospecificity of hydride transfer to the nicotinamide ring is conserved over a vast variety of species. This has been taken as evidence for a highly conserved evolutionary property and hence of great importance in catalysis. Why is this so when these structures reveal that the only difference of A- and B-side specificity in terms of coenzyme conformation is a difference in the position of the carboxamide group due to a rotation of 180° of the nicotinamide ring around the glycosidic bond? An answer to this question was provided by a comparison of the arrangement of catalytic groups around the reacting carbon of the substrate in A-specific LDH and B-specific GAPDH (Garavito et al., 1977a). Three groups of similar function were considered, (i) the acid-base catalysts His 195 and His 176 respectively, (ii) C-4 of the nicotinamide and (iii) substrate binders, in this case Arg 171 and Cys 149, respectively. These three groups should be roughly positioned in three of the tetrahedral directions around the reacting carbon. It was found that when the first two corresponding groups were superimposed the substrate binding groups were in opposite directions. It was also found that the positions of the substrate binders were such that the substrates avoided the carboxamide groups of the coenzyme. This is illustrated in Figure 20 . In other words, a change to A-side specificity for GAPDH would require rearrangement in space of the substrate binding groups. This would involve several simultaneous mutations which is an unlikely event.

LADH is somewhat different because secondary alcohols such as cyclohexanol are substrates which means that both directions from the reacting carbon must be available for substrate binding. As a consequence there is no specific substrate binder but instead a wide hydrophobic barrel into which the remaining groups of the substrate bind. This observation can be added to the list of other indications (Brändén, 1977) that LADH is not designed for small substrate molecules like ethanol but rather for large ones containing secondary alcohol groups. On the other hand if the position of the methyl group of ethanol is used as the substrate binder in the known stereospecificity for hydride transfer from ethanol (1-pro-R) of A-specific LADH the same orientation of catalytic groups as in A-specific LDH is found (Argos <u>et al</u>., 1978).

The second important observation from these comparisons concerns similarities of groups of unknown function in the vicinity of the active site. It was found that Arg 109 in LDH, Arg 231 in GAPDH and His 51 in LADH are positioned roughly in the same place relative to the reacting carbon and the nicotinamide group. Similar positions of these three polar groups strongly indicate similarities in function which might be involved in productive substrate binding.

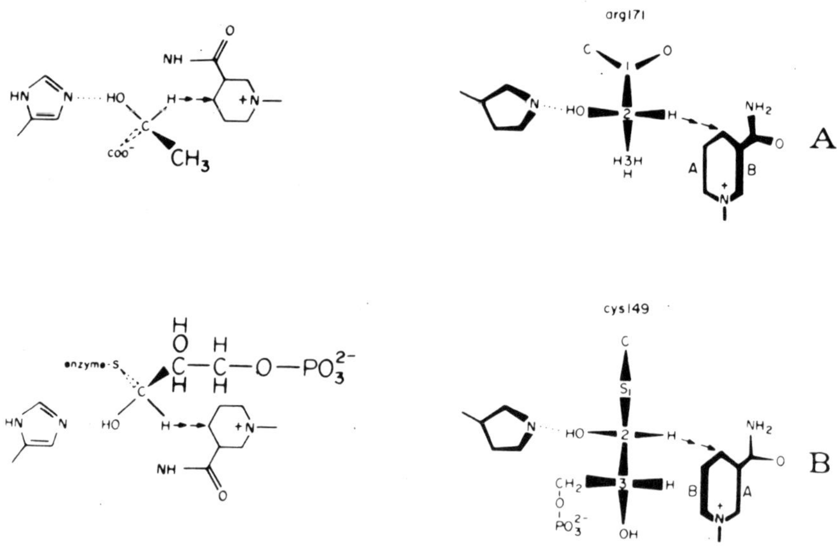

Fig 20. <u>Diagrammatic representation of the relationship between substrate binding positions and the orientation of the carboxamide group of the nicotinamide ring in a) LDH and b) GAPDH (Garavito et al., 1977a).</u>

ACKNOWLEDGEMENTS

This work was supported by grants from the Swedish Natural Science Research Council.

REFERENCES

Abdallah, M.A., Biellmann, J.-F., Nordström, B. & Brändén, C.-I. (1975) Eur. J. Biochem. 50, 475-481.
Adams, M.J., Ford, G.C., Koekoek, R., Lentz, P.J. Jr., McPherson, A. Jr., Rossmann, M.G., Smiley, I.E., Schevitz, R.W. & Wonacott, A.J. (1970) Nature (London) 227, 1098-1103.
Adams, M.J., Liljas, A. & Rossmann, M.G. (1973) J. Mol. Biol. 76, 519-531.
Adams, M.J., Helliwell, J.R. & Bugg, C.E. (1977) J. Mol. Biol. 112, 183-197.
Argos, P., Garavito, R.M., Eventoff, W., Rossmann, M.G. & Brändén, C.-I. (1978) J. Mol. Biol. 126, 141-158.
Banaszak, L.J. & Bradshaw, R.A. (1975) in The Enzymes (Boyer, P.D. ed.) 3rd ed. Vol XI, pp 369-396, Academic Press, New York.
Banner, D.W., Bloomer, A.C., Petsko, G.A., Phillips, D.C., Pogson, I., Wilson, A., Corran, P.H., Furth, A.J. Milman, J.D., Offord, R.E., Priddle, J.D. & Waley, S.G. (1975) Nature (London) 255, 609-614.
Bernhard, S.A. & MacQuarrie, R.A. (1973) J. Mol. Biol. 74, 73-78.
Biesecker, G., Harris, J.I., Thierry, J.-C., Walker, J.E. & Wonacott, A.J. (1977) Nature (London) 266, 328-333.
Bignetti, E., Rossi, G.-L. & Zeppezauer, E. (1979) FEBS Letters (in press).
Blackwell, L.F. & Hardman, M.J. (1975) Eur. J. Biochem. 55, 611-615.
Blake, C.C.F. & Evans, P.R. (1974) J. Mol. Biol. 84, 585-601.
Boiwe, T. & Brändén, C.-I. (1977) Eur. J. Biochem. 77, 173-179.
Brändén, C.-I. & Rossmann, M.G. (1970) in Pyridine Nucleotide Dependent Dehydrogenases (Sund, H. ed.) pp 133-134, Springer Verlag, Berlin.
Brändén, C.-I., Eklund, H., Nordström, B., Boiwe, T., Söderlund, G., Zeppezauer, E., Ohlsson, I. & Åkeson, Å. (1973) Proc. Natl. Acad. Sci. (USA) 70, 2439-2442.
Brändén, C.-I., Jörnvall, H., Eklund, H. & Furugren, B. (1975) in The Enzymes (Boyer, P.D. ed.) 3rd ed. Vol XI, pp 103-190, Academic Press, New York.
Brändén, C.-I. (1977) in Alcohol and Aldehyde Metabolizing Systems (Thurman, R.G., Williamson, J.R., Drott, H.R. & Chance, B. eds.) Vol 2, pp 17-32, Academic Press, New York.
Brooks, R.L., Shore, J.D. & Gutfreund, H. (1972) J. Biol. Chem. 247, 2382-2383.

Bryant, T.N., Watson, H.C. & Wendell, P.L. (1974) Nature (London) 247, 14-17.
Buehner, M., Ford, G.C., Moras, D., Olsen, K.W. & Rossmann, M.G. (1973) Proc. Natl. Acad. Sci. (USA) 70, 3052-3054.
Buehner, M., Ford, G.C., Moras, D., Olsen, K.W. & Rossmann, M.G. (1974a) J. Mol. Biol. 82, 563-585.
Buehner, M., Ford, G.C., Moras, D., Olsen, K.W. & Rossmann, M.G. (1974b) J. Mol. Biol. 90, 25-49.
Campbell, J.W., Watson, H.C. & Hodgson, G.I. (1974) Nature (London) 250, 301-305.
Chandrasekhar, K., McPherson, A. Jr., Adams, M.J. & Rossmann, M.G. (1973) J. Mol. Biol. 76, 503-518.
Dunn, M.F. (1975) in Structure and Bonding Vol 23, pp 61-122, Springer Verlag, Berlin and New York.
Dunn, M.F., Biellmann, J.-F. & Branlant, G. (1975) Biochemistry 14, 3176-3182.
Dutler, H. (1977) in Pyridine Nucleotide Dependent Dehydrogenases (Sund, H. ed.) pp 339-350, Walter de Gruyter, Berlin and New York.
Dworschack, R.T., Tarr, G. & Plapp, B.V. (1975) Biochemistry 14, 200-203.
Dworschack, R.T. & Plapp, B.V. (1977) Biochemistry 16, 2716-2725.
Einarsson, R., Eklund, H., Zeppezauer, E., Boiwe, T. & Brändén, C.-I. (1974) Eur. J. Biochem. 49, 41-47.
Eklund, H. & Brändén, C.-I. (1979) J. Biol. Chem. in press.
Eklund, H., Nordström, B., Zeppezauer, E., Söderlund, G., Ohlsson, I., Boiwe, T. & Brändén, C.-I. (1974) FEBS Letters 44, 200-204.
Eklund, H., Nordström, B., Zeppezauer, E., Söderlund, G., Ohlsson, I., Boiwe, T., Söderberg, B.-O., Tapia, O., Brändén, C.-I. & Åkeson, Å. (1976) J. Mol. Biol. 102, 27-59.
Eventoff, W., Hackert, M.L. & Rossmann, M.G. (1975) J. Mol. Biol. 98, 249-258.
Eventoff, W., Rossmann, M.G., Taylor, S.S., Torff, H.-J., Meyer, H., Keil, W. & Kiltz, H.-H. (1977) Proc. Natl. Acad. Sci. (USA) 74, 2677-2681.
Fersht, A. (1977) Enzyme Structure and Mechanism, Freeman, Reading and San Francisco.
Garavito, R.M., Rossmann, M.G., Argos, P. & Eventoff, W. (1977a) Biochemistry 16, 5065-5071.
Garavito, R.M., Berger, D. & Rossmann, M.G. (1977b) Biochemistry 16, 4393-4398.
Hackert, M.L., Ford, G.C. & Rossmann, M.G. (1973) J. Mol. Biol. 78, 665-673.
Harrigan, P.J. & Trentham, D.R. (1973) Biochem. J. 135, 695-703.
Harris, J.I. & Waters, M. (1976) in The Enzymes (Boyer, P.D. ed.) Vol XIII, pp 1-50, Academic Press, New York.
Hill, E., Tsernoglou, D., Webb, L. & Banaszak, L.J. (1972) J. Mol. Biol. 72, 577-591.
Hol, W.G.J., van Duijnen, P.T. & Berendsen, H.J.C. (1978) Nature (London) 273, 443-446.

Holbrook, J.J., Liljas, A., Steindel, S.J. & Rossmann, M.G. (1975) in The Enzymes (Boyer, P.D. ed.), 3rd ed., Vol XI, pp 191-292, Academic Press, New York.
Jörnvall, H., Woenckhaus, C. & Johnscher, G. (1975) Eur. J. Biochem. 53, 71-78.
Jörnvall, H. (1977) Eur. J. Biochem. 72, 443-452.
Jörnvall, H., Eklund, H. & Brändén, C.-I. (1978) J. Biol. Chem. 253, 8414-8419.
Klinman, J.P. (1976) Biochemistry 15, 2018-2026.
Kvassman, J. & Pettersson, G. (1978) Eur. J. Biochem. 87, 417-427.
Lindberg, M., Larsson, P.-O. & Mosbach, K. (1973) Eur. J. Biochem. 40, 187-193.
Matthews, D.A., Alden, R.A., Bolin, J.T., Filman, D.J., Freer, S.T., Hamlin, R., Hol, W.G.J., Kisliuk, R.L., Pastore, E.J., Plante, L.T., Xuong, N. & Kraut, J. (1978) J. Biol. Chem. 253, 6946-6954.
McFarland, J.T. & Chu, Y.-H. (1975) Biochemistry 14, 1140-1146.
Moras, D., Olsen, K.W., Sabesan, M.N., Buehner, M., Ford, G.C. & Rossmann, M.G. (1975) J. Biol. Chem. 250, 9137-9162.
Musick, W.D.L. & Rossmann, M.G. (1979) J. Biol. Chem. in press.
Ohlsson, I., Nordström, B. & Brändén, C.-I. (1974) J. Mol. Biol. 89, 339-354.
Olsen, K.W., Garavito, R.M., Sabesan, M.N. & Rossmann, M.G. (1976a) J. Mol. Biol. 107, 571-576.
Olsen, K.W., Garavito, R.M., Sabesan, M.N. & Rossmann, M.G. (1976b) J. Mol. Biol. 107, 577-584.
Orsi, B.A. & Cleland, W.W. (1972) Biochemistry 11, 102-109.
Parker, D.M. & Holbrook, J.J. (1977) in Pyridine Nucleotide Dependent Dehydrogenases (Sund, H. ed.) pp 485-501, Walter de Gruyter, Berlin and New York.
Plapp, B.V., Brooks, R.L. & Shore, J.D. (1973) J. Biol. Chem. 248, 3470-3475.
Plapp, B.V., Eklund, H. & Brändén, C.-I. (1978a) J. Mol. Biol. 122, 23-32.
Plapp, B.V., Zeppezauer, E. & Brändén, C.-I. (1978b) J. Mol. Biol. 119, 451-453.
Popják, G. (1970) in The Enzymes (Boyer, P.D. ed.) 3rd ed., Vol II, pp 115-215, Academic Press, New York.
Racker, E. & Krimsky, I. (1952) J. Biol. Chem. 198, 731-743.
Rossmann, M.G., Moras, D. & Olsen, K.W. (1974) Nature (London) 250, 194-199.
Rossmann, M.G., Liljas, A., Brändén, C.-I. & Banaszak, L.J. (1975) in The Enzymes (Boyer, P.D. ed.) 3rd ed., Vol XI, pp 61-102, Academic Press, New York.
Saenger, W., Reddy, B.S., Mühlegger, K. & Weimann, G. (1977) Nature (London) 267, 225-229.
Samama, J.-P., Zeppezauer, E., Biellmann, J.-F. & Brändén, C.-I. (1977) Eur. J. Biochem. 81, 403-409.
Schulz, G.E., Elzinga, M., Marx, F. & Schirmer, R.H. (1974) Nature (London) 250, 120-123.

Shore, J.D., Gutfreund, H., Brooks, R.L., Santiago, D. & Santiago, P. (1974) Biochemistry 13, 4185-4191.
Sloan, D.L., Young, J.M. & Mildvan, A.S. (1975) Biochemistry 14, 1998-2008.
Stallcup, W.B. & Koshland, E.E., Jr. (1973) J. Mol. Biol. 80, 41-62.
Suhadolnik, R.J., Lennon, M.B., Uematsu, T. Monahon, J.E. & Baur, R. (1977) J. Biol. Chem. 252, 4125-4133.
Tapia, O., Lamborelle, C. & Johannin, G. (1979) Biochem. Biophys. Res. Comm. in the press.
Theorell, H. & McKinley-McKee (1961) Acta Chem. Scand. 15, 1811-1833.
Wassarman, P.M. & Lentz, P.J., Jr. (1971) J. Mol. Biol. 60, 509-522.
Watson, H.C., Duée, E. & Mercer, W.D. (1972) Nature New Biol. 240, 130-139.
White, J.L., Hackert, M.L., Buehner, M., Adams, M.J., Ford, G.C., Lentz, P.J., Jr., Smiley, I.E., Steindel, S.J. & Rossmann, M.G. (1976) J. Mol. Biol. 102, 759-779
Wills, C. & Jörnvall, H. (1979) Nature (London) in the press.
Woenckhaus, C. & Jeck, R. (1971) Hoppe-Seyler's Z. Physiol. Chem. 352, 1417-1425.
Wonacott, A.J. & Biesecker, G. (1977) in Pyridine Nucleotide-Dependent Dehydrogenases (Sund, H. ed.) pp 140-156, Walter de Gruyter, Berlin and New York.
Zeppezauer, E., Jörnvall, H. & Ohlsson, I. (1975) Eur. J. Biochem. 58, 95-104.

THE SPECIFICITY OF DEHYDROGENASES

Jonathan Jeffery

CONTENTS

SUMMARY

The specificity of dehydrogenases for coenzyme (and coenzyme analogues), and substrate (and substrate analogues) is discussed in relation to structure, function, and evolution.

Examples concern compounds that have very different structures, reactions that play widely differing roles in the life of the organism, and organisms of greatly differing types. The examples illustrate general points of interest and importance.

1. INTRODUCTION

Formally, a substrate binds to an enzyme because the Gibbs free energy of the enzyme-substrate complex in solution is lower than that of the enzyme and substrate separately in solution. The contribution of a particular group in the substrate to the binding energy will depend upon the circumstances of the case. A CH_2 group might contribute 11 kJ/mol (2.6 kcal/mol), an OH group forming one hydrogen bond about 12 kJ/mol (2.9 kcal/mol), and the interaction of an $-NH_3^+$ and a $-CO_2^-$ (a so-called salt bridge) might provide 15 kJ/mol (3.6 kcal/mol). These figures are merely for illustration, but show that when there is appropriate complementarity between substrate and enzyme, the binding energy may be considerable.

It might seem, on first thoughts, that to achieve specificity the substrate binding site of the enzyme should be complementary to the substrate, and bind it as strongly as possible. But an enzyme must not merely bind the substrate, it must catalyse the reaction and release the products. The requirement for optimal catalysis is that binding should be best for the transition state. This carries the implication that transition state analogues may be bound better than the substrate (a kind of inevitable non-specificity), and transition state analogues that bind much more strongly than the substrates are known for some enzymes (Wolfenden, 1972; Lienhard, 1973).

Jencks (1975) developed the idea that energy of interaction between the substrate and the enzyme (" the intrinsic binding energy") may be larger than it is generally thought to be, and that part of it may be utilized to provide the driving force for catalysis, the observed binding energy representing only what is left over from this utilization. It follows from this concept that introduction into a substrate of a group that provides additional binding energy will not necessarily increase the observed binding; the additional intrinsic binding energy may be utilized to cause a local destabilization, conformation change, more-productive binding, or greater loss of entropy in the enzyme-substrate complex, so increasing the reaction rate rather than the substrate binding.

Many enzymic reactions can be represented by an equation of the form of Eq. (1).

$$v = [\text{free enzyme}] \times [\text{free substrate}] \frac{k_{cat}}{K_m} \qquad (1)$$

(The term "free enzyme" does not mean only apoenzyme, but would include holoenzyme when appropriate.) Fersht (1974, 1977) pointed out that a high value of v will result from a high value of k_{cat}/K_m (which requires the enzyme to be complementary to the transition state of the substrate), and a high "free enzyme" concentration (which is achieved when K_m is high in comparison with the concentration of free substrate). It may sometimes be the case that the organism would obtain a selective advantage from a high v, either because a high reaction rate is itself advantageous, or because it allows less enzyme to suffice. In such cases, evolution should favour increasing k_{cat}/K_m and increasing K_m. The maximum value that k_{cat}/K_m could achieve would be the rate constant for the diffusion-controlled encounter of the enzyme and substrate, which is about $10^9\ s^{-1}\ dm^3\ mol^{-1}$. The useful upper limit for K_m is effectively set by the similarity of the structure of the substrate and of the transition state, which presumably makes it inevitable that beyond a certain point, any further change in the structure of the binding site that increases K_m will decrease k_{cat}/K_m. However, increase in K_m above about four times the concentration of free substrate gives only slight benefit, so this aspect need not be unduly restrictive.

The highest possible v may not always be what is most advantageous to an organism. It may be that advantage would lie in having v relatively unaffected by changes in substrate concentration, for example, so that evolution might favour a low K_m. More generally, if regulation is of critical advantage, structures allowing the operation of allosteric effects favourable for control may be selected rather than structures providing only increased catalytic efficiency.

The concept of complementarity between enzyme and transition state (Pauling, 1948; Wolfenden, 1969) is not without some resemblance to older "lock and key" theories (Fischer, 1894). A further development was the suggestion that change in the shape of the enzyme induced by substrate binding may itself create the fully active site (Koshland, 1963). Changes in conformation in horse liver alcohol dehydrogenase that result from NAD^+ binding (Samama _et al_., 1977), and are facilitated by the presence of ethanol (Shore _et al_., 1974) have been described. A conformation change in lactate dehydrogenase when lactate bound to the enzyme-NAD^+ complex was also proposed (Parker & Holbrook, 1977). These changes, which are further discussed by Brändén & Eklund (this book, chapter 2), are closely associated with the mechanism of reaction.

2. SPECIFICITY FOR NAD AND NADP

In vivo, the roles of NADH (in energy production) and

NADPH (in synthetic reactions) are somewhat different, and the ratio NAD^+ / NADH differs from the ratio $NADP^+$ / NADPH . Most dehydrogenases show a marked but incomplete specificity for NAD or NADP, which in some cases is easily rationalized in terms of cellular function. Muscle lactate dehydrogenase and yeast alcohol dehydrogenase, for example, show the considerable specificity for NAD that one would expect from their metabolic role. High specificity of glucose-6-phosphate dehydrogenase for $NADP^+$ is often quoted, and related to the production of NADPH. The specificity of glucose-6-phosphate dehydrogenase does, however, depend upon the source. The enzyme from *Leuconostoc mesenteroides* is active with either NAD^+ or $NADP^+$ (DeMoss et al., 1953), and a glucose-6-phosphate dehydrogenase from *Acetobacter xylinum* is specific for NAD^+ (Benziman & Mazover, 1973). Glucose-6-phosphate dehydrogenases from quite a large number of sources have been studied to some extent, and provide interesting comparisons (see Levy, 1979).

The interactions that give rise to the specificity of dogfish lactate dehydrogenase and horse liver alcohol dehydrogenase for NAD were compared briefly by Dalziel (1975), following the work of Adams *et al*., (1973) and Brändén *et al*. (1973,1975). (Extensive summaries were given by Holbrook *et al*., 1975, and Brändén *et al*., 1975; and, a table that allows comparison also with glyceraldehyde-3-phosphate dehydrogenase is given by Brändén and Eklund, this book, chapter 2). In both these enzymes, the 2'-hydroxyl of the adenosine-ribose is hydrogen bonded to an aspartic acid (asp-53 and asp-223, respectively). The phosphate group in NADP would prevent formation of this bond, steric hindrance would arise from the larger size of the phosphate group, and there would be charge repulsions between the phosphate and aspartic residue. This seems to provide such an adequate basis for specificity for NAD^+ and NADH, that it is perhaps a little surprising that $NADP^+$ and NADPH do serve as coenzymes (Fawcett & Kaplan, 1962; Dalziel & Dickinson, 1965a).

The binding of NADPH to dihydrofolate reductase (from *Lactobacillus casei*) (Matthews *et al*., 1978) is somewhat different both from the binding of NAD to the enzymes discussed above, and from the binding of $NADP^+$ to 6-phosphogluconate dehydrogenase (Abdallah *et al*., 1979). In dihydrofolate reductase, interaction between an arginine residue (arg-43) and the 2'-phosphate of NADPH (Matthews *et al*., 1978) provides an explanation for specificity for the phosphorylated coenzyme. In the case of 6-phosphogluconate dehydrogenase, the coenzyme binds in a cleft on the outside surface of the subunit. The 2'-phosphate of the bound coenzyme is evidently packed somewhat unfavourably against the ribose ring, and it is not yet apparent why $NADP^+$ binds so much better than NAD^+ (Abdallah *et al*., 1979).

3. SPECIFICITY TOWARDS COENZYME ANALOGUES

Colowick *et al*. (1966) tabulated 64 analogues of NAD

of which 24 could serve instead of NAD in at least some dehydrogenase reactions. Coenzyme analogues were reviewed in relation to lactate dehydrogenase (Everse & Kaplan, 1973) and analogues used with alcohol dehydrogenase and lactate dehydrogenase were later discussed in the light of the structures of these enzymes (Brändén et al., 1975; Holbrook et al., 1975; Brändén & Eklund, this book, chapter 2). Marked differences in specificity for coenzyme analogues were found between different enzymes, a striking example being the case of the isobutyryl analogue of NAD^+, which with horse liver alcohol dehydrogenase gave 7.92 times the rate for NAD^+, but with rabbit muscle glyceraldehyde-3-phosphate dehydrogenase gave 0.01 times the rate for NAD^+ (Anderson & Kaplan, 1959). Differences have also been found between isozymes. For horse liver alcohol dehydrogenase, with cinnamyl alcohol as substrate and 2-methyl-NAD^+ as coenzyme, for example, a 10-fold difference in maximum velocity was found (Biellmann & Samama, 1974) between two isozymes (the isozymes IV and V of Lutstorf et al., 1970). In many cases, the ability of an analogue to serve as a coenzyme was associated with the presence at the 3 position of the pyridine ring of a single bond to a carbon atom to which an electronegative element, such as O, S, or N, was attached by a double bond (as, for example, in a thioamide or acetyl group). However, the 3-iodopyridine analogue of NAD^+ was active with liver alcohol dehydrogenase (Abdallah et al., 1976; see also Brändén & Eklund, this book, chapter 2). The 3-iodopyridine analogue of $NADP^+$, was not active with sheep liver 6-phosphogluconate dehydrogenase (Abdallah et al., 1979). Synthesis of a fluorescent analogue of NAD^+, which served as coenzyme for various dehydrogenases (Barrio et al., 1972) afforded interesting possibilities for future work.

In the case of dogfish lactate dehydrogenase, interactions between the enzyme and the AMP portion of NAD^+ apparently generated a binding site for the nicotinamide moiety (McPherson, 1970).

In the case of horse liver alcohol dehydrogenase, interactions involving the nicotinamide moiety evidently facilitate a conformational change involved in productive complex formation (Abdallah et al., 1976).

More generally, the ability of a holoenzyme E.NAD^+ to bind the reduced substrate rather than the oxidized substrate, and of the holoenzyme E.NADH to make the contrary distinction, must depend upon interactions with the nicotinamide moiety. The basis of the distinction between the 3-carboxamido-pyridinium and the 3-carboxamido-dihydropyridine could be the charge difference, or the shape difference, or both. Inhibition studies (Danenberg et al., 1978; Biellmann & Jung, 1971) indicated that interactions with horse liver alcohol dehydrogenase were similar for pyridine adenine dinucleotide and NAD^+ (same shape and same charge), and for benzene adenine dinucleotide and NADH (different shape and same charge), and were different for benzene adenine dinucleotide and NAD^+ (same shape and different charge). However, pyridine adenine dinucleotide did not serve as a coenzyme for horse liver alcohol dehydrogenase, and the binding and associated conformational changes were shown to be of some subtlety (Abdallah *et al.*, 1976), so this limited evidence should be interpreted with caution. Nevertheless, the importance of the charge would be consonant with evidence for lactate dehydrogenase, in which his-138 and glu-140 may provide a charge-sensing mechanism associated with a conformational change (see Holbrook *et al.*, 1975).

The continuing interest in coenzyme analogues (Woenckhaus, 1974; Woenckhaus & Jeck, 1977), reveals some shift of emphasis to the use of analogues with reactive groups for affinity labelling (Fisher *et al.*, 1973; Biellmann *et al.*, 1974a; Dietz *et al.*, 1977; Guillory & Jeng, 1977; Biellmann *et al.*, 1979).

The use of solid-supported coenzyme analogues as affinity media for the purification of dehydrogenases is an extremely attractive prospect. If sufficient work were put into devizing the optimal affinity columns, highly selective purification could, in principle, then be achieved very easily. In practice, the cost of such developments may often be prohibitive, though no doubt favourable cases can be found, and a few useful media are available commercially.

4. STEREOSPECIFICITY OF HYDROGEN TRANSFER WITH RESPECT TO COENZYME

The 4-positions of NADH and NADPH are prochiral centres. Dehydrogenases show specificity for the 4-*pro*-*R* position (A-side) or the 4-*pro*-*S* position (B-side) (Fisher *et al.*, 1953), and this specificity(unlike the specificity for NAD or NADP) is usually represented as being absolute. (A useful compilation of 130 enzymes was given by You *et al.*, 1978). In many cases, however, it is not clear that the specificity of oxidoreductase reactions was, in fact, complete, though the findings allowed that it might have been (e.g. Krackow *et al.*, 1963). Flavin mononucleotide oxidoreductase from *Beneckea harveyi* catalyses the transfer of hydrogen from NADH to FMN. Riboflavin and 5-deazariboflavin can serve in place of FMN, and with these as hydrogen acceptors the hydrogen appeared to be transferred mainly but not entirely from the 4-pro-R position (Fisher & Walsh, 1974). In a footnote, the authors commented that the ^{3}H-NADH used may not have been chirally pure, or that there may have been "nonenzymatic deproportionation" between oxidized and reduced nicotinamide coenzymes. Orcinol hydroxylase (EC 1.14.13.6) transferred hydrogen mainly from the 4-*pro*-*R* position of NADH when orcinol was substrate, but the transfer was evidently less specific when resorcinol or *m*-cresol was added instead of orcinol, and there appeared to be a large isotope effect so that the steric course was thought to be influenced by the identity of the isotope being transferred (Ribbons et al., 1972).

The stereospecificity of hydrogen transfer with respect to coenzyme is evidently determined by the structure of the enzyme. In dogfish lactate dehydrogenase, horse liver alcohol dehydrogenase, and lobster muscle glyceraldehyde-3-phosphate dehydrogenase, hydrogen bonds with the carboxamide group of NAD^+ "anchor" the nicotinamide ring in a particular orientation (see Brändén & Eklund, this book, chapter 2, Table 3 and related text), aligning the 4-pro-R position (A-side) for hydrogen transfer in the first two cases, and the 4-pro-S position (B-side) for hydrogen transfer in the case of the glyceraldehyde-3-phosphate dehydrogenase. The steric course was shown to be the same with the thioamide analogue, the acetyl analogue, and the nitrile analogue of NAD^+ for horse liver and yeast alcohol dehydrogenase (both 4-pro-R specific) and glutamate dehydrogenase (4-pro-S specific) (Biellmann *et al.*, 1974 b). Hydrogen bond formation with the carboxamide of NAD^+ cannot, therefore, be the sole basis of the stereospecificity. In the case of horse liver alcohol dehydrogenase, steric hindrance in which val-203 may be important probably prevents binding of the coenzyme analogues with the 4-pro-S position facing the active site zinc (see Brändén *et al.*, 1975). Regarding "absolute" differences, it should be borne in mind that every 17.6 kJ/mol (4.2 kcal/mol) difference in activation energy between the 4-pro-R reaction and the 4-pro-S reaction could give up to about 1000-fold difference in rate, and there are limits (which depend upon the technique used) to what can be distinguished.

5. EMPIRICAL GENERALIZATIONS CONCERNING THE STEREOSPECIFICITY OF HYDRIDE TRANSFER WITH RESPECT TO COENZYME

Vennesland (1961) on the basis of 40 reactions of which the steric course was then known, made a number of generalizations. Although she stated that they were "by no means final", most have stood the test of time remarkably well. There have, of course, been developments (Bentley, 1970; Popjak, 1970; Vennesland, 1974; Alizade & Brendel, 1975; You *et al.*, 1978).

When an enzyme utilizes a range of substrates, or utilizes NAD and NADP, the steric course with respect to the coenzyme does not vary. The implication is presumably that no substrate causes the nicotinamide ring of either coenzyme to turn over, and no enzyme has an alternative binding site for some substrates on the other side of the bound coenzyme. This does not seem surprising. However, for most of the empirical generalizations, no explanation of this kind is apparent. For example, the steric course of a particular reaction with respect to coenzyme is usually the same whatever the source of the enzyme, (except for some dehydrogenases involved in electron transport processes). The exceptions are varied, but not numerous. The transhydrogenases (EC 1.6.1.1) from *Pseudomonas aeruginosa* (Louie & Kaplan, 1970) and

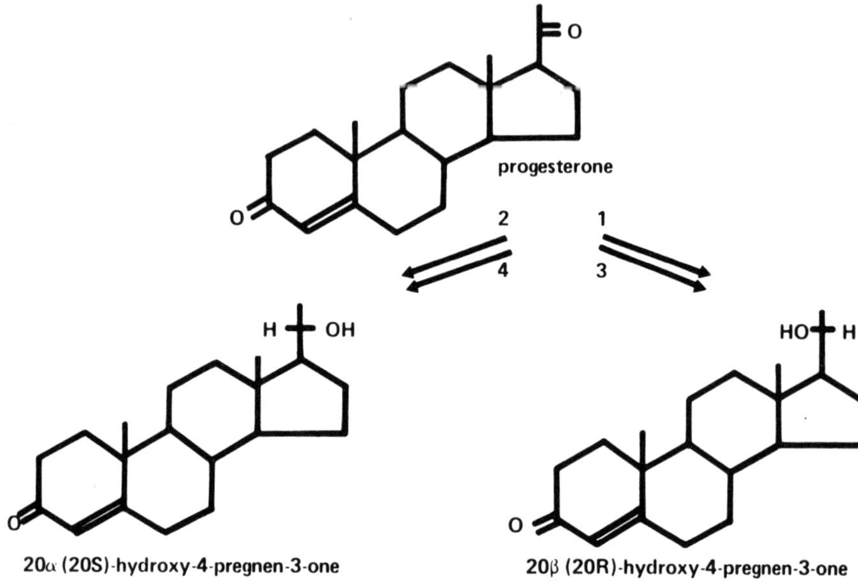

Fig. 1 The four possible courses of enzymic reduction of the 20-oxo group of progesterone to a 20-alcohol.

(1) transfer of the 4-pro-R hydrogen of the coenzyme to the 20α-position; this is not known.

(2) transfer of the 4-pro-R hydrogen of the coenzyme to the 20β-position; enzymes from rat ovary and pig testis do this.

(3) transfer of the 4-pro-S hydrogen of the coenzyme to the 20α-position; an enzyme from Streptomyces hydrogenans does this.

(4) transfer of the 4-pro-S hydrogen of the coenzyme to the 20β-position; an enzyme from human placenta does this.

Chemical reduction of the 20-oxo group with BH_4^- or AlH_4^- gives mainly the 20β-alcohol.

Azotobacter vinelandii (Middleditch & Chung, 1971) are 4-pro-S specific for NADH and NADPH; the bovine heart mitochondrial enzyme is 4-pro-S specific for NADPH, and 4-pro-R specific for NADH (Hoek et al., 1974). Nitrate reductase (EC 1.6.6.2) from Chlorella vulgaris is 4-pro-R specific (Guerrero & Vennesland, 1975), whereas that from Candida utilis (EC 1.6.6.3) is 4-pro-S specific (Davies et al., 1972). The 2-oxopantoyl-lactone reductase (EC 1.1.1.168) of Saccharomyces cerevisiae (4-pro-R specific for NADPH) and Escherichia coli (4-pro-S specific for NADPH) (Wilken et al., 1975) will constitute another exception if the product (2-hydroxypantoyl lactone) has the same chirality in both cases; this appears not to have been established. There are also enzymes (EC 1.1.1.2) with similar though not identical substrate specificities, which transfer hydrogen from NADPH to various aldehydes and ketones, using in many cases the 4-pro-R hydrogen (Flynn et al., 1975; Feldman et al., 1977), but the 4-pro-S hydrogen in the case of an enzyme from rabbit kidney cortex (Culp & McMahon, 1968). Ribitol oxidation by ribitol dehydrogenase (EC 1.1.1.56) from Klebsiella pneumoniae (4-pro-S specific for NAD^+) and by sorbitol dehydrogenase (EC 1.1.1.14) from sheep liver (4-pro-R specific for NAD^+) is a more recently-found exception (Alizade et al., 1976). Other exceptions are evident from Figs. 1-3. In the great majority of cases, however, the steric course with respect to the coenzyme is the same whatever the source of the enzyme.

This suggests the possibility that there might be some compelling steric requirement for a particular dehydrogenase reaction, so that in almost every case only enzymes using a particular side of the coenzyme could be evolved. The reduction of pyruvate with L-lactate dehydrogenase (EC 1.1.1.27) by transfer of the 4-pro-R hydrogen to give L-lactate (S-lactate), and the reduction of pyruvate with D-lactate dehydrogenase (EC 1.1.1.28) by transfer of 4-pro-R hydrogen to give D-lactate (R-lactate) suggest no such requirement. However, this would be a poor case to test such a proposition. No lactate dehydrogenase is known that transfers the 4-pro-S hydrogen to pyruvate to give either L-lactate or D-lactate. Moreover, pyruvate is prochiral but symmetrical, and itself imposes on the steric course of reaction only such constraints as arise from its interaction with a chiral reagent; the chemical reduction of the oxo group of pyruvate with non-asymmetric reagents gives racemic lactate. Some dehydrogenase reactions for which, on the basis of analogous chemical reactions, steric influences could be expected to be important, are summarized in Figs. 1-3. These reactions give no indication that any steric requirement has limited evolution to the use of a particular side of the coenzyme.

Two other empirical generalizations are that the oxidation of an alcohol carbon atom adjacent to a carboxyl group occurs with transfer of hydrogen to the 4-pro-R position of the coenzyme,(Vennesland, 1961) and that dehydrogenases that act on phosphorylated carbohydrate derivatives are 4-pro-S specific (Vennesland, 1974). There are very few exceptions to either of these two generalizations. An exception to the

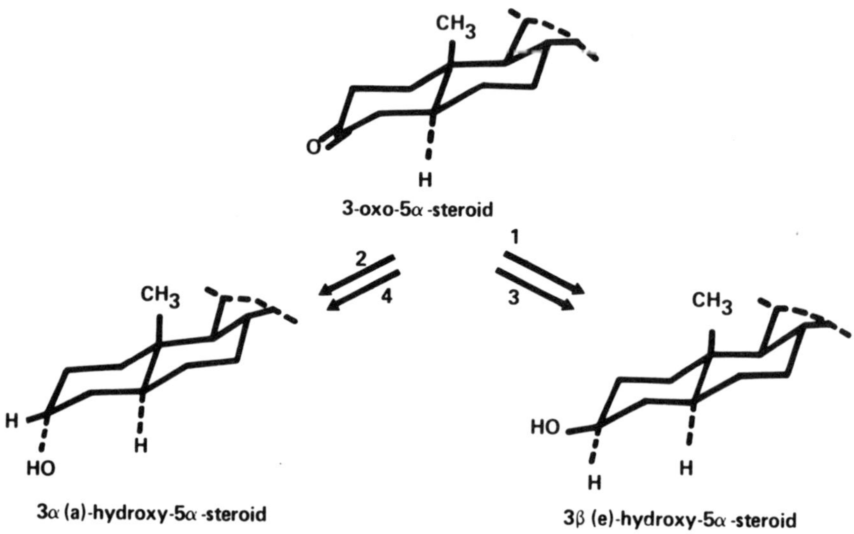

Fig. 2 The four possible courses of enzymic reduction of the 3-oxo group to a 3-alcohol in a 5α-steroid are:

(1) transfer of 4-pro-R hydrogen of the coenzyme to the 3α (a)-position; certain horse liver alcohol dehydrogenase isozymes do this.

(2) transfer of 4-pro-R hydrogen of the coenzyme to the 3β (e)position; a rat liver enzyme does this.

(3) transfer of 4-pro-S hydrogen of the coenzyme to the 3α (a)-position; an enzyme from Pseudomonas testosteroni does this.

(4) transfer of 4-pro-S hydrogen of the coenzyme to the 3β (e)-position; an enzyme from Pseudomonas testosteroni and an enzyme from Streptomyces hydrogenans do this.

Chemical reduction with BH_4^- or AlH_4^- gives mainly the 3β (e)-alcohol.

former is 2-oxopantoate reductase (EC 1.1.1.169), which is 4-pro-S specific (Wilken *et al*., 1975). Phosphoglycerate dehydrogenase (EC 1.1.1.95) must be an exception to one or other of these generalizations, because they are mutually exclusive and it falls into both categories. In fact it is 4-pro-R specific (Slaughter & Davies, 1968; Winicov, 1975), and so is an exception to the latter generalization.

Another generalization of wide validity is that when the natural (physiological) substrate of a dehydrogenase is a non-phosphorylated organic compound having three or fewer carbon atoms, the reaction is 4-pro-R specific with respect to coenzyme (You *et al*., 1978). Of 17 enzymes of known specificity, that fall within this category, 16 are 4-*pro*-*R* specific, and one is 4-*pro*-*S* specific. This one exception is an enzyme from *Mucor javanicus* (EC 1.1.1.156) which is able to utilize quite large substrates, but is believed to serve the function *in vivo* of reducing dihydroxyacetone (Hochuli *et al*., 1977).

Because no reason related to steric requirements of the reactions is evident (see, for example, earlier discussion in this section), the possibility must be considered that the broad validity of the better of the empirical generalizations depends upon retention of the prochiral specificity of a common ancestor. Thus, a suggestion made by You *et al*., (1978) can be elaborated as follows. Prebiotic chemical processes may be thought likely to have produced small non-phosphorylated organic substances in greater abundance than more complex molecules. The first organisms seem likely to have had rudimentary metabolic systems able to utilize whatever substrates were relatively abundant at that time, that is, small non-phosphorylated organic substances. Most present-day dehydrogenases that have non-phosphorylated organic substrates with three or fewer carbon atoms as their natural (physiological) substrates are 4-pro-R specific. So the suggestion is that the 4-*pro*-*R* specific dehydrogenases evolved first, and the 4-*pro*-*S* specific dehydrogenases later. Such a view could be considered broadly consistent with the view of Rossmann *et al*. (1977) that the nucleotide binding domains of all dehydrogenases show a divergent evolutionary relationship, and that there has been convergence of catalytic residues at the active sites. The difficulty here, however, is that whether a particular dehydrogenase is 4-*pro*-*R* specific or 4-*pro*-*S* specific would in itself appear to make no difference whatsoever to the organism. (This is in contrast to specificity with respect to NAD or NADP, which might be expected to affect the organism; see Section 2). Insofar as dehydrogenases that fall within the validity of some of the empirical generalizations are related extremely remotely, it is not apparent why the prochiral specificity of hydrogen transfer to the coenzyme should have been conserved. It will be interesting to learn, as more structures are elucidated in sufficient detail, whether the orientation of the nicotinamide moiety is generally *anti* in the 4-*pro*-*R* specific dehydrogenases, and *syn* in those that are 4-*pro*-*S* specific (You *et al*., 1978).

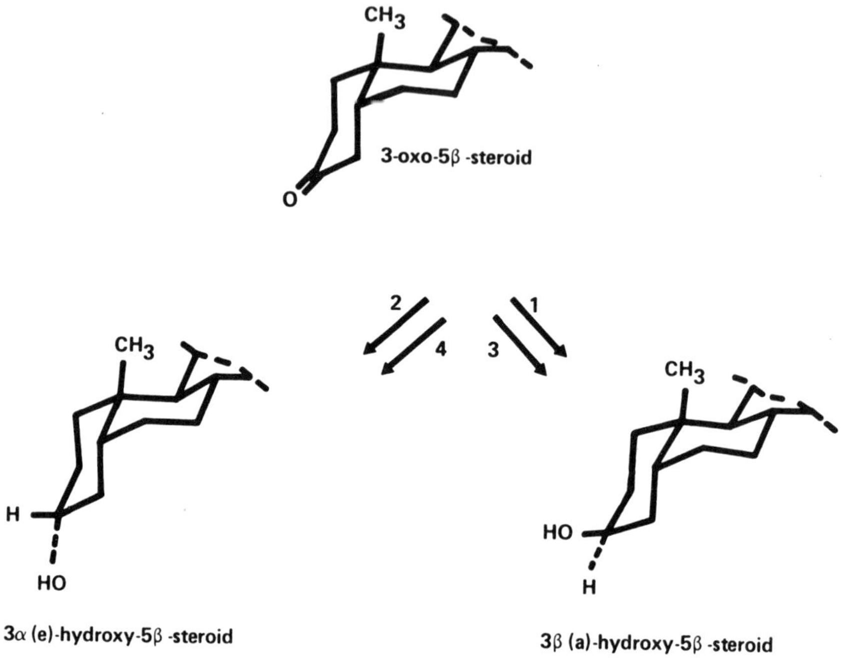

Fig. 3 The four possible courses of enzymic reduction of the 3-oxo group to the 3-alcohol in a 5β -steroid.

(1) transfer of the 4-pro-R hydrogen of the coenzyme to the 3α (e)-position; rat liver alcohol dehydrogenase does this.

(2) transfer of the 4-pro-R hydrogen of the coenzyme to the 3β (a)-position; a rat liver enzyme does this.

(3) transfer of the 4-pro-S hydrogen of the coenzyme to the 3α (e)-position; an enzyme from Pseudomonas testosteroni does this.

(4) transfer of the 4-pro-S hydrogen of the coenzyme to the 3β (a)-position; an enzyme from Pseudomonas testosteroni does this.

Chemical reduction with BH_4^- or AlH_4^- gives mainly the 3α (e)-alcohol.

This would allow the altered steric course to arise from rotation of the nicotinamide moiety through 180° about the nicotinamide glycosyl bond, without necessitating radical alteration of binding to the ADPR moiety (Fig. 4).

Several reaction sequences are known in which there are successive dehydrogenase reactions.

Examples are as follows:

1. ethanol, acetaldehyde, acetic acid
2. 5β-cholestane-3α, 7α, 12α, 26-tetrol; 3α, 7α, 12α-trihydroxy-5β-cholestan-26-al; 3α, 7α, 12α-trihydroxy-5β-cholestan-26-oic acid
3. D-xylulose, xylitol, L-xylulose
4. mesotartrate, oxaloglycollate, glycerate
5. D-glucose, sorbitol, D-fructose
6. D-mannitol-1-phosphate, D-fructose-6-phosphate, D-glucitol-6-phosphate
7. D-glucose-6-phosphate, 6-phosphoglucono-δ-lactone, (6-phosphogluconate), ribulose 5-phosphate
8. L-aspartate β-semialdehyde, L-β-aspartyl phosphate, L-homoserine
9. nitrate, nitrite, ammonia, hydroxylamine

In each of these nine sequences, the steric course of hydrogen transfer to the coenzyme is the same for the whole sequence. Sequences 1-5 are 4-*pro*-*R* specific, and sequences 6-9 are 4-*pro*-*S* specific (Davies *et al*., 1972; Fukuba, 1974; Do Nascimento & Davies, 1975; Alizade *et al*., 1974; Feldman *et al*., 1977). In a reaction sequence, the product of one enzyme is the substrate of the next. It is not difficult to imagine that in sequences like the above, a small change in structure affecting the substrate-binding site of the first dehydrogenase might produce a protein that catalysed the second reaction, and one can conceive of the evolution of the sequence by gene duplication and subsequent mutation. The consistent prochiral specificity would arise because this involved least change rather than because there was any particular requirement for it. This line of reasoning seems attractive because it suggests a basis for a unifying hypothesis to explain some of the empirical generalizations. It would, however, seem to require a close structural relationship between all the dehydrogenases of a particular sequence. There is insufficient evidence for an authoritative comment on this, but such close structural similarities are not evident at present.

6. SUBSTRATE SPECIFICITY

6.1 General

The most valuable single item of information for the understanding of the specificity of a dehydrogenase may be imagined to be the full 3-dimensional structure of the productive complex, such as the productive ternary complex of enzyme, coenzyme, and substrate. There are not many cases in which anything approaching this is available. Moreover, when

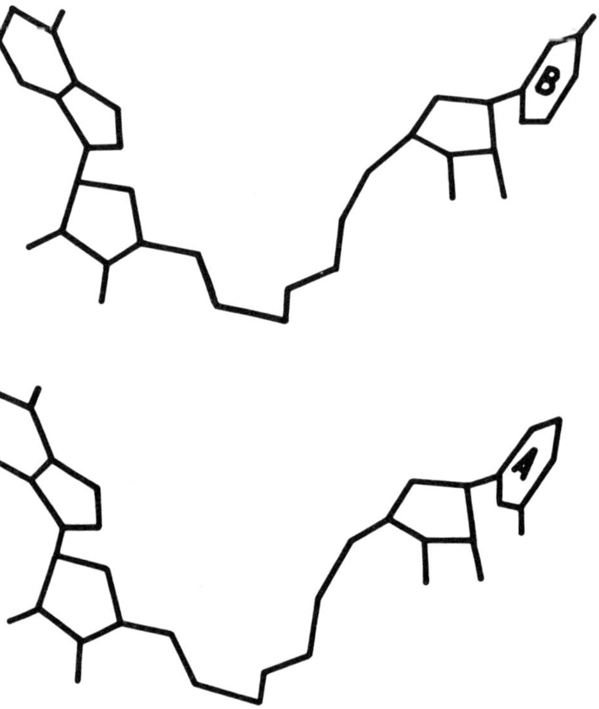

Fig. 4 An indication of the conformation that bound NAD^+ is thought to have in various dehydrogenases. The nicotinamide ring is shown on the right, with the side indicated facing upwards and to the left. In the 4-pro-R specific dehydrogenases, the conformation of the nicotinamide about the glycosidic bond is anti (top diagram), and in the 4-pro-S specific dehydrogenases it is syn (bottom diagram), according to evidence so far available.

coenzyme-induced or substrate-induced conformational changes occur (which seems commonly to be the case), even the structure of, say, the productive ternary complex can reveal only part of the story. One would wish to see the 3-dimensional structure of the apoenzyme, and of the binary complexes, for example. And one would wish to see all of these things for at least a selection of substrates as different in structure as possible, so that one might assess what adjustments were possible in the enzyme structure. Yet even this quite impracticable list of requirements would very probably not make it possible to say (without previous knowledge) which substrates would differ in reaction rate by a factor of a thousand, or whether most of the substrates would have about the same maximum velocity (because the rate at which the coenzyme left the binary complex was actually rate-limiting). To understand how the specificity comes about, knowledge of the structure is essential. Unfortunately, a very detailed knowledge of the structure is needed to give even a slight understanding of the mechanism of the specificity. It is the kinetics that tell us what the specificity in fact is.

6.2 Hopes and expectations

Well-planned studies usually centre either upon a known substrate, analogues of which(containing small structural modifications)are then tested; or, they centre upon compounds known to be present in the cell, one or other of which it is hoped will turn out to be an important natural substrate. There is always the possibility (particularly when, as is usually the case, the 3-dimensional structure of the enzyme is not known) that a compound markedly different from known substrates might be a good substrate, perhaps even the natural substrate. The following illustrates some of these points.

Aldehyde reductases from various tissues utilize aldehydes such as propionaldehyde and D-glyceraldehyde. 2-Hydroxybenzaldehyde, pyridine-3-carboxaldehyde, and 3-nitrobenzaldehyde are, in fact much better substrates for some aldehyde reductases (e.g. from brain, see Ris & von Wartburg, 1973).

Rabbit liver cytosol contains an enzyme designated 3-hydroxyhexobarbital dehydrogenase (Toki & Tsukamoto, 1964), which must obviously serve some function other than merely oxidizing 3-hydroxyhexobarbital.

17β-Hydroxy-5β-androstan-3-one (Fig. 5) is a good substrate for an aldehyde reductase from rat liver (Pietruszko & Chen, 1976). The product, namely 5β-androstane-3α, 17β-diol, is a good substrate for rabbit liver 3-hydroxyhexobarbital dehydrogenase, giving the corresponding 17-oxo compound (Fig. 5). (Takenoshita & Toki, 1978).

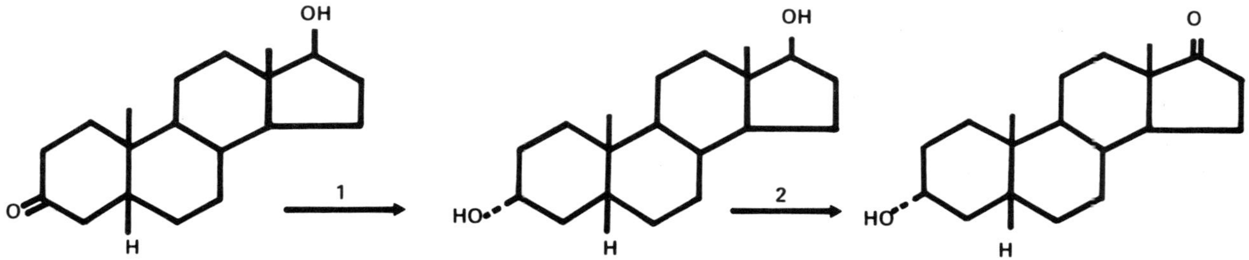

Fig. 5 Reaction 1 is catalysed by rat liver aldehyde reductase using NADPH or NADH. Reaction 2 is catalysed by rabbit liver 3-hydroxyhexobarbital dehydrogenase using $NADP^+$ or NAD^+. These reactions illustrate the point that some dehydrogenases utilize compounds considerably different from their previously known substrates, and catalyse reactions which they might not have been expected to catalyse (see text, Section 6.2, and 7.1 - 7.4 with Figs. 6 and 7)

7. STUDIES OF SUBSTRATE SPECIFICITY: SELECTED EXAMPLES

7.1 L-Lactate dehydrogenase (EC 1.1.1.27)

This enzyme catalyses the transfer of the 4-*pro*-*R* hydrogen from NADH to an α-keto acid substrate by *re* attack, and the reverse reaction, between NAD^+ and α(*S*)-hydroxy acid. The mechanism is discussed by Brändén & Eklund (this book, chapter 2). Of various α-keto acids, pyruvate had the highest value of V/K_m (Meister, 1950; 1952; Holbrook & Stinson, 1973). From models of the structure deduced from X-ray crystallography, it seems that a gap between arginine-101 and glycine-196 allows the "tail" of larger substrates to be accommodated, though in most cases the "tail" generates interactions unfavourable for reaction. There are marked differences between isozymes in specificity for α-keto acids such as α-ketoglutarate, 2-oxobutyrate, and 2-oxovalerate (Schatz & Segal, 1969; Hawtrey & Goldberg, 1973). The alkyl group (methyl in pyruvate) is not essential; glyoxalate is reduced to glycollate in the presence of NADH (Meister, 1952; Romano & Cerra, 1969; Warren, 1970). Presumably it is the non-hydrated oxo form of glyoxalate that serves as substrate, as is the case for pyruvate (Pocker *et al.*, 1969; Hegazi & Meany, 1972; Tienhaara & Meany, 1973).

$$\mathrm{NADH} + \underset{\displaystyle CO_2^-}{\overset{\displaystyle H}{C}} = O + H^+ \rightleftharpoons \mathrm{NAD}^+ + H - \underset{\displaystyle CO_2^-}{\overset{\displaystyle H}{C}} - OH$$

Glyoxalate in aqueous solution exists mainly as the hydrate (Debus, 1904), and the hydrate bears a resemblance to lactate (a hydroxyl replacing the methyl group) rather than pyruvate. Presumably it is for this reason that glyoxalate serves as a substrate in the presence of NAD^+, giving oxalate (Sawaki *et al.*, 1967;

$$\mathrm{NAD}^+ + HO-\underset{\displaystyle CO_2^-}{\overset{\displaystyle H}{C}}-OH \longrightarrow \underset{\displaystyle CO_2^-}{CO_2H} + \mathrm{NADH} + H^+$$

Duncan & Tipton, 1969; Warren, 1970). This reaction is virtually irreversible because at pH values at which the enzyme is active, the concentration of half-ionized oxalic acid (which would be the substrate for the reverse reaction) is negligible.

7.2 Alcohol dehydrogenase (EC 1.1.1.1) as an aldehyde dehydrogenase (cf. EC 1.2.1.3)

Exactly analogous to the oxidation of glyoxalate to oxalate by lactate dehydrogenase (see above) is the oxidation of aldehydes to the corresponding acids by liver alcohol dehydrogenase. Acetaldehyde, which is about 55%

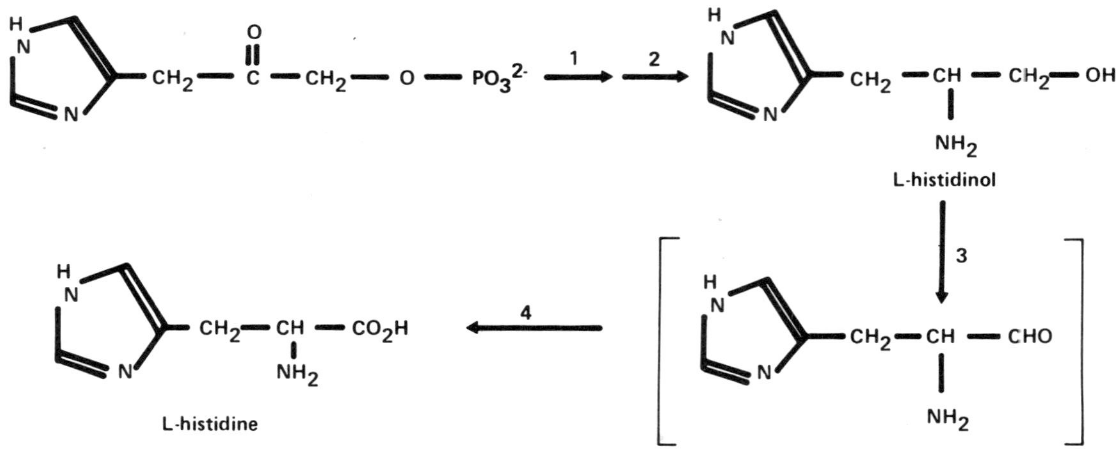

Fig. 6 Dutler's group has shown that horse liver alcohol dehydrogenase with NAD^+ can catalyse the oxidation of L-histidinol to L-histidine (reactions 3 and 4, above), and has suggested that liver alcohol dehydrogenase with pyridoxal or pyridoxal phosphate might possibly also catalyse reactions 1 (transamination) and 2 (hydrolysis of phosphate ester) (Ambar et al., 1979).

hydrated in aqueous solution, is oxidized to acetic acid (Dalziel & Dickinson, 1965b); formaldehyde, which is almost entirely hydrated, is oxidized to formic acid, and reacts faster than acetaldehyde (Abeles & Lee, 1960). In the case of formaldehyde, a further phenomenon occurs in reaction mixtures in which the reaction is allowed to proceed for some time. The enzyme-catalysed reaction between NAD^+ and (hydrated) formaldehyde gives NADH and formic acid; the enzyme then catalyses the reaction of NADH and(non-hydrated)formaldehyde to give NAD^+ and methanol; these alternating oxidations and reductions of formaldehyde are facilitated by the circumstance that the slow step of coenzyme dissociation from the enzyme-coenzyme binary complex need not occur. The methanol and (non-hydrated) formaldehyde will then react chemically (non-enzymically) to form some methyl hemiacetal. Methyl hemiacetal is an excellent substrate for alcohol dehydrogenase and NAD^+, giving methyl formate (Gupta, 1970).

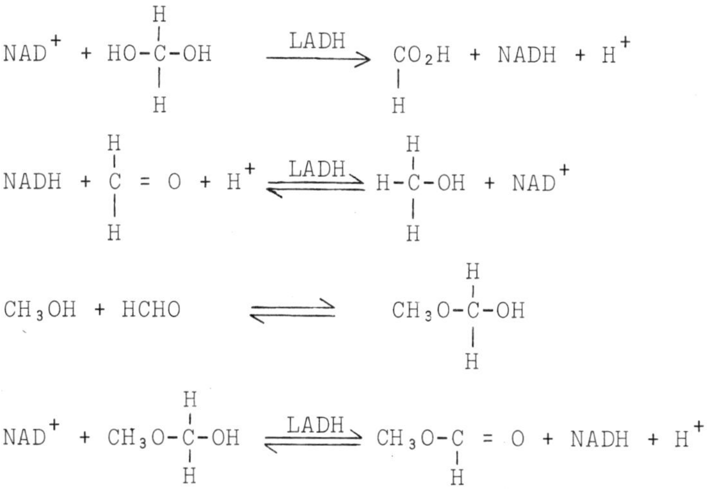

Regarding the formation of the hemiacetal in the general case of an aldehyde RCHO, the equation would be:

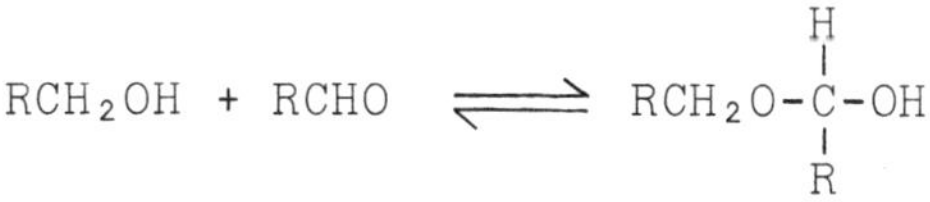

The hemiacetal would be asymmetric at C-1, and the chemical reaction would give both epimers of hemiacetal (in equal amounts unless R were asymmetric).

Octanal was a good substrate for horse liver alcohol dehydrogenase and NAD^+, having a K_m value about the same as acetaldehyde, but a V value 72 times that of acetaldehyde (Hinson & Neal, 1972). Presumably dissociation of NADH from the binary complex, E.NADH, is not rate-limiting for the oxidation of acetaldehyde to acetic acid as it is for the oxidation of ethanol to acetaldehyde.

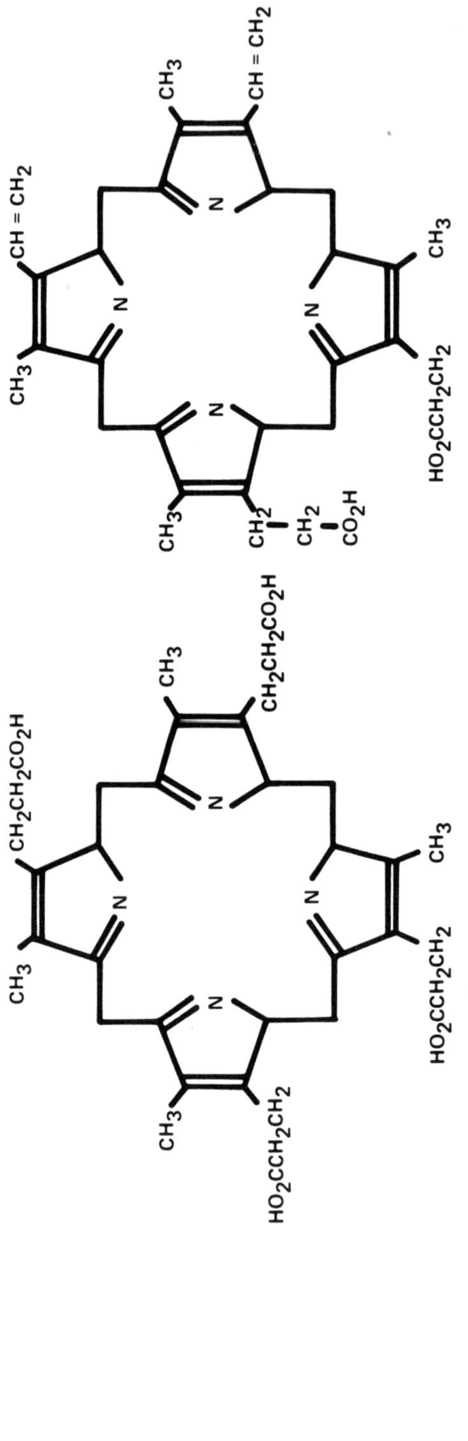

coproporphyrinogen III

protoporphyrinogen IX

Fig. 7 Dutler's group put forward evidence that liver alcohol dehydrogenase can participate with NAD^+ and pyridoxal phosphate in the oxidative decarboxylation reaction of porphyrin synthesis, in which propionic acid moieties are converted into vinyl groups (Ambar et al., 1979).

7.3 Substrate specificity of alcohol dehydrogenases and their function in vivo.

In cells such as yeasts, which produce acetaldehyde from pyruvate, the function of alcohol dehydrogenase is regarded as being the reduction of acetaldehyde and regeneration of NAD^+.

The function of liver alcohol dehydrogenase may be the oxidation of ethanol and other alcohols absorbed from the gastro-intestinal tract (Krebs & Perkins, 1970). However, other functions may also be possible, such as the oxidation of the 26-hydroxy group during bile acid synthesis (for example in 5β-cholestane-3α, 7α, 12α, 26-tetrol) (Waller et al., 1965; Okuda & Takigawa, 1970), the oxidation of the omega-hydroxy group in omega-hydroxy fatty acids (Björkhem et al., 1973; 1974), the interconversion of retinol and retinal (Bliss, 1951; Reynier, 1969), the reduction of D-glyceraldehyde (during D-fructose metabolism) (Holtzer et al., 1955), the oxidation of allylic alcohols such as 3-methyl-2-buten-1-ol, geraniol (Christophe & Popjack, 1961) and farnesol (Waller, 1965) formed from isoprenyl and polyisoprenyl phosphates. 3-Methyl-2-buten-1-ol is, in fact, an extremely good substrate for human liver alcohol dehydrogenase (K_m 4.5×10^{-6}M, V 13, V/K_m 2.9×10^6 compared with K_m 400×10^{-6}M, V 10.7, V/K_m 0.027×10^6 for ethanol) (Pietruszko et al., 1973), though a very poor substrate for the yeast enzyme (K_m 17200×10^{-6}M, V 43, V/K_m 2500, compared with ethanol K_m 17800×10^{-6}M, V 6776, V/K_m 381000) (Pietruszko et al., 1973). Differences in the specificity of alcohol dehydrogenases from different sources, and between different isozymes are well known. It seems reasonable to suppose that these confer some advantage upon the organism. Alcohol dehydrogenases from mammals, yeast, and bacteria appear to have evolved from a common ancient origin, diverging greatly, and conserving tertiary structure more than primary structure (Eklund et al., 1976b; Jörnvall, 1977, and this book, chapter 4). Prelog (1964) commented (in another connection) that the reduction of a carbonyl group is one of the commonest steps in the biogenesis of natural products and that a multipurpose oxidoreductase would be "very economical".

7.4 Liver alcohol dehydrogenase as a histidinol dehydrogenase (EC 1.1.1.23) and as a coproporphyrinogen oxidase

Evidence was recently put forward (Ambar et al., 1979) that horse liver alcohol dehydrogenase will catalyse the oxidation of histidinol to histidine via a tightly-bound histidinal intermediate (Fig. 6). The preceding steps in histidine biosynthesis are transamination of 3-(4-imidazolyl)-2-oxopropanol phosphate to histidinol phosphate, and hydrolysis of the histidinol phosphate to histidinol (Fig. 6). It was thought possible that liver alcohol dehydrogenase might also participate in these reactions. Pyridoxal and pyridoxal phosphate bind to an active site lysine (lys-228) (Sogin & Plapp, 1975) which is what might be expected for a transaminase,

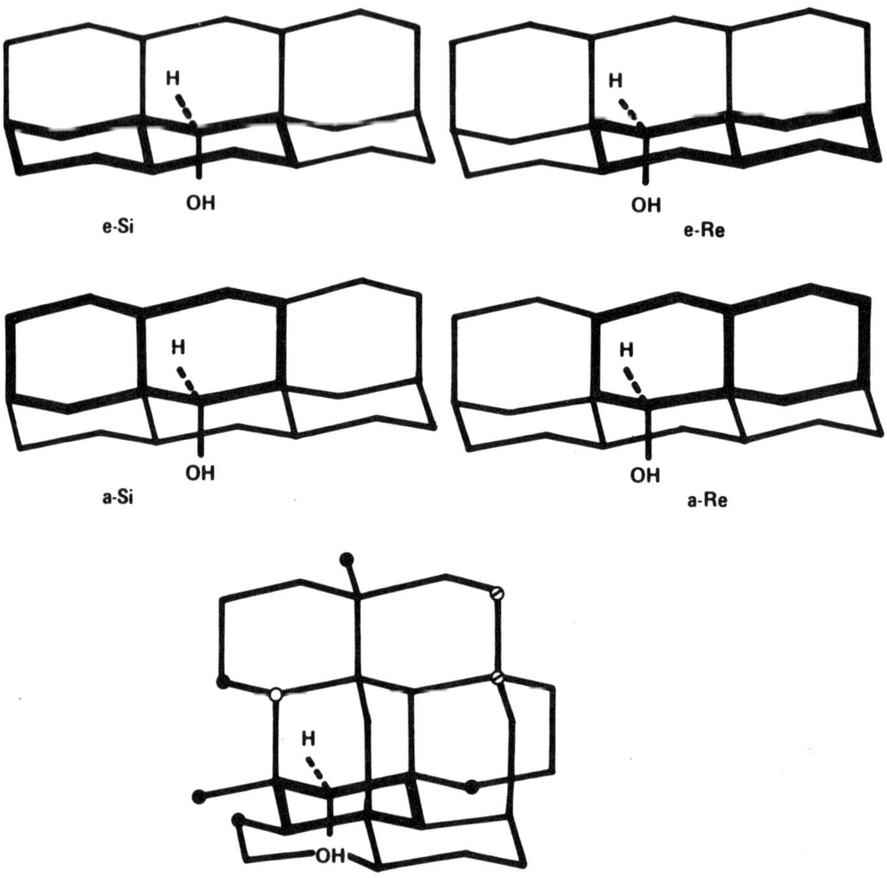

Fig. 8 The specificity of a dehydrogenase can often be represented by a characteristic diamond lattice section. The top four diagrams illustrate four types of stereospecificity. The top left diagram represents hydride attack from the Si side of the carbonyl along an equatorial direction, giving the R(a) alcohol (Dutler et al., 1977). Horse liver alcohol dehydrogenase has e-Re specificity (top right), and an extended lattice section representing its specificity (Irwin & Jones, 1976) is shown (bottom diagram). If any part of the substrate occupies one of the lattice positions indicated, the reaction will be retarded to a greater (●), intermediate (⊘), or lesser (○) extent.

and this binding does not destroy the dehydrogenase activity (it actually increases the activity towards ethanol).

Evidence has also been put forward (Ambar _et al._, 1979) that horse liver alcohol dehydrogenase can participate (with NAD^+ and pyridoxal phosphate) in the oxidative decarboxylation reaction of porphyrin synthesis in which propionic acid moieties are converted into vinyl groups. This is an unusual type of reaction. No evidence was put forward regarding alcohol dehydrogenases from plants, but the possibility obviously suggests itself that if they too can participate in such a reaction, their function may be related to chlorophyll synthesis.

Insufficient evidence is available for the significance of these findings to be properly assessed at present. However, the findings do suggest the possibility that liver alcohol dehydrogenase may not be, as some may have thought, a simple oxidoreductase with an unnecessarily large active site for ethanol, but may be a multifunctional enzyme catalysing diverse reactions with exquisite economy of structure. Perhaps different forms (of the isozyme type) perform the different functions with greatly differing facility; this could simplify the problems of control. More evidence is needed before further discussion is appropriate.

7.5 Definitive descriptions of specificity

It will be clear from the above discussion (especially Sections 7.2, 7.3, and 7.4) that it may be difficult to define the specificity of a dehydrogenase. For attempts to characterize the specificity, the advantage of using relatively rigid substrates, the shapes of which are quite clearly defined, is at once apparent. Alicyclic compounds have proved particularly useful. Prelog, Dutler, and co-workers studied enzymes from _Curvularia falcata_, _Mucor javanicus_, and pig liver using decalones, and decalinols. For each substrate, the preferred alcohol could be characterized as of _R_ or _S_ configuration, and axial or equatorial with respect to the ring. All three of these enzymes were 4-_pro_-_S_ specific for NADP. The enzyme from _Curvularia falcata_ favoured the equatorial alcohol of _R_ configuration, the enzyme from pig liver favoured the axial alcohol of _R_ configuration, and the enzyme from _Mucor javanicus_ favoured the axial alcohol of _S_ configuration. These specificities could be represented in each case by a characteristic diamond lattice section (Prelog, 1964) (Fig. 8). Use of this concept (characteristic diamond lattice section) enables the stereospecificity of a dehydrogenase to be represented with economy and precision. Moreover, the diamond lattice could be extended in all directions around the characteristic section, and occupancy of particular lattice positions by particular groups (methyl carbon, hydroxyl oxygen, etc.) could be related to substrate behaviour in attempts to produce a more general description of the specificity.

Cyclohexanone was a good substrate for horse liver alcohol dehydrogenase. Using alkylcyclohexanones, the influence of alkyl substituents was compared, and from the kinetic

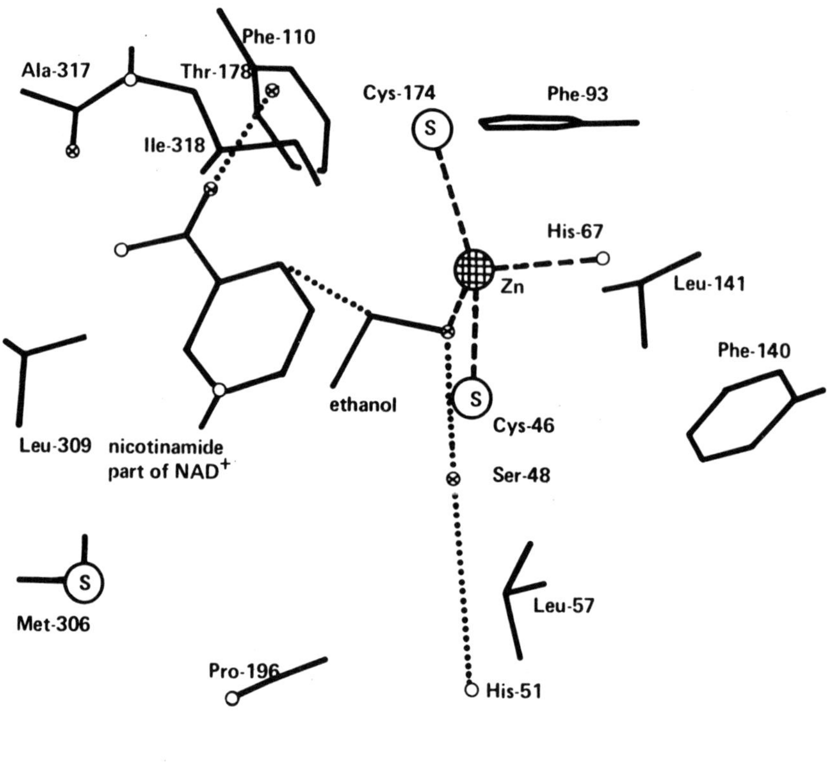

Fig. 9 The small ethanol molecule is bound in the large active site of horse liver alcohol dehydrogenase so that specific transfer of hydrogen occurs from the 1-pro-R position of the ethanol to the 4-pro-R position of NAD^+. The figure indicates how the ethanol is oriented, according to an analysis of kinetic and structural information by Dutler and Branden (1979) (see Section 7.5)

findings (Good *et al.*, 1979) together with conformational analysis of the substrates, X-ray evidence of the structure of the enzyme (Eklund et al., 1976a) and reasonable assumptions regarding the arrangement of the reacting atoms, a model was constructed of the enzyme-NAD^+-alkylcyclohexanone complex (Dutler & Brändén, 1979). The position that ethanol can be expected to occupy (built into this model) is illustrated (Fig. 9). Four interesting aspects of specificity were shown by this study. First, the high stereospecificity for the ethanol oxidation appeared to result from the unfavourable interaction that would have occurred with the methyl group of the ethanol (C-2) if the ethanol had been positioned for transfer of the 1-*pro-S* hydrogen. Second, isopropanol, in which a methyl group occupies this position, would be expected to be, and is, a very much worse substrate than ethanol. Third, the reason that cyclohexanol was quite a good substrate was evidently that the C-4 end of the cyclohexanol ring created interactions with the surrounding hydrophobic residues that were sufficient to overcome the unfavourable interactions elsewhere. Fourth, the arrangement of the residues making a hydrophobic "belt" and a hydrophilic "bottom" to the active site pocket meant that the positioning of any group close to the hydrophobic/hydrophilic interface would be particularly critical; this is an interesting structural basis for substrate alignment.

7.6 Miscellaneous aspects of general interest concerning dehydrogenase specificity

In a sense, the specificity of every dehydrogenase poses a unique problem. However, powerful circumstances militate against the kind of isolated uniqueness that would make the findings for one dehydrogenase of no relevance to other dehydrogenases. There is both a general expectation (Dobzhansky *et al.*, 1977) and some evidence (Rossmann *et al.*, 1977) that dehydrogenases are related to one another (and to certain other proteins). Half of the problem, namely the binding of the nicotinamide coenzyme and provision for hydride transfer to and from C-4 of the nicotinamide moiety, poses essentially similar problems in every case. Substrates are divisible (very roughly) into categories according to characteristics such as whether they are large, small, rigid, flexible, hydrophobic, hydrophilic, etc.; the common properties of substrates of any one such category impose certain common constraints on what properties would be possible for the binding site. The essentially similar requirements for the dehydrogenase reaction necessitate alignment of the reacting groups in a closely similar way. The purpose of this section is to present miscellaneous aspects of specificity that are of some general interest.

Hydrogen dehydrogenase (EC 1.12.1.2) catalyses the reaction:

$H_2 + NAD^+ \rightleftharpoons NADH + H^+$

The hydrogenase from Alcaligenes eutrophus H16 requires no other cofactor or electron carrier (Schleicher and Simon, 1977), and is 4-pro-S specific with respect to coenzyme. The substrate (hydrogen) is uniquely small; there is no product, other than NADH and H^+. The structure of this enzyme should be most interesting.

Transhydrogenases (EC 1.6.1.1) were obtained from Pseudomonas aeruginosa and Azotobacter vinelandii, and found to be different, but closely similar (van den Broek & Veeger, 1970; Chung, 1970). The former consisted of subunits of molecular weight about 43,000, assembled, in groups of about 20, into units (molecular weight about 900,000) which formed filamentous aggregates of non-uniform length (Cohen & Kaplan, 1970; Louie et al., 1972). The transhydrogenase from Azotobacter vinelandii consisted of a single polypeptide chain of molecular weight about 54,000 (Wermuth & Kaplan, 1976; Höjeberg et al., 1976), assembled, in groups of about 8, into units (molecular weight about 421,000) which associated to form threads and rods (van den Broek et al., 1971b; Voordouw et al., 1979). They catalysed the reaction:

$NAD^+ + NADPH \rightleftharpoons NADH + NADP^+$

The hydrogen was transferred from the 4-pro-S position of one nicotinamide ring to the 4-pro-S position of the other (San Pietro et al., 1955; Louie & Kaplan, 1970; Middleditch et al., 1972). Various analogues (of the kind referred to in Section 3) served in place of NAD^+ or $NADP^+$, and the specificity of the enzymes from the Pseudomonas and the Azotobacter were not identical; for example, the 3-acetyl analogue of NAD^+ served with the former, but not the latter (Shuster & Kaplan, 1955; van den Broek et al., 1971a). The reaction was influenced by Ca^{2+} and by 2′-AMP (Kaplan et al., 1953; Louie et al., 1972; Rydström et al., 1973; van den Broek et al., 1971a; Chung, 1970). Separate binding sites for hydrogen donor and hydrogen acceptor were thought likely, at least for the Azotobacter enzyme (Chung, 1970; van den Broek et al., 1971a), but some doubt remained, especially in the case of the Psueodomonas enzyme (Cohen & Kaplan, 1970; Louie & Kaplan, 1970; Hoek et al., 1974). The enzymes were flavoproteins (Cohen & Kaplan, 1970; van den Broek et al., 1971a). It would certainly be interesting to know the structure and mechanism in sufficient detail to be able to compare coenzyme binding and catalysis in these enzymes with events in dehydrogenases that have one binding site for nicotinamide coenzyme and one for substrate.

Pig liver contains a glucose dehydrogenase (EC 1.1.1.47) that utilizes either NAD^+ or $NADP^+$. Particularly with $NADP^+$ as coenzyme, glucose-6-phosphate also served as substrate (Thompson & Carper, 1970). The activity towards glucose-6-phosphate (but not towards glucose) was reported to be abolished by EDTA, and partial restoration of this activity was found to be possible by subsequent addition of excess Mg^{2+}

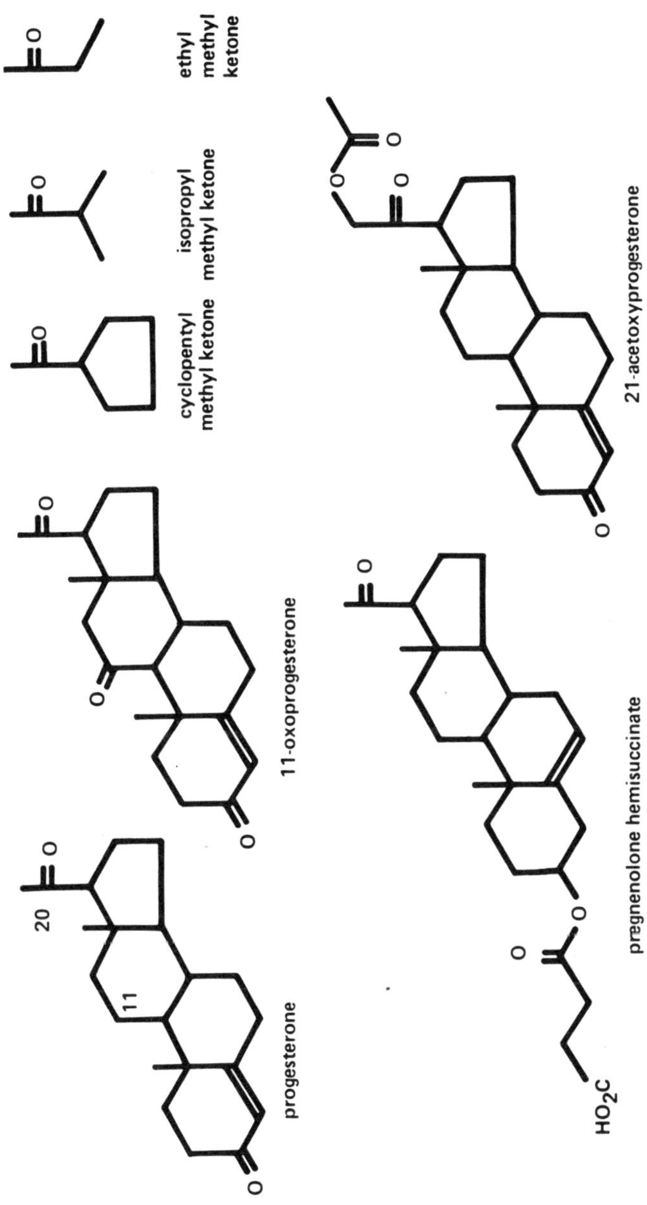

Fig. 10 20β-Hydroxysteroid dehydrogenase from Streptomyces hydrogenans utilizes various steroids and some analogous compounds (see text, and references therein)

(Carper et al., 1976).

Glucose-6-phosphate dehydrogenases (EC 1.1.1.49) transfer the 1α-hydrogen of β-D-glucopyranose-6-phosphate (Colowick & Goldberg, 1963; Salas et al., 1965; Smith & Beutler, 1966; Mazover & Benziman, 1973; Lessmann et al., 1975; Malaisse et al., 1976) to the 4-pro-S position of the coenzyme (Stern & Vennesland, 1960; Davies et al., 1972; Lessmann et al., 1975; Arnold et al., 1976). The $NADP^+$ dependent activity is often increased by the presence of Mg^{2+} (Kornberg, 1950; Glaser & Brown, 1955; Levy, 1963; Yoshida, 1967; Domagk et al., 1969), though the enzyme from most sources has no true requirement for Mg^{2+}. Other cations (bivalent and univalent) can also cause activation in some cases (Glock & McLean, 1953; Glaser & Brown, 1955; Jagganathan et al., 1956; Marquet, 1963). Mg^{2+} may increase tetramer formation (Bonsignore et al., 1971), or the equilibrium between dimeric forms (Levy et al., 1966), but the precise reasons for the effects of cations are not known. Glucose is a poor substrate for many glucose-6-phosphate dehydrogenases (e.g. Colowick & Goldberg, 1963; Salas et al., 1965). The activity towards glucose is increased by bicarbonate, sulphate, and phosphate, though these anions decrease the activity towards glucose-6-phosphate (Anderson & Nordlie, 1968; Anderson et al., 1968; Horne et al., 1970). There is evidently an anion binding site, occupation of which is favourable for enzyme activity, and for which the 6-phosphate group of the substrate (if present) and the other anion (bicarbonate, sulphate, or phosphate) compete. It would certainly be interesting to know the structural basis of these effects in both glucose dehydrogenase and glucose-6-phosphate dehydrogenase, and to be able to compare the mechanism of these enzymes.

Glucose-6-phosphate dehydrogenases from various tissues of mammals were inhibited by certain steroids (Mark & Banks, 1960; McKerns & Bell, 1960; McKerns & Kaleita, 1960; Marks et al., 1961; Levy, 1961; McKerns, 1963; Nielson & Warren, 1965; Benes et al., 1970; Menzel et al., 1970; Benes & Oertel, 1971; Lopes & Rene, 1973). Detailed studies of the enzyme from rat mammary gland (Raineri & Levy, 1970) led to the view that this enzyme had a hydrophobic pocket approximately 1.5 nm × 0.8 nm × 0.6 nm, with one side virtually planar. Almost the whole steroid ring system was required for tight binding, but for inhibition there was also a requirement for a carbonyl group at C-17 (androstane series) or C-20 (pregnane series). Most of the kinetic evidence suggested that the steroid bound only to the ternary complex, or to the enzyme-NADPH complex. The most potent steroid inhibitors had K_i values less than micromolar (Raineri & Levy, 1970), and there was some evidence for steroid binding in vivo (Stechenska et al., 1975), but there is really no evidence for mediation of any important regulatory effect in this way. Could the steroid-binding be considered as not yet important, but a candidate to become a regulatory mechanism after mutations have produced further structural change? Perhaps it would be too wild a fantasy to wonder whether other mutations could make the steroid-binding site become the substrate-binding site of a 17- or 20-hydroxysteroid dehydrogenase; and whether a hydroxysteroid dehydrogenase may even have arisen in this way (from a duplicate gene) at some time past?

The 20β-hydroxysteroid dehydrogenase (EC 1.1.1.53)from Streptomyces hydrogenans transfers the 4-*pro*-S hydrogen from NADH (Gibb & Jeffery, 1973) to the 20α-position of substrates such as progesterone (White & Jeffery, 1973b) (Fig.10). The highest V/K_m value was for progesterone and the highest V value was for 11-oxoprogesterone (White & Jeffery,1974a). Analysis of findings for a carefully chosen series of about forty compounds (White & Jeffery, 1972; 1973a; 1974a,b; 1975; 1977) could be consistent with a partly flexible binding site ("induced fit") as previously suggested (see, for example, White & Jeffery, 1974a,b), but it is now suggested that they could also be consistent with an open cavity type of site, to which the substrate bound with β-face to the enzyme, making favourable hydrophobic interactions across the β-face of rings B and C, and with ring A possibly projecting out of the open cavity. The small molecules that reacted slowly(e.g. cyclopentyl methyl ketone) or not at all (e.g. ethyl methyl ketone) were evidently bound weakly (White & Jeffery, 1977), and, in the absence of further evidence, it seems possible to explain this either by supposing that they were too small to induce a good fit, or that they did not make sufficiently extensive favourable contacts to locate them properly in the open cavity. Put in another way, they may have had insufficient "intrinsic binding energy" to provide the driving force for catalysis (cf. Section 1), or they may have formed (Michaelis) complexes in which the entropy of activation was insufficiently "frozen out".

The 15-hydroxyprostaglandin dehydrogenase of pig lung (EC 1.1.1.141)(Ånggård & Samuelsson, 1966) transferred hydrogen from the 15-position of prostaglandinsthat had a (15*S*)-15-hydroxy configuration (that is, the well-known, natural, active prostaglandins) to NAD^+ (Nakano *et al*.,1969; Shio *et al*.,1970). Transfer was to the 4-*pro*-S position in the cases of similar enzymes from human placenta (Saintot *et al*., 1976) and pig kidney (Tai, 1976), and apparently various rat tissues including lung (Tai, 1976), but this point may not have been proved for the pig lung enzyme. The prostaglandins have a somewhat different structure (Fig.11) from any of the substrates for dehydrogenases discussed so far, and are of great interest because of their biological activities. Their oxidation by the 15-hydroxyprostaglandin dehydrogenase is of particular interest because it is a major inactivation mechanism in vivo. More than a hundred compounds have been tested as substrates or inhibitors. (Vonkeman *et al*., 1969; Nakano *et al*., 1969; Ohno *et al*., 1978). Prostaglandin E_1 was one of the best substrates (K_m 0.007 mM, V 100, V/K_m 14286), 11-deoxyprostaglandin E_1 (K_m 0.0057mM, V 107, V/K_m 18772) being slightly better, and 11-deoxyprostaglandin I_2 methyl ester being the best substrate (K_m 0.0024 mM, V 99, V/K_m 41250). That the large structural change from the E to the I series did not prevent reaction is striking (Fig.11). Comparison of 11-deoxyprostaglandin E_1 with prostaglandin A_1 (K_m 0.014 mM, V 86, V/K_m 6143) indicates some

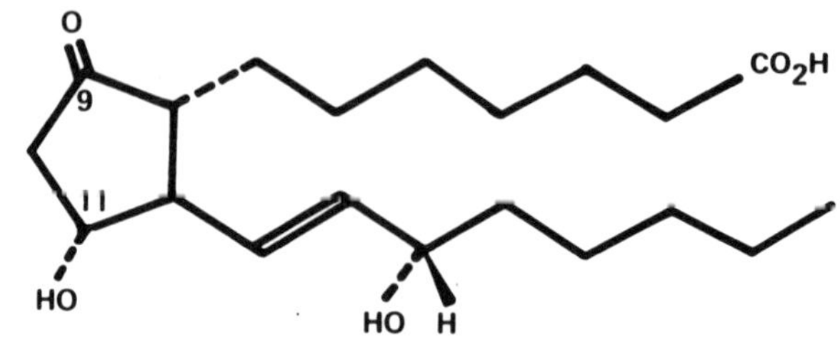

prostaglandin E_1

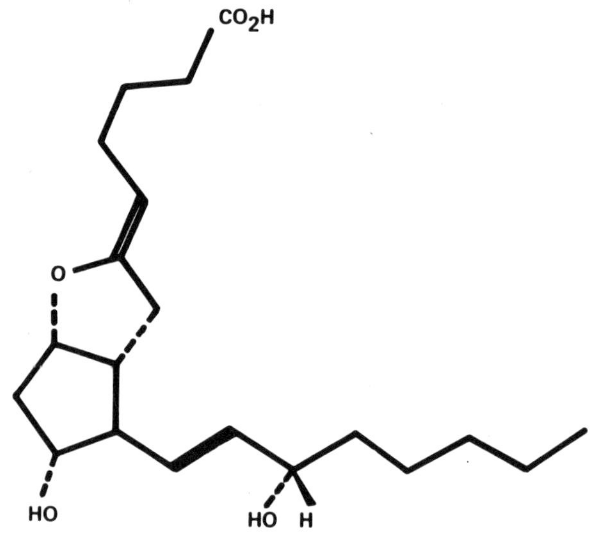

prostaglandin I_2

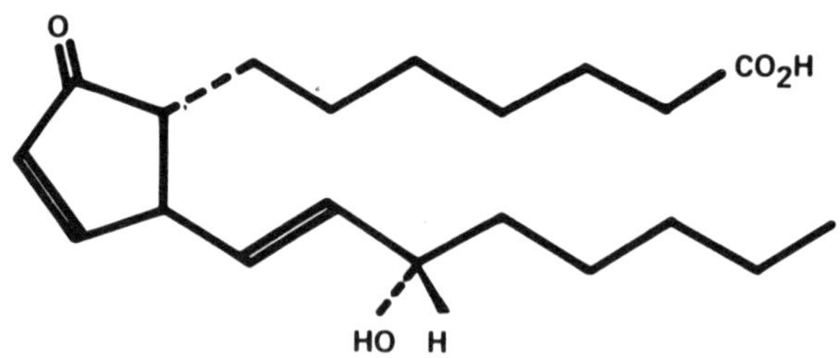

prostaglandin A_1

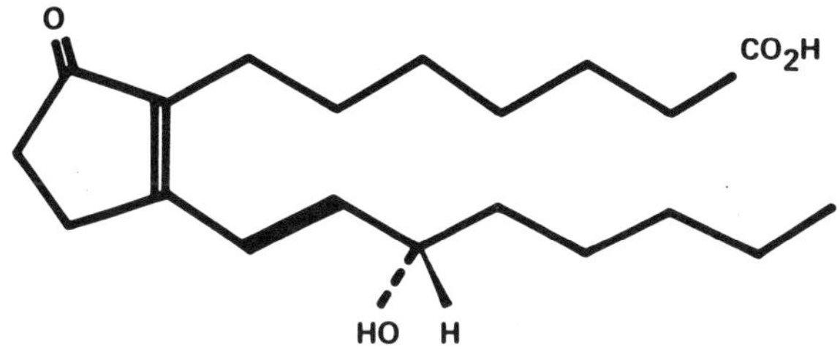

prostaglandin B_1

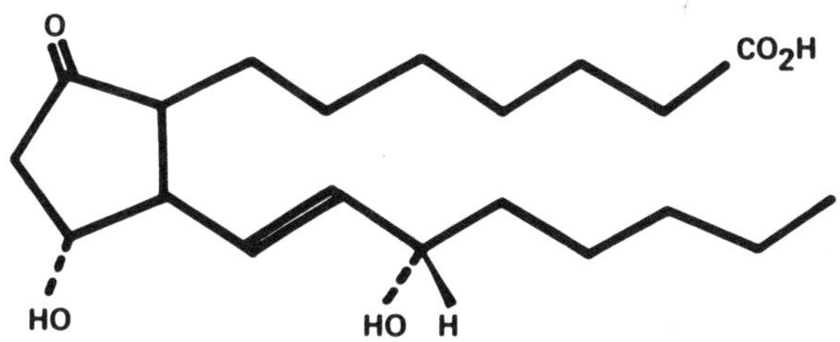

8-epi-prostaglandin E_1

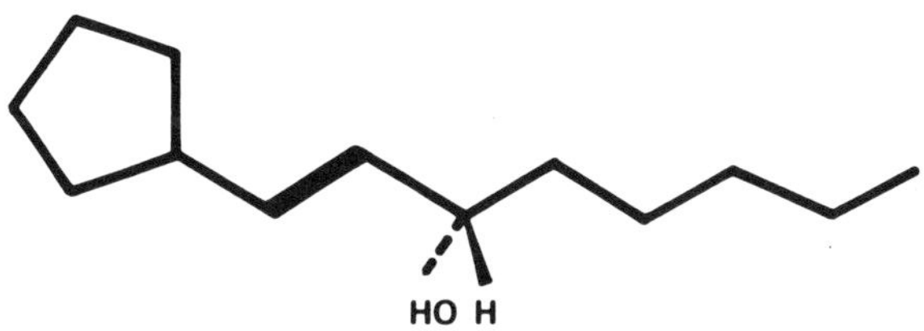

3S - trans-1-cyclopentyloct-1-en-3-ol

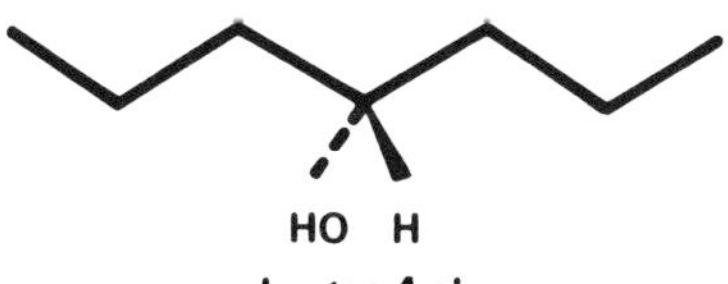

heptan-4-ol

Fig. 11 15-Hydroxyprostaglandin dehydrogenase from pig lung utilizes various prostaglandins and some analogous compounds (see text)

sensitivity to the structure of the cyclopentane ring. This is confirmed by the fact that prostaglandin B_1 was not a substrate at all. 8-epi-Prostaglandin E_1 was not a particularly bad substrate (K_m 0.052 mM, V 93, V/K_m 1788) although this structural change alters the overall shape of the molecule considerably. The one charged group in the molecule (the terminal carboxyl) was of little importance; prostaglandin E_1 methyl ester, K_m 0.0086 mM, V 100, V/K_m 11628; prostaglandin I_2, K_m 0.012 mM,V 91, V/K_m 7583; prostaglandin I_2 methyl ester, K_m 0.010, V 96, V/K_m 9600. The findings might be compatible with an open slot type of binding site. It would be helpful to know whether some much simpler compounds, such as the appropriate isomer of 1-cyclopentyloct-1-en-3-ol, and even heptan-4-ol were substrates (Fig. 11). Interest has understandably centred upon biologically active structures that were poor substrates. If simpler compounds serve, it should be possible to formulate a characteristic diamond lattice section for this enzyme. Substrate analogues tested as inhibitors had inhibition constants in the micromolar to millimolar range. They would scarcely cause inhibition at concentrations similar to prostaglandin concentrations in vivo. A more systematic knowledge of the specificity would probably enable more potent inhibitors to be devised; whether that would be a worthwhile therapeutic aim is another matter.

8. CONCLUSIONS

It is a commonplace that enzymes are much more specific than non-enzymic catalysts. Specificity must be a main consideration in the description of any enzymic reaction.

Since information became available on both the structure and the specificity of a few dehydrogenases, the structural basis of the mechanism has become clearer in these cases. But many problems remain, and the number of cases for which detailed structural and kinetic information is available is a very small proportion of the total number of important dehydrogenases.

The need to combine structural and kinetic studies scarcely requires emphasis. From kinetic studies with horse liver alcohol dehydrogenase, primary alcohols such as benzyl alcohol were known to be good substrates, and imidazole was known to be a potent inhibitor. This provided the logic for seeking a naturally-occurring substance that contained both an imidazole moiety and a primary alcohol group. Histidinol was such a compound, and was shown to be a substrate (Ambar et al., 1979; see Section 7.4). From the structure of the enzyme, it appeared a little surprising that a small molecule, such as ethanol, should be the substrate. The binding site seemed suited to a large, fairly flat molecule, such as a porphyrin, though no oxo/alcohol conversion with such a substance seemed to be required. Considerations of this kind provided the logic for looking for the catalysis of some other reaction using this kind of substrate. The interesting outcome of this approach was briefly discussed (Ambar *et al.*, 1979; see Section 7.4). Whether or not it turns out that liver alcohol dehydrogenase plays any important role in the synthesis of histidine or porphyrins *in vivo*, these studies have surely been an intellectually stimulating development.

Many biochemists now suspect that liver alcohol dehydrogenase is rather freakish; they feel sure that other dehydrogenases will have structures more obviously suited to their accepted function. This may turn out to be the case, but it is not many years since alcohol dehydrogenase was almost the only dehydrogenase that had been studied in any real detail, and at that time it was often spoken of confidently as if it were a prototype for all dehydrogenases.

Other examples of "improbable" specificity, also from mammals, were mentioned (Section 6.2). A somewhat different example would be the glycerol dehydrogenase of *Mucor javanicus*, which showed marked stereospecificity, and for which (*9S*)-*trans*-1, 4-decalindione was an excellent substrate, though the function *in vivo* was thought to be the reduction of dihydroxyacetone to glycerol (that is, a small symmetrical substrate to a small symmetrical product) (Dutler *et al.*, 1977; Hochuli *et al.*, 1977).

The specificity of a dehydrogenase could influence the survival fitness of an organism, and is a property upon which natural selection may operate. Small changes in enzyme structure could greatly alter specificity, and it is interesting that at present dehydrogenases appear to resemble one another in tertiary structure more closely than in primary structure. It must again be noted, however, that few dehydrogenases have so far been examined in detail, and that in some of the respects in which they resemble one another they also resemble other enzymes (see, for example, Blake, 1975), so the significance of these resemblances is uncertain.

The better empirical generalizations regarding prochiral specificity with respect to coenzyme (see Section 5) are sufficiently widely valid to require explanation. It is not clear what proper explanations may be.

The use of dehydrogenases as analytical reagents is well established. A principal advantage is high specificity, but incomplete specificity is often tolerable in practice provided that the specificity is known and there is good reason to believe that alternative substrates are unlikely to be present in substantial quantities. The simplicity of an assay using a dehydrogenase often makes it attractive even when a satisfactory non-enzymic assay is possible, for example, by GLC of a derivative. In favourable cases, two enzymes of different specificity may be obtainable, as for example, dihydroxyacetone reductase and glyceraldehyde reductase, which, in principle, could both be used to assay glycerol. Dehydrogenases are of unrivalled analytical value for the assay of a particular enantiomorph when the reacting alcohol group is at a chiral centre. An example would be L-lactate and D-lactate. In this case, a dehydrogenase is available to assay either enantiomorph. The sensitivity of an assay using a dehydrogenase can be made very high either by incorporating a "cycling system" for NADH

(Änggård *et al.*, 1969; Kato *et al.*, 1973) or a diaphorase-resazurin system (Ohno *et al.*, 1972), and fluorimetry.

Assay of the dehydrogenase rather than the substrate may be required. For this purpose, a knowledge of the specificity of the enzyme to be assayed, and of other enzymes likely to be present is needed, to allow a favourable choice of test substrate. If, for example, β-hydroxypyruvate were used to assay D-glycerate dehydrogenase in a liver preparation, its reduction by this enzyme would not be distinguished from its reduction by L-lactate dehydrogenase, which would be likely to be present, and for which it is also a substrate. If the specimen to be tested happens to absorb close to 340 nm, it may be convenient, when the specificity of the enzyme allows, to use the thioamide analogue of the nicotinamide coenzyme, and follow the reaction at 400 nm. By careful choice of substrates it is sometimes possible to assay a particular isozyme; *trans*-hex-2-en-1-ol has been regarded as a useful substrate for distinguishing certain alcohol dehydrogenase isozymes (Holmes *et al.*, 1978).

In contrast to the widespread use of dehydrogenases for analytical purposes, their use on a preparative scale has not been extensive. Studies during the past few years using horse liver alcohol dehydrogenase (Irwin & Jones, 1976; 1977a,b; Irwin *et al.*, 1978) made clear the potential value of dehydrogenases in small-scale synthesis (grams). An interesting feature of this work was that it did not merely make use of the known course of reaction for known substrates, but applied the characteristic diamond lattice section for the enzyme (i.e. the conceptualized specificity) to potential substrates, the behaviour of which could be predicted in this way, but was not previously known.

For the future, it seems that appropriate studies on the specificity of dehydrogenases have much to contribute to the understanding of mechanism, and the relationship of structure and function. Analytical applications for which some knowledge of specificity is needed will continue to be important. Preparative use of dehydrogenases is likely to increase, making use, in some cases, of concepts such as the characteristic diamond lattice section.

It is self-evident that specificity studies should be done, whenever possible, in conjunction with structural studies. The necessary facilities and expertise are difficult to assemble and retain in a single laboratory. Multicentre collaboration is therefore likely to play an important role in future developments.

9. REFERENCES

Abdallah, M.A., Biellmann, J.F., Samama, J.P. & Wrixon, A.D. (1976) Eur. J. Biochem. 64, 351-360.

Abdallah, M.A., Adams, M.J., Archibald, I.G., Biellmann, J.F., Helliwell, J.R. & Jenkins, S.E. (1979) Eur. J. Biochem. 98, 121-130.

Abeles, R.H. & Lee, H.A. Jr. (1960) J. Biol. Chem. 235 1499-1503.

Adams, M.J., Buehner, M., Chandrasekhar, K., Ford, G.C., Hackert, M.L., Liljas, A., Rossmann, M.G., Smiley, I.E., Allison, W.S., Everse, J., Kaplan, N.O. & Taylor, S.S. (1973) Proc. Natl. Acad. Sci. (U.S.A.), 70, 1968-1972.

Alizade, M.A. & Brendel, K. (1975) Naturwissenschaften, 62, 346-348.

Alizade, M.A., Bressler, R., & Brendel, K. (1974) Biochim. Biophys. Acta, 370, 354-357.

Alizade, M.A., Gaede, K., & Brendel, K. (1976) Hoppe-Seyler's Z. Physiol. Chem. 357, 1163-1169.

Ambar, A., Good, P. & Dutler, H. (1979) Communicated at FEBS Special Meeting on Enzymes, Dubrovnik, but not printed in the papers of the meeting.

Anderson, B.M. & Kaplan, N.O. (1959) J. Biol. Chem. 234, 1226-1232.

Anderson, W.B. & Nordlie, R.C. (1968) Biochemistry, 7, 1479-1485.

Anderson, W.B., Horne, R.N., & Nordlie, R.C. (1968) Biochemistry, 7, 3997-4004.

Änggård, E. & Samuelsson, B. (1966) Arkiv för Kemi, 25, 293-300.

Änggård, E., Matschinsky, F.M. & Samuelsson, B. (1969) Science, N.Y., 163, 479-480.

Arnold, L.J., You, K., Allison, W.S. & Kaplan, N.O. (1976) Biochemistry, 15, 4844-4849.

Barrio, J.R., Secrist, J.A. & Leonard, N.J. (1972) Proc. Natl. Acad. Sci. (U.S.A.), 69, 2039-2042.

Benes, P. & Oertel, G.W. (1971) J. Steroid Biochem., 2, 289-292.

Benes, P., Menzel, P., & Oertel, G.W. (1970) J. Steroid Biochem. 1, 291-293.

Bentley, R. (1970) In "Molecular asymmetry in biology" Vol. 2, pp. 1-89, Academic Press, New York.

Benziman, M. & Mazover, A. (1973) J. Biol. Chem. 248, 1603-1608.

Biellmann, J.F. & Jung, M.J. (1971) Eur. J. Biochem. 19, 130-134.

Biellmann, J.F. & Samama J.P. (1974) FEBS Letters, 38, 175-178.

Biellmann, J.F., Branlant, G., Foucaud, B.Y. & Jung, M.J. (1974a) FEBS Letters, 40, 29-32.

Biellmann, J.F., Hirth, C.G., Jung, M.J., Rosenheimer, N. & Wrixon, A.D. (1974b) Eur. J. Biochem., 41, 517-524.

Biellmann, J.F., Goulas, P.R., Nicolas, J.C., Descomps, B. & Crastes de Paulet, A. (1979) Eur. J. Biochem. 99, 81-84.

Björkhem, I., Jörnvall, H., Zeppezauer, E. (1973) Biochem. Biophys. Res. Commun., 52, 413-420.
Björkhem, I., Jörnvall, H. & Åkeson, A. (1974) Biochem. Biophys. Res. Commun., 57, 870-875.
Blake, C.C.F. (1975) Essays in Biochemistry, 11, 37-79.
Bliss, A.F. (1951) Arch. Biochem. Biophys., 31, 197-204.
Bonsignore, A., Lorenzoni, I., Cancedda, R., Cosulich, M.E., & De Flora, A. (1971) Biochem. Biophys. Res. Commun. 42, 159-165.
Brändén, C.I., Eklund, H., Nordström, B., Boiwe, T., Söderlund, G., Zeppezauer, E., Ohlsson, I. & Åkeson, A.(1973) Proc. Natl. Acad. Sci. (U.S.A.), 70, 2439-2442.
Brändén, C.I., Jörnvall, H., Eklund, H. & Furugren, B. (1975) In "The enzymes" (Boyer, P.D., ed.) 3rd edn., Vol. 11, pp. 103-190, Academic Press, New York.
Carper, W.R., Toews, M.L., Thompson, R.E. & Buess, C.M. (1976) Arch. Biochem. Biophys. 175, 312-320.
Chandrasekhar, K., McPherson, A. Jr., Adams, M.J. & Rossmann, M.G. (1973) J. Mol. Biol., 76, 503-518.
Christophe, J. & Popjack, G. (1961) J. Lipid. Res. 2, 244-257.
Chung, A.E. (1970) J. Bacteriol., 102, 438-447.
Cohen, P.T. & Kaplan, N.O. (1970) J. Biol. Chem. 245, 2825-2836.
Colowick, S.P. & Goldberg, E.B. (1963) Bull. Res. Counc. Israel, 11A4, 373.
Colowick, S.P., van Eys, J., & Park, J.H. (1966) In "Comprehensive biochemistry" (Florkin, M. & Stotz, E.H., eds.),Vol. 14, pp. 1-98, Elsevier, Amsterdam.
Culp, H.W. & McMahon, R.E. (1968) J. Biol. Chem. 243, 848-852.
Dalziel, K. (1975) Phil. Trans. R. Soc. Lond. B272, 109-122.
Dalziel, K. & Dickinson, F.M. (1965a) Biochem. J. 95, 311-320.
Dalziel, K. & Dickinson, F.M. (1965b) Nature (London), 206, 255-257.
Danenberg, P.V., Danenberg, K.D. & Cleland, W.W. (1978) J. Biol. Chem. 253, 5886-5887.
Davies, D.D., Teixeira, A. & Kenworthy, P. (1972) Biochem. J. 127, 335-343.
Debus, H. (1904) J. Chem. Soc., 85, 1382-1403.
DeMoss, R.D., Gunsalus, I.C. & Bard, R.C. (1953) J. Bacteriol. 66, 10-16.
Dietz, G., Woenckhaus, C., Jaenicke, R. & Schuster, I. (1977) Z. Naturforsch., 32c, 85-92.
Dobzhansky, T., Ayala, F.J., Stebbins, G.L. & Valentine, J.W. (1977) In "Evolution" pp. 77-94, Freeman, San Francisco.
Domagk, G.F., Domschke, W., & Engel, H.J. (1969) Hoppe-Seyler's Z. Physiol. Chem. 350, 1242-1246.
Do Nascimento, K.H. & Davies, D.D. (1975) Biochem. J., 149, 553-557.
Duncan, R.J.S. & Tipton, K.F. (1969) Eur. J. Biochem. 11, 58-61.
Dutler, H. & Brändén, C.I. (1979) In preparation.

Dutler, H., van der Baan, J.L., Hochuli, E., Kis, Z., Taylor, K.E. & Prelog, V. (1977) Eur. J. Biochem. 75, 423-432.
Eklund, H., Nordström, B., Zeppezauer, E., Söderlund, G., Ohlsson, I., Boiwe, T., Söderberg, B.O., Tapia, O., Brändén, C.I., & Åkeson, Å. (1976a) J. Mol. Biol., 102, 27-59.
Eklund, H., Brändén, C.I. & Jörnvall, H. (1976b) J. Mol. Biol. 102, 61-73.
Everse, J. & Kaplan, N.O. (1973) Adv. Enzymol., 37, 61-133.
Fawcett, C.P. & Kaplan, N.O. (1962) J. Biol. Chem. 237, 1709-1715,
Feldman, H.B., Szczepanik, P.A., Havre, P., Corrall, R.J.M., Yu, L.C., Rodman, H.M., Rosner, B.A., Klein, P.D., & Landau, B.R. (1977) Biochim. Biophys. Acta 480,14-20
Fersht, A.R. (1974) Proc. R. Soc. Lond. B187, 397-407.
Fersht, A.R. (1977) In "Enzyme structure and mechanism" 2nd edn., pp. 244-273, Freeman, San Francisco.
Fischer, E. (1894) Ber. dt. chem. Ges., 27, 2985-2993.
Fisher, J. & Walsh, C. (1974) J. Am. Chem. Soc., 96, 4345-4346.
Fisher, H.F., Conn, E.E., Vennesland, B. & Westheimer, F.H. (1953) J. Biol. Chem. 202, 687-697.
Fisher, T.L., Vercellotti, S.V., & Anderson, B.M. (1973) J. Biol. Chem. 248, 4293-4299.
Flynn, T.G., Shires, J. & Walton, D.J. (1975) J. Biol. Chem. 250, 2933-2940.
Fukuba, R. (1974) Biochim. Biophys. Acta, 341, 48-55.
Gibb, W. & Jeffery, J. (1973) Eur. J. Biochem. 34, 395-400.
Glaser, L. & Brown, D.H. (1955) J. Biol. Chem. 216, 67-79.
Glock, G.E. & McLean, P. (1953) Biochim. Biophys. Acta, 12, 590.
Good, P., Helmchen, R.E., Carson, M. & Dutler, H. (1979) Eur. J. Biochem. submitted for publication.
Guerrero, M. & Vennesland, B. (1975) FEBS Letters, 51, 284-286.
Guillory, R.J. & Jeng, S.J. (1977) In "Methods in enzymology" (Jakoby, W.B. & Wilchek, M., eds.) Vol. 46, pp.259-289, Academic Press, New York.
Gupta, N.K. (1970) Arch. Biochem. Biophys. 141, 632-640.
Hawtrey, C.O. & Goldberg, E. (1973) J. Exp. Zool., 174, 451-461.
Hegazi, M. & Meany, J.E. (1972) J. Phys. Chem. 76, 3121-3124.
Hinson, J.A. & Neal, R.A. (1972) J. Biol. Chem. 247, 7106-7107.
Hochuli, E., Taylor, K.E., & Dutler, H. (1977) Eur. J. Biochem. 75, 433-439.
Hoek, J.B., Rydström, J. & Höjeberg, B. (1974) Biochim. Biophys. Acta, 333, 237-245.
Höjeberg, B., Brodelius, P., Rydström, J. & Mosbach, K. (1976) Eur. J. Biochem. 66, 467-475.
Holbrook, J.H. & Stinson, R.A. (1973) Biochem. J. 131, 739-748.
Holbrook, J.J., Liljas, A., Steindel, S.J. & Rossmann, M.G. (1975) In "The enzymes" (Boyer, P.D., ed.) 3rd edn., Vol. 11, pp. 191-292. Academic Press, New York.
Holmes, R.S., Jones, J.T. & Peters, J. (1978) J. Exper. Zoology, 206, 279-288.

Holtzer, H., Schneider, S. & Lange, K. (1955) Angew, Chem., 67, 276.
Horne,R.N., Anderson, W.B. & Nordlie, R.C. (1970) Biochemistry, 9, 610-616.
Irwin, A.J. & Jones, J.B. (1976) J. Am. Chem. Soc., 98, 8476-8482.
Irwin, A.J. & Jones, J.B. (1977a) J. Am. Chem. Soc., 99, 556-561.
Irwin, A.J. & Jones, J.B. (1977b) J. Am. Chem. Soc. 99, 1625-1630.
Irwin, A.J., Lok, K.P., Huang, K.W.C., & Jones, J.B. (1978) J. Chem. Soc. Perkin I., 1636-1642.
Jagannathan, V., Rangachari, P.N. & Damodaran, M. (1956) Biochem. J., 64, 477-481.
Jencks, W.P. (1975) Adv. Enzymol., 43, 219-410.
Jörnvall, H. (1977) In "Pyridine nucleotide-dependent dehydrogenases" (Sund, H., ed.) pp. 62-68, De Gruyter, Berlin.
Kaplan, N.O., Colowick, S.P., Neufeld, E.F., & Ciotti, M.M. (1953) J. Biol. Chem. 205, 17-29.
Kato, T., Berger, S.J., Carter, J.A. & Lowry, O.H. (1973) Anal. Biochem., 53, 86-97.
Kornberg, A. (1950) J. Biol. Chem. 182, 805-813.
Koshland, D.E. Jr. (1963) Cold Spring Harbor Symp. quant. Biol., 28, 473-482.
Krakow, G., Ludowieg, J., Mather, J.H., Normore, W.M., Tosi, L., Udaka, S., & Vennesland, B. (1963) Biochemistry, 2, 1009-1014.
Krebs, H.A. & Perkins, J.R. (1970) Biochem. J., 118, 635-644.
Lessmann, D., Schmiz, K.L., & Kurz, G. (1975) Eur. J. Biochem. 59, 545-559.
Levy, H.R. (1961) Biochem. Biophys. Res. Commun., 6, 49-53.
Levy, H.R. (1963) J. Biol. Chem. 238, 775-784.
Levy, H.R. (1979) Adv. Enzymol. 48, 97-192.
Levy, H.R., Raineri, R., & Nevaldine, B.H. (1966) J. Biol. Chem. 241, 2181-2187.
Lienhard, G.E. (1973) Science, N.Y., 180, 149-154.
Lopez, S.A. & Rene, A. (1973) Proc. Soc. Exp. Biol. Med., 142, 258-261.
Louie, D.D. & Kaplan, N.O. (1970) J. Biol. Chem. 245, 5691-5698.
Louie, D.D., Kaplan, N.O., & Lean, J.D. (1972) J. Mol. Biol. 70, 651-664.
Lutstorf, U.M., Schurch, P.M. & von Wartburg, J.P. (1970) Eur. J. Biochem. 17, 497-508.
McKerns, K.W. (1963) Biochim. Biophys. Acta, 71, 710-718.
McKerns, K.W. & Bell, P.H. (1960) Rec. Prog. Hormone Res., 16, 97-117.
McKerns, K.W. & Kaleita, E. (1960) Biochem. Biophys. Res. Commun., 2, 344-348.
McPherson, A. Jr. (1970) J. Mol. Biol. 51, 39-46.
Malaisse, W.J., Sener, A., Koser, M. & Herchuelz, A. (1976) J. Biol. Chem. 251, 5936-5943.
Marks, P.A. & Banks, J. (1960) Proc. Natl. Acad. Sci. (U.S.A.), 46, 447-452.
Marks, P.A., Szeinberg, A. & Banks, J.(1961) J.Biol.Chem.,236, 10-17.

Marquet, M. (1963) C.R. Acad. Sci. Paris, 256, 4997-5000.
Matthews, D.A., Alden, R.A., Bolin, J.T., Filman, D.J., Freer, S.T., Hamlin, R., Hol, W.G.J., Kisliuk, R.L., Pastore, E.J., Plante, L.T., Xuong, N-h., & Kraut, J. (1978) J. Biol. Chem. 253, 6946-6954.
Mazover, A. & Benziman, M. (1973) FEBS Letters, 36, 203-206.
Meister, A. (1950) J. Biol. Chem. 184, 117-129.
Meister, A. (1952) J. Biol. Chem. 197, 309-317.
Menzel, E., Menzel, P., & Oertel, G.W. (1970) Hoppe-Seyler's Z. Physiol. Chem., 351, 759-760.
Middleditch, L.E. & Chung, A.E. (1971) Arch. Biochem. Biophys. 146, 449-453.
Middleditch, L.E., Atchison, R.W. & Chung, A.E. (1972) J. Biol. Chem. 247, 6802-6809.
Nakano, J., Änggård, E. & Samuelsson, B. (1969) Eur. J. Biochem., 11, 386-389.
Nielson, M.H. & Warren, J.C. (1965) Biochim. Biophys. Acta, 97, 532-541.
Ohno, H., Miyamoto, T., Nishibori, S. & Hirata, F. (1972) Yakugaku Zasshi, 92, 1111.
Ohno, H., Morikawa, Y. & Hirata, F. (1978) J. Biochem.,Tokyo, 84, 1485-1494.
Okuda, K. & Takigawa, N. (1970) Biochim. Biophys. Acta, 220, 141-148.
Parker, D.M. & Holbrook, J.J.(1977) In "Pyridine nucleotide dependent dehydrogenases" (Sund, H., ed.) pp. 485-501, De Gruyter, Berlin.
Pauling, L. (1948) Nature (London), 161, 707-709.
Pietruszko, R. & Chen, F.F. (1976) Biochem. Pharmacol. 25, 2721-2725.
Pietruszko, R., Crawford, K. & Lester, D. (1973) Arch. Biochem. Biophys. 159, 50-60.
Pocker, Y., Meany, J.E., Nist, B.J. & Zadorojny, C. (1969) J. Phys. Chem. 73, 2879-2882.
Popjack, G. (1970) In "The enzymes" (Boyer, P.D., ed.) 3rd edn., Vol. 2, pp. 115-215, Academic Press,New York.
Prelog, V. (1964) Pure Appl. Chem. 9, 119-130.
Raineri, R. & Levy, H.R. (1970) Biochemistry, 9, 2233-2243.
Reynier, M. (1969) Acta Chem. Scand., 23, 1119-1129.
Ribbons, D.W., Ohta, Y. & Higgins, I.J. (1972) In "The molecular basis of electron transport", (Schultz, J. and Cameron, B.F., eds.) Proc. Miami Winter Symposia, Vol. 4, pp. 251-271, Academic Press, New York.
Ris, M.M. & von Wartburg, J.P. (1973) Eur. J. Biochem. 37, 69-77.
Romano, M. & Cerra, M. (1969) Biochim. Biophys. Acta, 177, 421-426.
Rossmann, M.G., Garavito, R.M. & Evantoff, W.(1977) In "Pyridine nucleotide-dependent dehydrogenases" (Sund, H., ed.) pp. 3-28, De Gruyter, Berlin.
Rydström, J., Hoek, J.B., & Höjeberg, B. (1973) Biochem. Biophys. Res. Commun., 52, 421-429.

Saintot, M., Thaler-Dao, H., Descomps,B. & Crastes de Paulet, A. (1976) FEBS Letters, 72, 238-242.
Salas, M., Vinuela, E., & Sols, A. (1965) J. Biol. Chem. 240, 561-568.
Samama, J.P., Zeppezauer, E., Biellmann, J.F. & Brändén, C.I. (1977) Eur. J. Biochem. 81, 403-409.
San Pietro, A., Kaplan, N.O. & Colowick, S.P. (1955) J. Biol. Chem. 212, 941-952.
Sawaki, S., Hattori, H. & Yamada, K. (1967) J. Biochem. Tokyo, 62, 263-268.
Schatz, L. & Segal, H.L. (1969) J. Biol. Chem. 244, 4393-4397.
Schleicher, E. & Simon, H. (1977) FEBS Letters, 74, 269-271.
Shio, H., Anderson, N.H., & Corey, E.J. (1970) Experientia, 26, 355-357.
Shore, J.D., Gutfreund, H., Brooks, R.L., Santiago, D. & Santiago, P. (1974) Biochemistry, 13, 4185-4191.
Shuster, L. & Kaplan, N.O.(1955) J. Biol. Chem. 215, 183-194.
Slaughter, J.C. & Davies, D.D. (1968) Biochem. J. 109, 743-748.
Smith, J.E. & Beutler, E. (1966) Proc. Soc. Exptl. Biol. Medicine, 122, 671-673.
Sogin, D.C. & Plapp, B.V. (1975) J. Biol. Chem. 250, 205-210.
Stechenska, M.S., Russanov, E.M., & Vassileva-Popova, J.G. (1975) FEBS Letters, 49, 297-300.
Stern, B.K. & Vennesland, B. (1960) J. Biol. Chem. 235, 205-208.
Tai, H.H. (1976) Biochemistry, 15, 4586-4592.
Takenoshita, R. & Toki, S. (1978) Biochem. Pharmacol., 27, 989-993.
Thompson, R.E. & Carper, W.R. (1970) Biochim. Biophys. Acta, 198, 397-406.
Tienhaara, R. & Meany, J.E. (1973) Biochemistry, 12, 2067-2070.
Toki, K. & Tsukamoto, H. (1964) J. Biochem., Tokyo, 55, 142-147.
van den Broek, H.W.J. & Veeger, C. (1970) In "Pyridine nucleotide dependent dehydrogenases" (H.Sund, ed.) p.335 , Springer, Berlin.
van den Broek, H.W.J., Santema, J.S., Wassink, J.H. & Veeger, C. (1971a) Eur. J. Biochem. 24, 31-45.
van den Broek, H.W.J., van Breemen, J.F.L., van Bruggen, E.F.L. & Veeger, C. (1971b) Eur. J. Biochem. 24, 46-54.
Vennesland, B. (1961) Proc. Internat. Cong. Biochem. 5th(Moscow) Symposium 5, preprint no. 145, pp.1-7.
Vennesland, B. (1974) Topics in Current Chemistry, 48, 39-65.
Vonkeman, H., Nugteren, D.H. & Van Dorp, D.A. (1969) Biochim. Biophys. Acta, 187, 581-583.
Voordouw, G., Veeger, C., van Breemen, J.F.L. & van Bruggen, E.F. J. (1979) Eur. J. Biochem. 98, 447-454.
Waller, G.R. (1965) Nature (London), 207, 1389-1390.
Waller, G., Theorell, H. & Sjövall, J. (1965) Arch. Biochem. Biophys. 111, 671-684.
Warren, W.A. (1970) J. Biol. Chem. 245, 1675-1681.
Wermuth, B. & Kaplan, N.O. (1976) Arch. Biochem. Biophys. 176, 136-143.

White, I.H. & Jeffery, J. (1972) Eur. J. Biochem. 25, 409-414.
White, I.H. & Jeffery, J. (1973a) Biochim. Biophys. Acta, 296, 604-614.
White, I.H. & Jeffery, J. (1973b) Biochem. Soc. Trans., 1, 767-769.
White, I.H. & Jeffery, J. (1974a) Biochem. J., 137, 349-354.
White, I.H. & Jeffery, J. (1974b) Biochem. Soc. Trans., 2, 984.
White, I.H. & Jeffery, J. (1975) Biochem. Soc. Trans., 2, 674-675.
White, I.H. & Jeffery, J. (1977) Biochem. Soc. Trans., 5, 723-724.
Wilken, D.R., King, H.L. Jr. & Dyar, R.E. (1975) J. Biol. Chem. 250, 2311-2314.
Winicov, I. (1975) Biochim. Biophys. Acta, 397, 288-293.
Woenckhaus, C.(1974) In "Topics in current chemistry" (Boschke, S.L. ed.) Vol. 52, pp.209-223, Springer, Berlin.
Woenckhaus, C. & Jeck, R. (1977) In "Methods in enzymology" (Jakoby, W.B. & Wilchek, M., eds.) Vol. 46, pp.249-258, Academic Press, New York.
Wolfenden, R. (1969) Nature (London), 223, 704-705.
Wolfenden, R. (1972) Accts. Chem. Res., 5, 10-18.
Yoshida, A. (1967) Biochem. Genet., 1, 81-99.
You, K., Arnold, L.J. Jr., Allison, W.S. & Kaplan, N.O. (1978) TIBS, 3, 265-268.

PRIMARY STRUCTURES OF DEHYDROGENASES. EVOLUTIONARY CHARACTERISTICS RELATED TO FUNCTIONAL ASPECTS; MODELS FOR ISOZYME DEVELOPMENTS AND ANCESTRAL CONNECTIONS

Hans Jörnvall

CONTENTS

SUMMARY

This chapter describes known characteristics of evolutionary changes in individual dehydrogenases, as well as possible relationships among this group of enzymes. Data from primary structures are correlated with those from other observations.

Variations in the amino acid sequences demonstrate functional properties, and can be interpreted in relation to conformational aspects, subunit arrangements and enzyme stabilities.

Different types of isozyme developments have occurred and show functional fixations at various levels. They define isozyme patterns of general significance in protein evolution.

Sequence similarities may be found between different segments. They are analyzed in relation to known conformations, subunit sizes, species divergence and genetic mechanisms. A wide-ranging evolutionary model is discussed relating dehydrogenases and some other oligomeric enzymes to a distant, frequently remodelled ancestral building unit of repetitive occurrence.

INTRODUCTION

The knowledge of primary structures is essential for most studies of protein properties, ranging from questions of catalysis to conformational and evolutionary interpretations. Regarding evolution, sequence studies can establish changes in a structure, relationships between different structures, isozyme developments, and occurrence of elements indicating duplications, fusions or other events. All sequence changes can be quantitated (Dayhoff, 1972), allowing many evolutionary properties of the proteins to be deduced. These include functional roles, such as for example substrate specificity of alcohol dehydrogenase (Jörnvall, 1977a).

In these respects, dehydrogenases reveal changes similar to those in other types of protein, but because they have large subunits structural investigations have still not been performed in many cases. Consequently, mutation rates and some other changes are not yet known in the same detail as they are for a few smaller proteins. Similarly, very distant relationships may now be just discerned but cannot be seen in fine resolution. However, in spite of these limitations, dehydrogenases do illustrate several types of general change -- and illustrate exceptionally well the evolution of isozymes and of variation in size and subunit arrangements. Various aspects of these features, observable from homologies in primary structures, are discussed in the first section below. They establish important properties of dehydrogenase evolution and suggest different patterns of functional roles.

In addition to these firm conclusions on individual types of dehydrogenase, this whole group of proteins may eventually resolve more difficult questions about distant relationships. These aspects concern structures within the subunits and between different types of dehydrogenase and other proteins. Observed similarities may illustrate wide-ranging events in protein evolution. Several of these have been discussed before (Rossmann et al., 1975; Jörnvall, 1977b) but may now be summarized and related to new observations. This is outlined in the second section below.

No attempt is made to cover all reports of dehydrogenase structures. Examples frequently relate to alcohol dehydrogenase or the other well-studied enzymes, but are selected to emphasize general properties that may link different functional aspects.

INDIVIDUAL ENZYME GROUPS

AMINO ACID EXCHANGES

Comparisons of homologous structures from different species indicate the rate of accumulation of amino acid exchanges during evolution of a protein. In order to obtain reliable values for the changes, many structures must be known. However, relatively few amino acid sequences of dehydrogenases have yet been determined. Consequently, successive changes are difficult to trace, but a few conclusions are clear.

1. Rates of mutation acceptance during evolution.

This rate is often slow for dehydrogenases. They therefore occur comparatively low in lists of amino acid sequence variabilities of all presently investigated proteins (Dayhoff, 1972). This is consistent and probably reflects general properties of oligomeric enzymes. Large proteins with complex and conserved functions have slow evolutionary changes in other cases as well. However, this property of dehydrogenases means that glyceraldehyde 3-phosphate dehydrogenase and vertebrate glutamate dehydrogenase show far less species divergence than even the similar-sized protein chains or molecular aggregates in albumin and immunoglobulins.

2. Spread.

There is a considerable difference between known dehydrogenases. Thus, alcohol dehydrogenase is more variable in amino acid sequence than other dehydrogenases studied, as is seen in Table 1. For glutamate dehydrogenase values differ; consideration of only vertebrate structures indicates large conservations, but inclusion of fungal enzymes increases variations (Table 1). This difference is also coupled with other variations and is probably related to additional ancestral events of the type further discussed in the next section. For alcohol dehydrogenase, however, sequence variations are large both within vertebrates, and after inclusion of the enzyme from other sources.

The more constant dehydrogenases (Table 1) have fixed and defined functions in glycolysis or ammonia metabolism, and catalyze important physiological steps common to many organisms. The more variable alcohol dehydrogenase, however, does not have such a function conserved in all cases. The correlation between functional role and structural variability is therefore noticed also within the group of dehydrogenases.

3. Variations.

The extent of amino acid exchanges along the subunits varies. This is also coupled to functional roles, but again there appears to be a difference between dehydrogenases. The more constant glyceraldehyde 3-phosphate dehydrogenase has a highly conserved primary structure around the active site cysteine residue (Harris & Waters, 1976). No similarly long conserved region has been noticed for alcohol dehydrogenase (Jörnvall, 1977b). On the contrary, this enzyme has isozyme differences close to the zinc-liganding cysteine residues at the active site, and exchanges affecting even catalytically potentially important residues (Jörnvall, 1977c). The *Drosophila* alcohol dehydrogenase in particular appears so different that a conserved catalytic mechanism has not been taken for granted (Schwartz & Jörnvall, 1976).

Table 1. Rates of mutation acceptance.

Rates are in accepted point mutations per 100 residues and 100 million years, after corrections for superimposed mutations (Dayhoff, 1972). Values are approximate, since only few structures are known, but appear consistent with different sources, except for glutamate dehydrogenase, where inclusion of the *Neurospora* $NADP^+$-specific enzyme results in the value within parentheses.

Enzyme	Rate	
Glyceraldehyde 3-phosphate dehydrogenase	4	
Lactate dehydrogenase	5	
Glutamate dehydrogenase	2	(28)
Alcohol dehydrogenase	23	

4. Correlations with tertiary structures

The two types of domain in dehydrogenases (Rossmann et al., 1975) have no significant differences in evolutionary rates as judged from amino acid sequence variations in glyceraldehyde 3-phosphate (Olsen et al., 1975) or alcohol (Jörnvall, 1977b) dehydrogenases. Internal regions are conserved as in other proteins. These distributions are especially well noticed when the comparatively few amino acid sequence identities in distantly related variants are considered. This is shown in Fig. 1, where identities between the yeast and horse liver alcohol dehydrogenases have been outlined in relation to the conformation of the horse protein, determined by Eklund et al. (1976b).

A particular feature, clearly demonstrated in alcohol dehydrogenase, but presumably of wide-spread occurrence also in other homologous structures, is the presence of compensated exchanges (Eklund et al., 1976a; Jörnvall et al., 1978). These

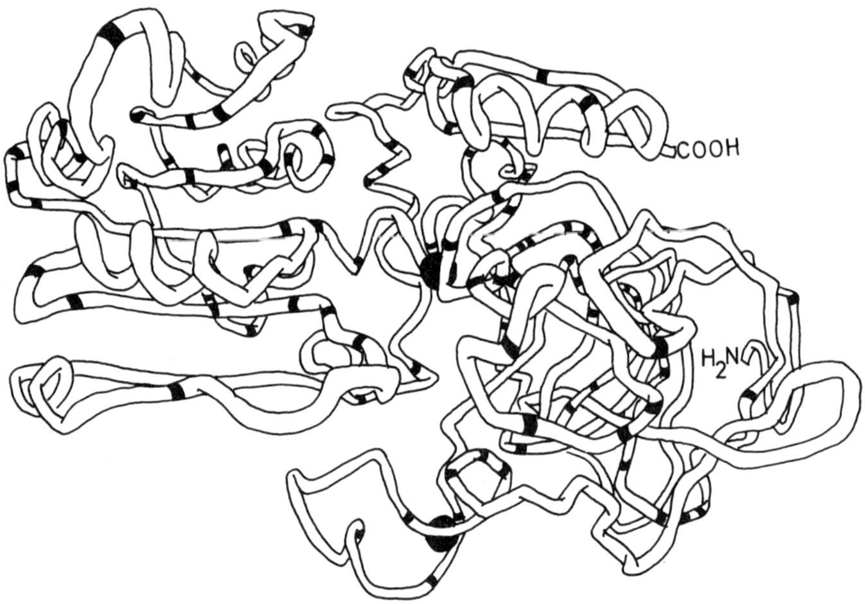

Fig. 1. Distribution of identical residues in amino acid sequences of distantly related alcohol dehydrogenases.

Amino acid sequence identities in the yeast and horse liver enzymes are illustrated black in the conformation of the subunit of the horse protein, determined by Eklund et al. (1976b).

allow comparatively unaltered packing in spite of extensive differences. Glycine is the most conserved type of residue in distantly related forms of dehydrogenase (Jörnvall, 1977b) as in other proteins, further demonstrating the importance of strict space requirements.

5. Subunit arrangements.

Dehydrogenases represent oligomeric enzymes that have been studied in detail. Hence, the amino acid sequences in areas for subunit interactions are of special interest. Expectedly, these are found to be highly conserved in glyceraldehyde 3-phosphate dehydrogenases with their unaltered quaternary structures (Olsen et al., 1975) but not in alcohol dehydrogenases with their different subunit arrangements (Jörnvall et al., 1978).

In lactate dehydrogenase, both these types of pattern have been shown. The subunit interaction corresponding to a similar region in malate dehydrogenase, and therefore presumably the oldest, is more conserved in amino acid sequence than the others (Eventoff et al., 1977). This therefore demonstrates that traces of the variation in associations may still be observed, although the lactate dehydrogenases compared are now all tetramers.

6. Stability.

Apart from the above characteristics, some more special physicochemical aspects have also been studied in sequence variations of dehydrogenases.

Thermophilic glyceraldehyde 3-phosphate dehydrogenases, stable at 70-90°C, have been compared to the ordinary mesophilic enzymes (Harris & Walker, 1977). In addition, halophilic malate and glutamate dehydrogenases, active at 4 M salt concentrations, have been purified (Eisenberg et al., 1977). Differences in the thermophilic enzymes are surprisingly small, and appear to correspond to additions of only single stabilizing bonds, whereas the halophilic enzymes have considerable changes which involve charges (extra acidic residues, but losses of basic ones).

ISOZYME DEVELOPMENTS

Many dehydrogenases have isozymes due to the occurrence of more than one type of subunit in the oligomeric molecules. It is possible to discern various classes of multiple forms in different arrangements. Several of these were originally first demonstrated in dehydrogenases, and the correlations between catalytic and structural differences of isozymes are still best known in this group of proteins.

The different classes are discussed below. It will be noticed that the last two depend on essentially identical mechanisms, only functional effects in the protein molecules differ, explaining the apparent distinctions. The multiple forms demonstrate various roles of isozymes, show different structural patterns and form important steps in protein evolution.

1. Modified forms.

Although this multiplicity is generally not included among isozymes, it is best discussed in this connection. It is well illustrated with mammalian alcohol dehydrogenases, for which a group of sub-forms has been described for each parent isozyme (Brändén et al., 1975). Post-translational modifications of the molecules explain the variable sub-forms. Differential labeling early suggested that oxidation of thiol groups may be one mechanism. A recently demonstrated variation in the extent of acetylation of yeast alcohol dehydrogenase (Wills & Jörnvall, 1979a) may suggest another mechanism, and the existence of yet others is likely.

At least some of the modified forms may be of functional significance in the metabolism of these proteins. It is possible that they are formed in steps of protein catabolism. Thus, some mammalian forms change slowly with time (Brändén et al., 1975). The yeast acetylation variability has been noticed to occur during isozyme change-over in metabolic switches from anaerobic to aerobic states (Wills & Jörnvall, 1979a). In that system, however, it seems to be mainly the newly synthesized isozyme rather than the old type that is non-acetylated.

2. Allelic variations.

This type of variation illustrates point mutations, and is well-studied in many systems. Drosophila alcohol dehydrogenase has two common allelic forms, S and F (for slow and fast migration in electrophoresis). Other forms have been shown to differ by single amino acid exchanges of glycine for glutamic acid (Schwartz & Jörnvall, 1976) and of alanine for aspartic acid (Thatcher & Camfield, 1977). Strain differences with probably allelic explanations have also been shown in yeast for glyceraldehyde 3-phosphate dehydrogenase (Jones & Harris, 1972) and for alcohol dehydrogenase (Wills & Jörnvall, 1979a). In both instances, the differences are small, being exchanges of the branched-chain residues valine and isoleucine at one and two positions, respectively.

These population differences appear unimportant. However, those mutations which alter functional properties are of particular value both to elucidate molecular mechanisms and physiological roles of the enzymes.

Three such mutations have been characterized in alcohol dehydrogenase, one spontaneous mutation in man (Berger et al., 1974) and two artificial ones in yeast (Wills & Jörnvall, 1979b). The human atypical form indirectly affects coenzyme binding (Brändén et al., 1975). Population differences in this enzyme have extensive interest in relation to alcoholism, as demonstrated for another yet uncharacterized form (Li et al., 1977). The yeast mutations demonstrate directly the importance of an arginine residue for strong binding of the coenzyme's pyrophosphate (Wills & Jörnvall, 1979b), which supports data from crystallographic analyses and chemical modifications (Lange et al., 1975).

3. Hybridizable forms derived from duplicated genes.

This step in development of new functions is very well illustrated from primary sturctures. It is the explanation for the familiar occurrence of isozyme systems with two types of subunits, as in lactate and some alcohol dehydrogenases.

Amino acid sequences of both types of subunits of lactate dehydrogenase have been compared (Eventoff et al., 1977). The M and H forms differ more between each other in one species (24-25% in the investigated cases) than identical types of subunits between different species (about 10% in the same cases). Quantitative proof of earlier suggestions from observations of peptide maps has therefore been obtained. These results show that the duplicatory event which has led to the synthesis of M and H subunits is old, and that the two types of subunits became early locked in fixed functions. Consequently, much of the species variation has been added later, and is in all probability functionally less important. A schematic representation of the events in the evolution of lactate dehydrogenase, as deduced from presently known primary structures, is given in Fig. 2B.

Alcohol dehydrogenases show another pattern of isozyme divergence. In this case, the structural differences explaining the well-known ethanol/steroid-active subunits of the horse liver enzyme are not apparent in other species studied. Thus, although isozymes also occur in man, their structures appear to differ in other positions than in the horse. Furthermore, the rat has probably neither of these two types of differences (Jörnvall, 1977c).

Hence, in alcohol dehydrogenases, presently known primary structures suggest that different isozyme subunits within a species are more similar than subunits from different species. This is the opposite of the situation for lactate dehydrogenase. It appears that duplications have occurred repeatedly in the isozyme evolutions of alcohol dehydrogenases. The situation that best explains presently determined structures is schematically represented in Fig. 2A. The actual positions in alcohol dehydrogenase of the residues that have been exchanged in different isozymes (including investigated alleles) are shown in Fig. 3, which further demonstrates that none of the forms have known amino acid substitutions in common.

The isozyme patterns in Fig. 2A and B suggest evolutionary differences between dehydrogenases in the same manner as the rates of mutation acceptances (Table 1). In both properties, alcohol dehydrogenase is the most variable dehydrogenase as judged from presently known structures. Its isozyme patterns can therefore also be related to functional aspects, probably to the lack of a universal and fixed physiological role.

Even if consistent, the parallelism between rates of mutation acceptance and of duplicatory events is still remarkable, since the genetic mechanisms are different. In any event, variable structures from mutations have apparently also led to multiple structures from duplications. In alcohol dehydrogenase this is not confined to the mammalian enzymes, since comparatively recent duplications have also occurred in the evolution of the yeast enzyme, as indicated in Fig. 1A and shown by extensive homologies between yeast isozymes I and II (Wills & Jörnvall, 1979a). It is therefore possible that duplicatory rates may be equally characteristic properties of individual proteins as the mutational rates. Repeated duplications during evolution have also been studied experimentally (Rigby et al., 1974).

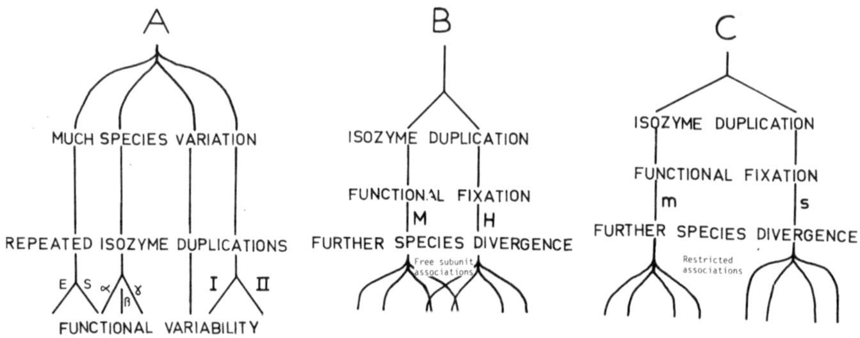

Fig. 2. Different patterns of isozyme development.

Scheme A fits present structures of alcohol dehydrogenases, and the designations at the bottom correspond to the various known isozyme subunits in, respectively, horse, man, rat and yeast. Scheme B fits present M and H structures of lactate dehydrogenases. Scheme C can presumably fit malate or aldehyde dehydrogenases, although structural proof is largely lacking. The letters m and s symbolize the different intracellular localizations (mitochondrial, soluble/cytoplasmic).

All relative scales, including time in the vertical dimension, are arbitrary. Functional co-variation, symbolized by the inter-locking species variations in B, may occur from other requirements than just common subunit associations. Further complexities may also be envisaged from different multiplicities in ancestral building units during isozyme developments. This probably applies to alcohol (and glutamate) dehydrogenases but is uncharacterized and is not represented in the figure. Future adjustments to the schemes are therefore likely but three principally different schemes have emerged so far.

In the isozymes thus far discussed, the amino acid exchanges affect catalytic properties. A few exchanges explain such differences in lactate dehydrogenase (Eventoff et al., 1977). In horse liver alcohol dehydrogenase one substitution appears to affect a loop that restricts the substrate pocket (Brändén, 1977). In both these isozyme systems, however, there is no evidence that exchanged residues affect areas for subunit interactions. The two types of subunit therefore associate in the same way, forming all possible hybrids in amounts proportionate to relative subunit concentrations.

4. Separated forms.

Isozymes that do not form hybrid oligomers are also common among dehydrogenases, although not yet studied in the same

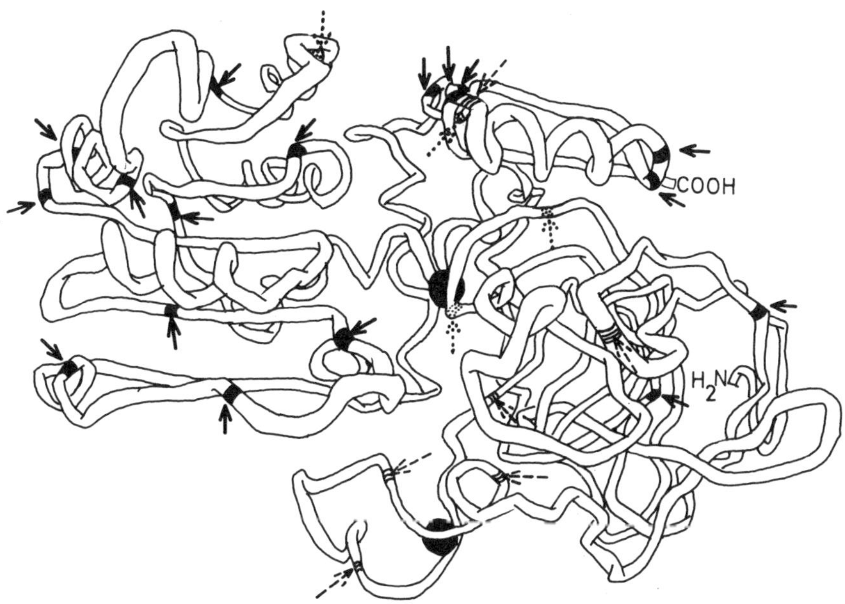

Fig. 3. Positions of known amino acid substitutions in isozyme-subunits in alcohol dehydrogenases.

Full black: differences between the two types of subunits in the yeast isozymes. Discontinuous black: differences between the two horse isozymes. Spotted black: differences among the human isozymes; those between yeast mutants are also shown spotted. Actual substitutions from Brändén et al. (1975) and from Wills & Jörnvall (1979a). All positions shown in relation to the subunit conformation of the horse enzyme, which was determined by Eklund et al (1976b).

detail as the previous classes. Malate and liver aldehyde dehydrogenases may illustrate this system. Both these enzymes occur in two forms with different intracellular localizations: cytoplasmic and mitochondrial. For each enzyme, the two forms have identical quaternary structures and similar sizes (2 subunits with molecular weights of about 33,000 for malate dehydrogenase, 4 with weights of about 52,000 for aldehyde dehydrogenase).

The two forms of each enzyme show other similarities apart from catalyzing the same reactions. Both forms of aldehyde dehydrogenase are inhibited by disulfiram (although with highly different sensitivities) indicating similarities at the active sites. They also have related amino acid compositions (Eckfeldt et al., 1976). The latter is true for the malate dehydrogenases, too (Banaszak & Bradshaw, 1975). Hence, it appears likely that for both enzymes the isozymes are synthesized from genes derived from duplications.

Subsequent mutational exchanges, however, have apparently been large. Although not yet proven, it is probable that they have affected the areas for subunit interactions to such an extent that non-identical subunits no longer fit together. The schematic representation of this evolution is shown in Fig. 2C. Similar mechanisms to those discussed above, gene duplications followed by mutational events, are therefore expected to form the explanation of the type of conceptually different isozymes where hybrids are not observed, and where subunits apparently may segregate in different cellular compartments.

Scheme C in Fig. 2 is structurally the least documented one. However, changes in the areas for subunit interactions according to this scheme have also been demonstrated in species variations of dehydrogenases, as noted above.

It should be emphasized that additional types of isozyme occur. This applies not only to other dehydrogenases but also to the enzymes previously mentioned, especially lactate dehydrogenase (Holbrook et al., 1975). Furthermore, only a few structures have yet been characterized. Present schemes in Fig. 2 may therefore be incomplete or may exaggerate some of the differences. For example, isozyme patterns of alcohol dehydrogenase in the mouse (Holmes et al., 1978) suggest the presence of a common gene multiplicity for this enzyme in mouse and man. However, numbers of structurally known duplications and their corresponding positional differences are still greater in alcohol dehydrogenase than in lactate dehydrogenase. In any event, Fig. 2 is not intended to establish details but to show the occurrence of principally different isozyme patterns along with schemes that fit presently analyzed structures.

For glutamate dehydrogenase from *Neurospora crassa*, another type of different forms has been investigated. Here, the primary structures of the NAD^+-specific and the $NADP^+$-specific forms are extremely different (Austen et al., 1977); the subunits also differ largely in size (116,000 versus 48,000) and in quaternary structure (tetramers versus hexamers). Mechanisms behind these patterns are unknown. A distinct proof for ancestral linkage between these forms is still lacking (Austen et al., 1977), but gene fusions or further duplications have probably occurred.

These patterns are more related to the questions of further relationships within dehydrogenases, and the tracing of evolution of constituent protein chains (next section).

SUMMARY OF COMPARISONS OF DIFFERENT FORMS

Even if the distinctions between isozyme classes are oversimplified, the structures suggest non-identical patterns of evolutionary lines. Different relative orders and rates of duplicatory events and mutations create various isozyme patterns. These are related to functional properties and reflect catalytic mechanisms, subunit arrangements and intracellular localizations, which can all be fixed at various levels. Minor and unimportant exchanges are noticed; small but functionally critical exchanges occur, and large variations causing non-hybridizable subunits or non-identical quaternary structures are also found. Indeed, the chemically and functionally most dissimilar isozymes and species variations almost simulate the evolution of "new" proteins.

A novel concept of parallelism in several different variables is noticed, that is, excess of mutational exchanges appears to be linked to acceptance of more duplicatory events, as illustrated with alcohol dehydrogenase. In addition, we will see below that internal organization in the corresponding gene has been submitted to changes.

Finally, the inter-relationships between isozymes should be noticed. The random subunit associations between M and H types of lactate dehydrogenases indicate that common restrictions operate in the respective species divergences. This covariation, symbolized in Fig. 2B, may form one of the explanations of the fact that the two types of subunit of lactate dehydrogenase appear to have roughly the same rate of mutation acceptance, although they have partly different physiological functions and organ distributions. It is possible that such covariations may be different in the C-type scheme, producing more dissimilar phylogenetic trees, as indicated in Fig. 2C. To what extent this possibility holds true remains to be seen. In any event, further isozyme studies can define additional relationships of functional significance.

INTER-ENZYME COMPARISONS

TYPE OF COMPARISONS

The structural homologies discussed so far have revealed properties for each enzyme group. Different types of change have developed, and functions have become fixed at various levels.

Sub-forms, isozymes and conceptually different enzymes can be related within each type of activity. But the individual structures and activities must also have evolved from earlier forms. Many attempts have been made to trace these relationships.

1. Background

One approach is to compare different dehydrogenase groups for common structural features that could indicate ancestral linkages. Another approach is to screen single subunits, or non-equivalent parts of any subunits, in order to detect evidence for ancestral building blocks.

Early comparisons were based on the fact that many dehydrogenases have reactive cysteine residues. It became evident, however, that similar reactivity does not necessarily imply identical function. In the three best studied dehydrogenases, these cysteine residues have completely different catalytic functions. Furthermore, reactive residues may vary with the experimental conditions or with the enzyme species (Twu et al., 1973; Woenckhaus et al., 1979). In this context, it may be noted that identities between different dehydrogenases have also been suggested from apparent coincidences of enzymatic properties (Holmes et al., 1978). If correct, this would of course illustrate a further example of relationships between different enzyme specificities and of developments of new functions.

The availability of complete primary structures made it possible to compare numerically identical positions in different dehydrogenases. On the other hand, crystallographically determined tertiary structures permitted amino acid sequence alignments based on conformational similarities. These similarities led to a scheme showing the evolution of the common coenzyme-binding domain in dehydrogenases, which has been summarized before (Rossmann et al., 1975). The variation in properties among alcohol dehydrogenases prompted new sequence comparisons. It was then found that sequence patterns could fit into a very wide-ranging evolutionary scheme, although distinctions are difficult because of the distant relationships (Jörnvall, 1977b).

In common, all studies suggest ancestral linkages between different dehydrogenases, and an evolution of present molecules from building units derived by gene duplications and other rearrangements. Several details are unclear and some suggestions appear to be partly conflicting. These difficulties mainly concern how far relationships may now be traced into other proteins, and which regions of present subunits are ancestrally linked.

The unsettled questions in these distinctions are inherent in the comparisons because the relationships are old and obscured by subsequent mutations, duplications, gaps and insertions. Furthermore, effects of sequence changes on folding are still incompletely known, complicating correlations between chemical, functional and crystallographic similarities. Finally, the suggested genetic mechanisms are little studied. These aspects will be discussed in relation to recent results, after a summary of the earlier observations.

2. Sequence similarities related to tertiary structures of dehydrogenases.

Lactate, alcohol, malate and glyceraldehyde 3-phosphate dehydrogenases are composed of two domains (Rossmann et al., 1975): one (the coenzyme-binding domain) is similar in these enzymes and has two sub-parts (mononucleotide-binding units), the other (the catalytic domain) is different in each enzyme, except the malate/lactate dehydrogenase pair. A common ancestral origin is accepted for the mononucleotide-binding units (Rossmann et al., 1975). Duplication of the gene for this part, followed by fusion with genes coding for different catalytic domains can explain the existence of all these dehydrogenases. Kinases and some other enzymes are also included in this evolutionary system (Rossmann et al., 1975).

Based on these similarities, corresponding amino acid sequences can be aligned, although gaps and irregularities have to be introduced in the mononucleotide-binding units from different dehydrogenases. Other dehydrogenases, without crystallographically determined conformations, also have parts that fit into this scheme (Smith et al., 1975; Otto, 1977), supporting the conclusions. In these amino acid alignments, three features are inherent:

a/ The sequence similarities align parts of different dehydrogenases. The similar regions differ numerically and are in the C-terminal half of alcohol dehydrogenase but in slightly shifted N-terminal halves of the other enzymes (Fig. 3 in Rossmann et al., 1977). Numerically unshifted sequence similarities between alcohol and the other dehydrogenases are therefore unexplained by known relationships in conformation.

b/ Sequence similarities between widely separated regions within single subunits are not expected for any of these dehydrogenases from the observed domain relationships. The two mononucleotide-binding units that have been related are adjacent in one domain.

c/ The sequence alignments are based on conformational and functional similarities. The amino acids therefore form another way of expressing these similarities. Consequently, in the sequence homologies observed, coincidences need not all indicate genetic relationships, but can also to some extent reflect common folding or binding requirements.

3. Sequence similarities related to all regions of dehydrogenases.

Similarities early reported among alcohol, lactate, glyceraldehyde 3-phosphate and glutamate dehydrogenases have been summarized before (Rossmann et al., 1975). The early observations showed low degrees of identity or affected short regions and did not consider presently known conformational properties. Some of these similarities have been questioned on statistical (Williams & Wilkins, 1974) or conformational (Eventoff & Rossmann, 1975) grounds. More extensive comparisons, performed in relation to alcohol dehydrogenase show considerably higher similarities, which have been presented in detail (Jörnvall, 1977b). Sequence patterns have also been summarized graphically (Jörnvall, 1977d). The main results of these studies, and the evolutionary interpretations are:

a/ Segments. Sequence similarities, unrelated to conformational patterns, may be traced between different dehydrogenases and within individual subunits. In common, all these sequences affect segments of the molecules, with a slight over-representation in the N-terminal parts. Together, they cover about half of the subunits of alcohol and glyceraldehyde 3-phosphate dehydrogenases in a complex fashion. Some reported similarities also affect lactate dehydrogenase and even aldolase.

b/ Identities. In 21 segment similarities, ranging in size up to 57 residues, shifts of from 6 to 193 residues occur. In 10 segments, each with a size of about 20 residues, the average identity is 47%. Two examples are given in Fig. 4, and are further discussed below.

c/ Significance. Questions inherent in this type of comparisons concern the significance of the sequence similarities. Statistical estimates vary both with the segments and with the methods of calculation, but several comparisons appear to give probability values for random occurrence below 1%. Segments in the comparisons also resemble each other at least to the same extent as known species variations.

d/ Scheme. An evolutionary scheme based on these observations suggests distant ancestral building units, comprising about 80 residues. The units occur repetitively and may have irregularities of up to about 30 residues. Entire chains of the common dehydrogenases studied may be explained by four such repetitive units, derived from duplications and modified by insertions/deletions or partial duplications. This scheme can explain all sequence coincidences and observed crystallographic similarities. The repeating unit corresponds to the mono-nucleotide binding unit in size and evolution but is not suggested to be limited to only the coenzyme-binding domain.

e/ Other enzymes. Sequence similarities are not restricted to dehydrogenases but may include other oligomeric enzymes, as shown for aldolase (Jörnvall, 1977b). This is a further extension of the similarities with kinases, suggested from tertiary structures (Rossmann et al., 1975).

f/ Multiplicity. The repeated similarities appear compatible with observed size classes for these enzymes. Alcohol dehydrogenases have two subunit ranges: 25,000 and 35,000-40,000. These could correspond to a difference in multiplicity of building units (3- and 4-fold, respectively). The same size ranges and two further ones, of 55,000 and 115,000 that could correspond to additional multiplicities, occur in all investigated dehydrogenases. In fact, all glycolytic and many oligomeric enzymes apparently have these ranges of subunit size.

In summary, if it holds true this scheme based on all the reports of possible homologies in primary structures may have far-reaching evolutionary consequences. It can extend the conclusions from the conformational interpretations on dehydrogenases and kinases to entire subunits and to further protein groups. It may also demonstrate new relationships between sequence and folding.

SIGNIFICANCE IN SEQUENCE SIMILARITIES

1. Distant relationships.

Statistical problems in tracing distant relationships were noted in relation to the segment comparisons above. Already the established dehydrogenase connections have variations large enough to approach values expected for chance similarities. Thus, in one dehydrogenase, alcohol dehydrogenase, the whole horse and yeast enzymes have low sequence homologies (25% identity), which in some regions become completely invisible (9% identity, i.e. as expected from chance). The species alignments also include frequent gaps, long insertions/deletions and variations in subunit size. Segment comparisons are expected to reflect older events in shorter sequences with further irregularities! It is perhaps more surprising that some good segment similarities have been found, rather than that a few appear inconsistent and unacceptable.

2. Examples.

A distinction should be made between two groups of reported sequence similarities. One group contains many of the early comparisons quoted above. These similarities between single pairs are small or short. For example, one comparison contains individual pairs with only 3 identities in a sequence of 23 residues (Bennett, 1974). Another comparison contains segments down to 6 residues (Engel, 1973). Consequently, these similarities have been questioned. Naturally, they may still be true but further discussion is difficult. In this group, however, it is clear that the questions regard the actual degree of identity.

The other group of sequence similarities is the one summarized above. It has larger identities without involving exceptionally short segments. In this group, segment similarities can be equally as impressive as species or conformational similarities. This is illustrated in Fig. 4, where four different pairs of comparable length and degree of sequence similarity are shown. They have been selected in the following way:

Pair A shows the most similar pieces of this size in the sequence comparisons of the coenzyme binding domains of alcohol dehydrogenases from horse (top) and yeast (bottom). This alignment thus compares maximally equivalent parts in a variation of a conserved conformation. The similarity definitely reflects a genetic significance, and is derived from a common ancestral origin.

Pair B shows a maximal similarity between regions of rabbit muscle glycerophosphate dehydrogenase (top) and dogfish muscle lactate dehydrogenase (bottom) according to Otto (1977). It fits excellently into relationships from known tertiary structures, and it has been suggested to indicate the highest degree of family connection of all investigated dehydrogenases (Otto, 1977).

Pair C shows a similarity between two slightly shifted regions of horse liver alcohol dehydrogenase. It corresponds to one of the suggested partial duplications, causing irregularities in multiplicity. It has a 29-residue shift and is in the border between repetitions of the supposed building unit. It includes the inter-domain connection (Eklund et al., 1976b) in the tertiary structure. It can also be associated with a sequence coincidence in the preceding border region between the supposed units (at positions around 100 in the enzyme, according to Table 2 in Jörnvall, 1977b).

Pair D shows a similarity between horse liver steroid-active alcohol dehydrogenase (top) and rabbit muscle aldolase (bottom). This is an example of the extension of sequence similarities to further oligomeric enzymes in the glycolytic pathway. It should be noticed that both segments are almost unshifted, and close to each protein's N-terminus. No further similarity of this size has been reported in relation to these enzymes, but an explanation for the greater similarity in the N-terminal regions of possibly distantly related structures is inherent in the fact that insertions/deletions increasingly may disturb order from a common starting point.

Thus, of the four pairs in Fig. 4, the first is an undisputed sequence variation, the second corresponds to an accepted similarity in tertiary structure, the third to the possible shifts of multiplicity in an evolutionary scheme, and the fourth to an extension to further proteins. Pairs A/B cannot be considered more significant than C/D from the similarities in Fig. 4.

It is clear that the actual identities noticed from the repetitive segments are sufficient in relation to the established connections. The crucial point is therefore if the size of the segments compared is sufficient. Considering the suggested irregularities of the repeated unit, the established frequency of insertions/deletions in individual dehydrogenases, and the extremely distant relationships searched for, the sizes of the similarities do not seem unreasonable.

In summary, Fig. 4 demonstrates the types of similarities that apply, the difficulties in distinctions, and the crucial point in present comparisons. Provided several coincidences with 20-odd residues, as in Fig. 4, are enough, and the known enzyme variations are representative, an extension of the dehydrogenase evolution to further families and to generally repetitive units is significant.

Pair	Sequence	Identity
A	178 Thr-Gly-Tyr-Gly-Ser-Ala-Val-Lys-Val-Ala-Lys-Val-Thr-Gln-Gly-Ser-Thr-Cys-Ala-Val-Phe-Gly-Leu- -Gly-Gly-Val-Gly 204	11/28
	157 Thr-Val-Tyr- -Lys-Ala-Leu-Lys-Ser-Ala-Asn-Leu-Met-Ala-Gly-His-Trp-Val-Ala-Ile-Ser-Gly-Ala-Ala-Gly-Gly-Leu-Gly 183	
B	176 Val-Ala-Val-Pro-Asp-Val-Val-Lys-Ala-Ala-Ala-Asp-Ala-Asp-Ile-Leu-Ile-Phe-Val-Val-Pro-His-Gln-Phe-Ile-Gly-Lys-Ile 203	6/28
	122 Phe-Ile-Ile-Pro-Asn-Ile-Val-Lys-His-Ser-Pro-Asn-Cys-Ile-Ile-Leu-Val-Val-Ser-Asn-Pro-Val-Asp-Val-Leu-Thr-Tyr-Val 148	
C	157 Val-Ala-Lys-Ile-Asp-Ala-Ala-Ser-Pro-Leu-Glu-Lys-Val-Cys-Leu-Ile-Gly-Cys-Gly-Phe-Ser-Thr-Gly-Tyr-Gly 181	9/25
	186 Val-Ala-Lys-Val-Thr-Gln-Gly-Ser-Thr-Cys-Ala-Val-Phe-Gly-Leu-Gly-Gly-Val-Gly-Leu-Ser-Val-Ile-Met-Gly 210	
D	12 Ala-Val-Leu-Trp-Glu-Gln-Lys-Lys-Pro-Phe-Ser- -Ile-Glu-Glu-Val-Glu-Val-Ala-Pro-Pro-Lys 32	11/22
	6 Ala-Leu-Thr-Pro-Glu-Gln-Lys-Lys-Glu-Leu-Ser-Asp-Ile-Ala-His-Arg-Ile-Val-Ala-Pro-Gly-Lys 27	

Fig. 4. Comparisons of four different sequence similarities.

Examples chosen to show the indistinguishable degree of sequence identity between species variations (A), conformational similarities (B), shifted segment similarities (C) and similarities between completely different enzymes (D). If anything, the unexpected similarity in D is the most impressive. Enzyme species and further details are given in the text.

CONFORMATIONAL CONSEQUENCES

1. Generally conserved tertiary structures.

Apart from the statistical problems, unknown aspects of the segment similarities concern the conformational consequences. It has been found that tertiary structures generally appear to be better conserved than primary structures in distant relationships (Richardson et al., 1976), allowing the clear demonstration of the common coenzyme-binding domain in dehydrogenases (Rossmann et al., 1975). Therefore, the tracing of repetitive segments throughout dehydrogenase subunits has been questioned, since it suggests sequence similarities where conformations are different. Is this possible?

One aspect of this question has been discussed before (Jörnvall, 1977b). It was then recalled that point mutations are known where single amino acid substitutions have large conformational effects. This applies to abnormal globins (Dayhoff, 1972), and probably to many inborn errors of metabolism, where unstable proteins can occur. In these cases, folding has apparently changed more than the amino acid sequence.

Another aspect of the question of the similarities in sequence versus folding concerns the mechanisms behind any conservation at all. It appears that a reason for constant tertiary structures is some functional requirement, as emphasized for all other evolutionary variations in this chapter. Functional requirements of a structure must dictate which conformations are acceptable. Consequently, generally conserved tertiary structures are likely to indicate common functional requirements. This also applies to the similarities in the coenzyme-binding domains of dehydrogenases.

2. Functional requirements.

The functional requirements are most stringent for the tertiary structure because it is the conformation that determines crucial properties. Primary structures can generally vary as long as they fit the conformation. This was shown in the previous section regarding the conservation of internal residues and glycine in dehydrogenases. Constant functional requirements can therefore explain why tertiary structures generally change less than primary structures.

However, it does not seem impossible that this normal relationship between tertiary and primary structures can be reversed. Thus, if functional requirements change during evolution, conformations may also be expected to do so, like all other properties. When this happens before all ancestral sequence characteristics have been lost, any remaining sequence resemblance in a new conformation would be due to the original connection. Therefore, repetitive sequence similarities in conformationally different domains of dehydrogenases cannot be excluded, although definite examples of this situation are still unknown (Richardson et al., 1976).

In summary, the conformational dissimilarities in dehydrogenases need not contradict the observed sequence patterns. On the contrary, they may serve to illustrate a further evolutionary coupling with functional properties.

GENETIC CONSEQUENCES, SIMILAR MECHANISMS IN OTHER PROTEIN SYSTEMS

The evolutionary events suggested from the sequence similarities of dehydrogenases, as well as from the domain relationships, require varied and repeated gene fusions. Other examples of such mechanisms have long been known in immunoglobulins, ferredoxins and the α_2-chain of haptoglobin (Dayhoff, 1972). Extensive duplications also occur in aminoacyl tRNA synthetases (Waterson & Konigsberg, 1974; Bruton et al., 1974; Koch et al., 1974). More complicated patterns, involving repeated elements, fusions and irregularities are found in plasminogen and prothrombin (Magnusson et al., 1976), and in albumin (McLachlan & Walker, 1977). Similarities between the gene for ovalbumin and two other, preceding genes (Royal et al., 1979), further demonstrate the occurrence of duplications. An apparent fusion between an oxidoreductase and another protein is also known. Dihydrofolate reductase has been suggested to be structurally related to the C-terminal fourth of the very long subunit of β-galactosidase, indicating a possible ancestral linkage (Hood et al., 1978).

On the genetic level, intervening sequences have been demonstrated, and splicing appears to be common in eucaryotic genes (Jeffreys & Flavell, 1977; Breathnach et al., 1977). It has also been suggested that an insertion in otherwise homologous proteins can be explained by a mutation that suddenly turns an adjacent normally non-coding DNA region into a coding state (Cohen et al., 1978). This could be one mechanism linking the chain length differences in proteins with genetic events. In this connection, the internal chain difference in yeast/horse liver alcohol dehydrogenases, involving a 41-residue long segment (Jörnvall, 1977b), may be of interest.

The genetic organizations behind the sequence patterns in dehydrogenases are yet unknown. It remains to be proven whether the internal differences or the sequence similarities have any correlation with special gene structures. However, the splicing systems are processes of great interest in dehydrogenase evolution. The frequent occurrence of repetitive elements, apparent fusions and irregularities in additional protein systems are also noticeable. They show that the sequence patterns observed in dehydrogenases are not limited to this protein group, but may demonstrate common mechanisms in protein evolution.

CONCLUSIONS AND FUTURE ASPECTS

Primary structures of dehydrogenases establish evolutionary properties for each enzyme group. Evolution is closely linked to functional aspects at various levels, and different stages of isozyme development are clearly seen. In addition, comparisons of the few primary structures presently available suggest the possibility of repeated common elements in large parts, if not entire regions of constituent subunits. This may reflect ancestral building units in an evolutionary system that can also include further glycolytic and oligomeric enzymes.

All these conclusions support similar observations from tertiary structures. Apparent inconsistencies between conformational and sequential similarities should probably not be overemphasized. They may simply indicate effects of extensive changes in functional requirements during evolution. Hence, two types of domain may have developed separately for a long time in spite of a possibly common building unit.

Genetic mechanisms required for all types of similarities include duplications and fusions, with different multiplicities and irregularities. The relative order of these events is unknown, and need not be the same in all the enzymes. The fusions of genes corresponding to whole domains, can therefore generally be later events as suggested from the crystallographically observed similarities. Hypothetically, it is possible that various combinations could lead to functional fixations at different levels in the evolution of the constituent protein chains, exactly as appears to be the case from primary structures for the more recent isozyme developments.

Further studies are obviously required to follow the evolutionary lines of dehydrogenases in greater detail. One future aspect concerns the relationships between primary and tertiary structures and their links to functional aspects. Similarities between proteins completely outside the dehydrogenase group are here also of interest. For example, the β-chain of haptoglobin has a clear sequence homology with serine proteases (Kurosky et al., 1974), but functions are different and conformational relationships unknown. Furthermore, immunoglobulin domains and the subunit of superoxide dismutase have similar folding patterns (Richardson et al., 1976). These and other groups, as well as dehydrogenases, may establish further relationships between sequence, conformation and function for proteins in general.

Other future aspects concern actual dehydrogenases. Additional isozymes, species variants and new enzymes need further investigation. Evolutionarily, dehydrogenases with widely different isozymes or variations in chain-lengths might be of special interest. This should establish the relationship between the "separated" isozymes and possibly show additional isozyme classes. It will also show which parts may be missing or added in subunits with shorter or longer protein chains, and how this is related to conformation and to genetic organization.

ACKNOWLEDGEMENTS

The author is grateful to Mr. Furugren for the drawing of Figs. 1 and 3, and to the Swedish Medical Research Council for support (project 13X-3532).

REFERENCES

Austen, B.M., Haberland, M.E., Nyc, J.F. & Smith, E.L. (1977) J. Biol. Chem. 252, 8142-8149.

Banaszak, L.J. & Bradshaw, R.A. (1975) The Enzymes, 3rd edn. (Boyer, P.D., ed.), vol. 11, pp. 369-396, Academic Press, New York.

Bennett, C.D. (1974) Nature 248, 67-68.

Berger, D., Berger, M. & von Wartburg, J.-P. (1974) Eur. J. Biochem. 50, 215-225.

Brändén, C.-I. (1977) in Alcohol and Aldehyde Metabolizing Systems (Thurman, R.G., Williamson, J.R., Drott, H.R. and Chance, B., eds.) vol. 2, pp. 17-32, Academic Press, New York.

Brändén, C.-I., Jörnvall, H., Eklund, H. & Furugren, B. (1975) The Enzymes, 3rd edn. (Boyer, P.D., ed.), vol. 11, pp. 103-190, Academic Press, New York.
Breathnach, R., Mandel, J.L. & Chambon, P. (1977) Nature 270, 314-319.
Bruton, C.J., Jakes, R. & Koch, G.L.E. (1974) FEBS Lett. 45, 26-28.
Cohen, L.H., Westerhuis, L.W., Jong, W.W. de & Bloemendal, H. (1978) Eur. J. Biochem. 89, 259-266.
Dayhoff, M.O. (1972) Atlas of Protein Sequence and Structure, vol. 5, National Biomedical Research Foundation, Silver Spring; (1973) supplement 1; (1976) supplement 2.
Eckfeldt, J., Mope, L., Takio, K. & Yonetani, T. (1976) J. Biol. Chem. 251, 236-240.
Eisenberg, H., Leicht, W., Mevarech, M. & Werber, M.M. (1977) in Pyridine Nucleotide-Dependent Dehydrogenases (Sund, H., ed.), pp. 368-380, Walter de Gruyter Co., Berlin and New York.
Eklund, H., Brändén, C.-I. & Jörnvall, H. (1976a) J. Mol. Biol. 102, 61-73.
Eklund, H., Nordström, B., Zeppezauer, E., Söderlund, G., Ohlsson, I., Boiwe, T., Söderberg, B.-O., Tapia, O., Brändén, C.-I. & Åkeson, Å. (1976b) J. Mol. Biol. 102, 27-59.
Engel, P.C. (1973) FEBS Lett. 33, 151-153.
Eventoff, W. & Rossmann, M.G. (1975) Crit. Rev. Biochem. 3, 111-140.
Eventoff, W., Rossmann, M.G., Taylor, S.S., Torff, H.-J., Meyer, H., Keil, W. & Kiltz, H.-H. (1977) Proc. Natl. Acad. Sci. USA 74, 2677-2681.
Harris, J.I. & Walker, J.E. (1977) in Pyridine Nucleotide-Dependent Dehydrogenases (Sund, H., ed.), pp. 43-61, Walter de Gruyter Co., Berlin and New York.
Harris, J.I. & Waters, M. (1976) The Enzymes 3rd edn. (Boyer, P. D., ed.) vol. 13, pp. 1-49, Academic Press, New York.
Holbrook, J.J., Liljas, A., Steindel, S.J. & Rossmann, M.G. (1975) The Enzymes, 3rd edn. (Boyer, P.D., ed.), vol. 11, pp. 191-292, Academic Press, New York.
Holmes, R.S., Jones, J.T., & Peters, J. (1978) J. Exper. Zoology 206, 279-288.
Hood, J.M., Fowler, A.V. & Zabin, I. (1978) Proc. Natl. Acad. Sci. USA 75, 113-116.
Jeffreys, A.J. & Flavell, R.A. (1977) Cell, 12, 1097-1108.
Jones, G.M.T. & Harris, J.I. (1972) FEBS Lett. 22, 185-189.
Jörnvall, H. (1977a) Biochem. Soc. Transact. 5, 636-639.
Jörnvall, H. (1977b) Eur. J. Biochem. 72, 443-452.
Jörnvall, H. (1977c) in Pyridine Nucleotide-Dependent Dehydrogenases (Sund, H., ed.), pp. 62-71, Walter de Gruyter & Co., Berlin and New York.
Jörnvall, H. (1977d) in Alcohol and Aldehyde Metabolizing Systems (Thurman, R.G., Williamson, J.R., Drott, H.R. and Chance, B., eds.) vol. 2, pp. 145-155, Academic Press, New York.
Jörnvall, H., Eklund, H. & Brändén, C.-I. (1978) J. Biol. Chem. 253, 8414-8419.

Koch, G.L.E., Boulanger, Y. & Hartley, B.S. (1974) Nature 249, 316-320.
Kurosky, A., Barnett, D.R., Rasco, M.A., Lee, T.-H. & Bowman, B. H. (1974) Biochem. Genet. 11, 279-293.
Lange, III, L.G., Riordan, J.F., Vallee, B.L. & Brändén, C.-I. (1975) Biochemistry 14, 3497-3502.
Li, T.-K., Bosron, W.F., Dafeldecker, W.P., Lange, L.G. & Vallee, B.L. (1977) Proc. Natl. Acad. Sci. USA 74, 4378-4381.
Magnusson, S., Sottrup-Jensen, L., Petersen, T.E., Dudek-Wojciechowska, G. & Claeys, H. (1976) in Proteolysis and Physiological Regulation (Ribbons, D.W. & Brew, K., eds.), pp. 203-238, Academic Press, New York.
McLachlan, A.D. & Walker, J.E. (1977) J. Mol. Biol. 112, 543-558.
Olsen, K.W., Moras, D., Rossmann, M.G. & Harris, J.I. (1975) J. Biol. Chem. 250, 9313-9321.
Otto, J. (1977) in Pyridine Nucleotide-Dependent Dehydrogenases (Sund, H., ed.), pp. 41-42, Walter de Gruyter & Co., Berlin and New York.
Richardson, J.S., Richardson, D.C., Thomas, K.A., Silverton, E. W. & Davies, D.R. (1976) J. Mol. Biol. 102, 221-235.
Rigby, P.W.J., Burleigh, jun., B.D. & Hartley, B.S. (1974) Nature 251, 200-204.
Rossmann, M.G., Liljas, A., Brändén, C.-I. & Banaszak, L.J. (1975) The Enzymes, 3rd edn. (Boyer, P.D., ed.), vol. 11, pp. 61-102, Academic Press, New York.
Rossmann, M.G., Garavito, R.M. & Eventoff, W. (1977) in Pyridine Nucleotide-Dependent Dehydrogenases (Sund, H., ed.), pp. 3-30, Walter de Gruyter Co., Berlin and New York.
Royal, A., Garapin, A., Cami, B., Perrin, F., Mandel, J.L., LeMeur, M., Brégégègre, F., Gannon, F., LePennec, J.P., Chambon, P. & Kourilsky, P. (1979) Nature 279, 125-132.
Schwartz, M.F. & Jörnvall, H. (1976) Eur. J. Biochem. 68, 159-168.
Smith, E.L., Austen, B.M., Blumenthal, K.M. & Nyc, J.F. (1975) The Enzymes, 3rd edn. (Boyer, P.D., ed.), vol. 11, pp. 293-367, Academic Press, New York.
Thatcher, D.R. & Camfield, R. (1977) Biochem. Soc. Transact. 5, 271-272.
Twu, J.-s., Chin, C.C.Q. & Wold, F. (1973) Biochemistry 12, 2856-2862.
Waterson, R.M. & Konigsberg, W.H. (1974) Proc. Natl. Acad. Sci. USA 71, 376-380.
Williams, J. & Wilkins, A.G. (1974) Nature 247, 556-557.
Wills, C. & Jörnvall, H. (1979a) Eur. J. Biochem. 99, 323-331.
Wills, C. & Jörnvall, H. (1979b) Nature 279, 734-736.
Woenckhaus, C., Jeck, R. & Jörnvall, H. (1979) Eur. J. Biochem. 93, 65-69.

PYRIDINE NUCLEOTIDE - DISULFIDE OXIDOREDUCTASES

Arne Holmgren

CONTENTS

I. INTRODUCTION

Electron transfer between pyridine nucleotides and disulfide compounds is catalyzed by three flavoproteins which are well characterized. Lipoamide dehydrogenase reoxidizes reduced lipoamide (lip-$(SH)_2$) by NAD^+. Glutathione reductase catalyzes reduction of glutathione (GSSG) by NADPH. Thioredoxin reductase catalyzes the reduction of oxidized thioredoxin (thioredoxin-S_2) by NADPH. These three enzymes resemble each other structurally and mechanistically. They have, in addition to FAD, a redox active disulfide which is functional in catalysis. The flavoproteins consist of two identical or near identical polypeptide chains, each with a functional cystine residue, and a molecule of FAD which is noncovalently linked.

Although the disulfide bonds in the substrates for the enzymes are chemically identical *per se*, the molecular structures of GSSG, lip-S_2 and thioredoxin-S_2 are widely different. It is thus not surprising that the physiological functions of the pyridine nucleotide-disulfide oxidoreductases are connected to many different areas of cellular metabolism. All three enzymes and their substrates are also widely distributed among procaryotes and eucaryotes.

An excellent review by Williams in the 3rd edition of *The Enzymes* covers most of the knowledge of these enzymes up to 1975 (Williams, 1976). The aim of this chapter is to review some recent results on the structure and functional properties of the enzymes.

The knowledge of the structure and mechanism of the pyridine nucleotide-disulfide oxidoreductases has taken a great step forward, through the recent elucidation of the three-dimensional structures of glutathione reductase and *E. coli* and phage T4 thioredoxin, which will be discussed here. A very rapid expansion of our knowledge in the field of metabolic functions for GSH and glutathione reductase and thioredoxin and thioredoxin reductase has taken place. The discovery of the glutaredoxin system is an example of a novel biochemical function for GSH and glutathione reductase in a reaction where the thioredoxin system has been implicated for a long time. Regulation of enzyme activities through thiol-disulfide exchange or "thiol redox control" is another example, to be discussed here, where both the glutathione and thioredoxin systems are of importance. The rapid development in the area of metabolic functions for the pyridine nucleotide-disulfide oxidoreductase systems makes it impossible to cover all of the extensive literature in this article. However, it illustrates the importance of these systems and the sulfhydryl and disulfide groups in biochemical reactions.

II. LIPOAMIDE DEHYDROGENASE

Lipoamide dehydrogenase was first isolated from the α-ketoglutarate complex (Massey, 1960). The enzyme catalyzes the reversible reaction:

$$NAD^+ + Lip\text{-}(SH)_2 \rightleftharpoons Lip\text{-}S_2 + NADH + H^+$$

Lipoamide dehydrogenase has been isolated from more than 30 species including E. coli and mammalian heart cells (see Williams, 1976, for a summary). The mammalian enzymes are of mitochondrial origin.

A. Metabolic functions

Lipoamide dehydrogenase (E.C. 1.6.4.3) is a component of the pyruvate dehydrogenase and α-ketoglutarate dehydrogenase multienzyme complexes involved in decarboxylation of pyruvate and α-ketoglutarate to yield acetyl-CoA and succinyl-CoA, respectively (Reed & Oliver, 1968; Reed, 1974; Perham, 1975). The pyruvate dehydrogenase complex of E. coli contains multiple copies of three different types of polypeptide chains, corresponding with three component enzyme activities. These are pyruvate decarboxylase (E1 or pyruvate-lipoate oxidoreductase), lipoate acetyl transferase (E2 or acetyl-CoA-dihydrolipoate S-acetyltransferase) and lipoamide dehydrogenase (E3 or NADH-lipoamide oxidoreductase). Together, they catalyze the oxidative decarboxylation of pyruvate via the following sequence of reactions:

$$\text{Pyruvate} + (\text{TPP})\text{-E1} \xrightarrow{Mg^{2+}} (\text{hydroxyethyl-TPP})\text{-E1} + CO_2 \quad (1)$$

$$(\text{hydroxyethyl-TPP})\text{-E1} + (\text{lipS}_2)\text{-E2} \rightleftharpoons$$

$$(\text{TPP})\text{-E1} + (\text{acetyl-S-lip-SH})\text{-E2} \quad (2)$$

$$(\text{acetyl-S-lip-SH})\text{-E2} + \text{CoA} \rightleftharpoons (\text{lip(SH)}_2)\text{-E2} + \text{acetyl-SCoA} \quad (3)$$

$$(\text{lip(SH)}_2)\text{-E2} + (\text{FAD})\text{-E3} + NAD^+ \rightleftharpoons$$

$$(\text{lipS}_2)\text{-E2} + (\text{FAD})\text{-E3} + \text{NADH} + H^+ \quad (4)$$

Overall reaction:

$$\text{Pyruvate} + \text{CoA} + NAD^+ \longrightarrow \text{acetyl-SCoA} + \text{NADH} + H^+ + CO_2$$

in which lipS_2 is an oxidized lipoic acid residue, lip(SH)_2 is a dihydrolipoic acid residue, and TPP is thiamine pyrophosphate. The substrate is carried by lipoic acid residues covalently attached to lysine residues on the E2 polypeptide chains (Reed, 1974). The pyruvate dehydrogenase complex of E. coli (Crookes strain) has a molecular weight of 4.6×10^6. The corresponding α-ketoglutarate complex has a molecular weight of 2.3×10^6.

Several studies on the molecular weight of the subunits of the three enzymes of pyruvate dehydrogenase have been published (Reed & Oliver, 1968; Perham & Thomas, 1971). Proteolytic degradation of the peptide chains is a likely explanation of differences noted between different authors (Perham & Thomas, 1971). The decarboxylase is 90,000 to 93,000, the lipoamide dehydrogenase is 56,000 and the transacetylase is 81,000 by sodium-dodecyl sulfate polyacrylamide gel electrophoresis (Perham & Thomas, 1971). Components with the lower molecular weights of the decarboxylase and transacetylase components reported in early papers (Reed & Oliver, 1968) can be produced by freezing and thawing pure enzyme.

Lipoamide dehydrogenase is rather easily dissociated from the α-keto acid dehydrogenase complexes. In *E. coli*, the enzyme from the pyruvate dehydrogenase and α-ketoglutarate complexes are identical in molecular weight, dimeric structure, electrophoretic and immunochemical properties, and they are functionally interchangeable (Pettit & Reed, 1967). Furthermore, an identical amino acid sequence of 17 residues surrounding the small intrachain disulfide loop of the active center in both proteins was determined (Brown & Perham, 1973) (see Table III). Mutants of *E. coli* lacking lipoamide dehydrogenase activity have been isolated (Guest & Creaghan, 1973) and it is now clear that the bacterial lipoamide dehydrogenase is the product of a single gene. The structural gene for the decarboxylase (E1) and lipoate acetyltransferase (E2) components of the pyruvate dehydrogenase complex, *aceE* and *aceF* respectively, are closely linked in the *leu* region of the *E. coli* linkage map, whereas the corresponding genes (*sucA* and *sucB*) for the E1 and E2 components of the α-ketoglutarate dehydrogenase component are closely linked in the *gal* region (Henning *et al.*, 1966; Herbert & Guest, 1969). The single lipoamide dehydrogenase (*lpd*) gene has been located next to the other two genes for the pyruvate dehydrogenase complex in the order, *aceE*-*aceF*-*lpd* (Guest & Creagham, 1974).

Perham and coworkers have recently drawn attention to this parsimonious use of structural genes for components of oligomeric structures (Perham *et al.*, 1978) of which there are several examples. Since lipoamide dehydrogenase fulfils a service function in the α-keto acid dehydrogenase multienzyme complexes, the economy of a single structural gene makes good sense. However, it poses complicated questions for the control of the biosynthesis of this common component of two different enzyme systems because the other structural genes are located in two independent operons. Expression of the *lpd* gene may be affected by the *ace* operon but can be independent of it (Langley & Guest, 1977). The mechanisms for the regulation of the gene are not understood.

It seems likely that a single gene also codes for the mammalian lipoamide dehydrogenase (see Williams, 1976) although several electrophoretically separable forms of the enzyme have been described. The degradation of the enzyme in handling the complex is a complication in analysis. However, it is possible that lipoamide dehydrogenase is modified in some way when it is inserted in the pyruvate and α-ketoglutarate complexes, respectively (Williams, 1976). The changes of the enzyme must be reversible because the enzyme shows essentially full activity upon reconstitution in cross-reactivity experiments.

Rapid intramolecular coupling of active sites in the *E. coli* pyruvate dehydrogenase complex has recently been suggested as a mechanism for catalytic rate enhancement in this multimeric structure (Danson *et al.*, 1978b; Collins & Reed, 1977). Examination of the fate of pyruvate decarboxylated with the pyruvate dehydrogenase complex in the absence of CoA (Reaction 1 and

2) showed that a single E1 dimer could bring about reductive acetylation of multiple copies of the E2 chains in the core complex (Danson *et al.*, 1978b). It was suggested that the lipoyllysine "swinging arms" can participate in two types of subunit interactions: each E1 dimer can interact directly with the lipoic acid residues of a small number of E2 chains, followed by the transfer of acetyl groups between the neighboring E2 chains in the complex. These internal reactions have been termed transacetylations. From studies of the rate of the intramolecular transacetylation reactions of the lipoic acid residues and the rate of the normal reaction mechanism of the pyruvate dehydrogenase complex it is clear that the transacetylation reaction is not rate-limiting. It seems likely that the rapid functional connection of active sites in a multienzyme complex will ensure that sequential reactions can be successfully coupled even under conditions of low substrate concentrations for the different steps. It is generally accepted that the intracellular concentrations of pyruvate and CoA are low in *E. coli* which would lead to problems of diffusion in the formation of acetyl-CoA from these two substrates catalyzed by three successive enzymes. The association of multiple copies of the three enzymes into the multienzyme complex and the functional connection of active sites would thus yield a rate enhancement and provide a plausible explanation for the unusual complexity of quaternary structure of the pyruvate dehydrogenase enzyme (Danson *et al.*, 1978a,b).

A different kind of site coupling in the electron and acetyl group transfer pathways has recently been detected in the *E. coli* pyruvate dehydrogenase complex by Frey *et al.* (1978). They have shown that only half of the lipoyl residues of the transacetylase are coupled to lipoamide dehydrogenase component and to NADH formation. This class of lipoyl groups is involved in the overall reaction of the pyruvate dehydrogenase complex. The other half of the lipoyl groups form a second class of groups, which are only coupled to the decarboxylase (E1) component but not to the lipoamide dehydrogenase (E3) enzyme. The latter class of lipoyl groups may have another, so far undefined, electron transfer function in *E. coli*. From the combined results regarding transacetylations (Danson *et al.*, 1978b), site coupling (Frey *et al.*, 1978) and measurements of the subunit structure (Reed *et al.*, 1975) the following model for *E. coli* pyruvate dehydrogenase has been suggested. The complex has E1:E2:E3 = 24:24:12 with a molecular weight of 4.6×10^6 and a total of 48 lipoyl groups. One class of 24 lipoyl groups has an unknown terminal acceptor, X, and is not linked to lipoamide dehydrogenase. In this connection the high reactivity of thioredoxin-S_2 and dihydrolipoate *in vitro* should be pointed out (Holmgren, A., 1979, unpublished work).

B. Enzyme activities of lipoamide dehydrogenase

When isolated, the dimeric enzyme catalyzes at least four separate reactions. These are: a) NADH - lipoamide - oxidoreduction, b) dihydrolipoamide - NAD - oxidoreduction,

c) reduction of thionicotinamide adenine dinucleotide (TNAD) by NADH (transhydrogenase assay) and d) 2,6-dichlorophenolindophenol reduction by NADH (diaphorase assay) (Koike & Hayakawa, 1970)

In the physiological direction (as in the pyruvate dehydrogenase complex) optimal reduction of NAD^+ is at pH 8.0. The extrapolated maximal rate of NAD^+ reduction at infinite dihydrolipoamide is around 42,000 mol NADH formed per min per mol enzyme. The corresponding rate of NADH oxidation at pH 8.0 at infinite substrate concentration of NAD^+ and oxidized lipoate is around 15,000. The rate of NADH oxidation is much higher at lower pH-values. Thus, at pH 6.5, in the presence of low NAD^+ and NADH, rates at infinite concentration of lipoamide of 160,000 mol NADH oxidized per min per mol enzyme have been calculated (Williams, 1976).

The oxidation of NADH, at low pH, shows a lag period which is abolished in the presence of NAD^+. This is evidence for the role of NAD^+ as modifier to protect the enzyme from overreduction by NADH to a catalytically inactive state, (see IIC), (Williams, 1976).

Steady state kinetic studies give parallel linear double reciprocal plots when one substrate is varied at several concentrations of the second substrate, consistent with a simple binary complex mechanism. The K_m-values for pig heart lipoamide dehydrogenase at pH 8.0 were: NAD^+ 0.40 mM, dihydrolipoamide 0.50 mM (Moore _et al._, 1978). Values for a rat liver enzyme were: NADH 0.062 mM and lipoamide 0.84 mM (Williams, 1976). The high K_m-value for lipoamide is consistent with the function of the enzyme in the multienzyme complexes where lipoic acid is protein bound and consequently present in high local concentration.

C. Catalytic mechanism

A reaction mechanism for lipoamide dehydrogenase was formulated already in 1961 by Massey and Veeger. The general outlines of the mechanism are shown in Fig. 1. A main feature

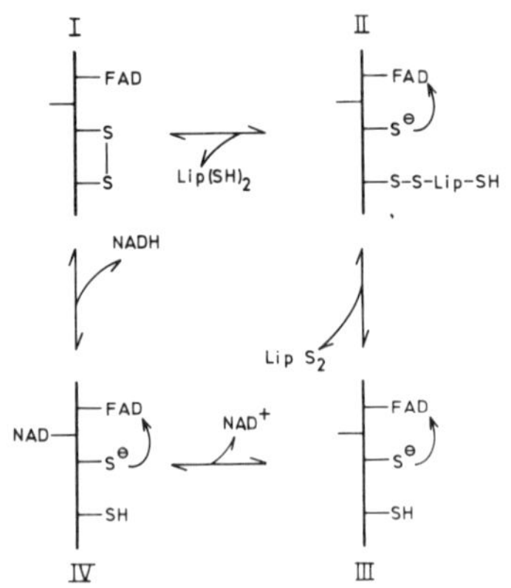

Fig. 1. Reaction mechanism of lipoamide dehydrogenase. I = oxidized enzyme; II = reduced enzyme lipoate complex; III = reduced enzyme, EH_2; IV = reduced enzyme NAD complex. Only one subunit is shown. (From Williams, 1976.)

of this mechanism is a 2 electron-reduced enzyme intermediate, EH_2, which is formed once in each catalytic cycle. The catalysis can be described by a ping-pong bi-bi mechanism in which the enzyme shuttles between the oxidized form, E, and the reduced form, EH_2. The overall catalytic mechanism can thus be divided into two parts consisting of an initial reduction of the enzyme by dihydrolipoamide followed by a second oxidation of the enzyme by NAD^+. The two electron-reduced enzyme, EH_2, can be produced *in vitro* by anerobic reduction with dihydrolipoamide or by one equivalent of dithionite (Williams, 1976). The reduced enzyme, EH_2, has diminished flavin absorbance which is shifted to shorter wave lengths when compared with the oxidized enzyme, E, which has a typical flavin spectrum (maximum at 455 nm with an extinction coefficient of 11.3 mM^{-1} cm^{-1} per FAD). In addition, EH_2 has a red color from a broad long wave length band with a maximum around 530 nm. This has been attributed to charge transfer between a thiolate anion formed on reduction of the active center disulfide, and the flavin, which remains oxidized (Kosower,1966).

Since the enzyme contains both FAD and an oxidation reduction active disulfide, it can accept 4 electrons (EH_4) under various forcing conditions (Massey & Veeger, 1961). Thus, the addition of 2 mol of dithionite per mol of FAD, produced EH_4 (Fig. 2) which has a spectrum of fully reduced flavin. The EH_4 component is considered to be an overreduced catalytically inactive species. This was shown by the absolute requirement for NAD^+ during catalysis of the NADH-lipoamide reductase reaction at pH 6.3; it was also shown that NAD^+ was required for catalytic activity in order to prevent the accumulation of EH_4 during turnover. The requirement for NAD^+ as an activator has been studied as a function of pH (Voetberg & Veeger, 1973) using lipoic acid as the electron acceptor, and by Matthews *et al.* (1976), using lipoamide as the electron acceptor. In both cases the degree of activation exerted by NAD^+ decreases with increasing pH. The explanation is that NAD^+ exerts its effect by influencing the equilibrium of the following reaction:

$$EH_2 + NADH + H^+ \rightleftharpoons EH_4 + NAD^+$$

and that the rate at which EH_4 is reoxidized by NAD^+ is very fast. The observed rate constant is 633 S^{-1} with 65 μM NAD^+ and 17 μM enzyme.

The oxidation-reduction potentials for the reduction to

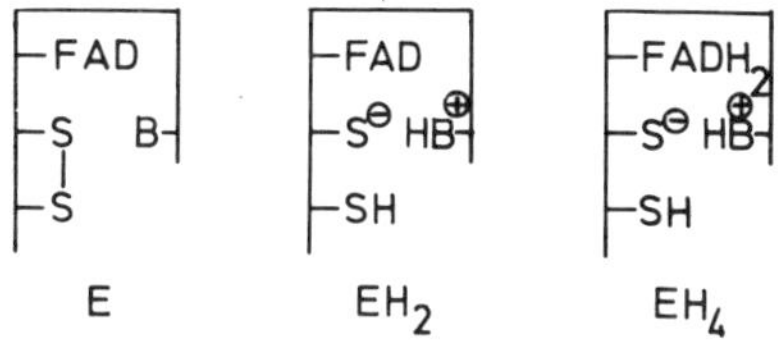

Fig. 2. A schematic representation of the three oxidation-reduction states of pig heart lipoamide dehydrogenase. B is a base. (From Thorpe & Williams, 1976a).

EH_2 and EH_4 forms of pig heart lipoamide dehydrogenase were recently determined (Matthews & Williams, 1976) from the equilibria with lipoamide and dihydrolipoamide. The potential for the couple oxidized lipoamide dehydrogenase/2 electron reduced lipoamide dehydrogenase is -0.280 V at pH 7 and $\Delta E/\Delta$ pH is -0.06 V in the pH range 5.5 to 7.6. The corresponding potential for the couple 2 electron reduced enzyme/4 electron reduced enzyme is -0.346 V at pH 7 and $\Delta E/\Delta$ pH is -0.06 V in the pH range 5.7 to 7.6. The spectrum of the EH_2 form of the enzyme is pH-dependent in the region at 530 nm and gives a lower absorbance at lower pH. These pH-dependent changes are consistent with the disappearance of the charge transfer complex at low pH. An amino acid side chain with an apparent pK of 5 and oxidized flavin seem to be involved. The suggestion was made that the amino acid side chain is a thiolate ion with an unusually low pK due to a stabilization by interaction with a protonated base as shown in Fig. 3.

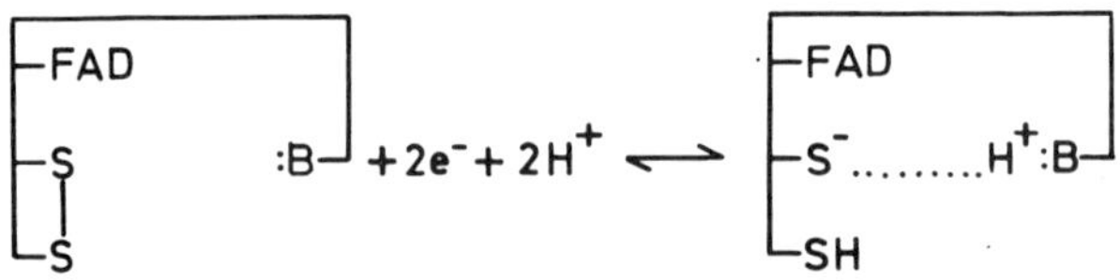

Fig. 3. Formation and stabilization of the 2 electron reduced lipoamide dehydrogenase by a protonated base.

The $\Delta E/\Delta pH$ for the E/EH_2 couple is consistent with 2 protons and 2 electrons, which must be added to E to form EH_2. Thiol base pairs of the type $S^-....BH^+$ are suggested in the mechanism of other pyridine nucleotide-disulfide oxidoreductases (Schulz et al., 1978). It has also been shown in papain, by modification of the enzyme by dibromopropanone, to form a covalent linkage between a histidine and the active site cysteine (Husain & Lowe, 1968). Polgar (1973) has measured the pK-values of 4.0 and 8.4 for the cysteine and histidine residues of papain, respectively. In lipoamide dehydrogenase, pK-values for a thiolate of 5 and of 8.2 for a histidine would explain the slope of $\Delta E/\Delta$ pH of -0.06 V for E/EH_2.

Kinetic data from rapid mixing experiments has provided further evidence for the participation of a base, probably an imidazolium side chain, in the active center of the enzyme (Matthews et al., 1977). It is proposed that this base deprotonates the substrate dihydrolipoamide and activates it for nucleophilic attack on the enzyme active center disulfide (see Fig. 4).

The rate of transfer of two reducing equivalents between dihydrolipoamide and the active center cystine of the enzyme by disulfide interchange is in fact independent of pH between 5.5 and 8.1. This thiol-disulfide interchange probably occurs by the formation of a mixed disulfide, which is likely to involve a nucleophilic attack on the disulfide by a thiolate anion. Yet, the macroscopic pK of the first thiolate of dihydrolipoamide is as high as 9.35. The activation mechanism in Fig. 4, would explain the high reactivity of lipoamide at pH 5.5. The

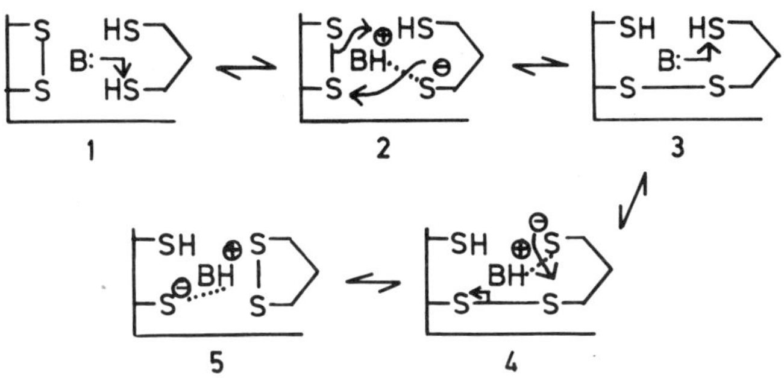

Fig. 4. Thiol/base ion pair in active center of lipoamide dehydrogenase in a mechanism for reduction of active center disulfide by dihydrolipoamide. (After Matthews et al., 1977.)

rate of alkylation of one of the active center thiols in EH_2 by iodoacetamide is also consistent with a thiolate/imidazolium ion pair (Thorpe & Williams, 1976a).

The two active site cysteine residues of pig heart lipoamide dehydrogenase, generated upon 2-electron reduction of the enzyme, have different reactivities towards alkylating reagents. One of the cysteine residues is at least 13-fold more active than its partner towards iodoacetamide at pH 7.6 under anerobic conditions (Thorpe & Williams, 1976a). Selective labeling of the reactive residue by [^{14}C]-iodoacetamide followed by complete alkylation of the enzyme and tryptic digestion plus sequence analysis was used to establish the position of the reactive residue in the active center sequence as follows:

```
                   CM                  SH
                   |                   |
-Thr-Leu-Gly-Gly-Thr-Cys-Leu-Asn-Val-Gly-Cys-Ile-Pro-Ser-Lys-
```

The monoalkylated lipoamide dehydrogenase is catalytically inactive towards oxidized or reduced lipoamide, whereas it is approximately two-fold more active as a transhydrogenase with NADH and acetylpyridine adenine dinucleotide as substrates. This result is consistent with independent binding and catalytic sites for NADH and dihydrolipoamide in the enzyme (Thorpe & Williams, 1976a).

The mechanism discussed so far ascribes different roles to the cysteine residues of the active site. One is thus proposed to participate in the transient formation of the mixed disulfide bond with lipoamide while the other interacts with the FAD forming the thiolate-flavin change transfer complex at the EH_2 level. Recently, evidence for a transitory covalent intermediate between a thiolate moiety and the C4a position of the isoalloxazine ring has been presented (Thorpe & Williams,1976b;

Moore *et al.*, 1978). Addition of one equivalent of NAD^+ to the monoalkylated EH_2 species induces a very rapid reversible spectral change which is consistent with the formation of a covalent adduct between FAD and the active site cysteine residue whose oxidation-reduction partner is selectively alkylated. A covalent flavin linkage in pig heart has been produced by reaction of the apolipoamide dehydrogenase with 8-Cl-FAD (Moore *et al.*, 1978). The likely structure of the adduct, as deduced from model spectra, is a cysteinyl residue covalently linked to position 8 of FAD, although this has yet to be proved by chemical analysis. The further analysis of the 8-Cl-flavin adduct of lipoamide dehydrogenase should help to establish the chemistry of the active site of this enzyme. The method of preparing the apoenzyme and the covalent FAD adduct should be applicable also to the closely related glutathione reductase, where a three-dimensional structure is already known (Schulz *et al.*, 1978), see also section III.

III. GLUTATHIONE REDUCTASE

Glutathione reductase catalyzes the reduction of glutathione disulfide (GSSG) by NADPH:

$$GSSG + NADPH + H^+ \longrightarrow 2\ GSH + NADP^+ \qquad (5)$$

This reaction is required for the maintenance of glutathione in the reduced form, GSH, which is the form used in most known biochemical reactions of glutathione (Meister, 1975).

Glutathione reductase (Asnis, 1955) has been isolated in pure form from many sources. These include *E. coli* (Williams & Arscott, 1971; Pigiet & Conley, 1977); rat liver (Carlberg & Mannervik, 1975), and human erythrocytes (Worthington & Rosemyer, 1975, 1976; Krohne-Ehrich *et al.*, 1977). The introduction of affinity chromatography on 2',5'-ADP-Sepharose 4B (Mannervik *et al.*, 1976; Brodelius *et al.*, 1974) has been an important advance in the purification procedures. The enzymes isolated are all dimeric flavoproteins of around 100,000 in molecular weight. The available information suggests that the essential catalytic properties of the enzyme are identical irrespective of the source of the enzyme.

The major break-through in the understanding of the structure and catalytic properties of glutathione reductase has come from the three-dimensional structure of the human erythrocyte enzyme at 3Å resolution (Schultz *et al.*, 1978); see section IIIC.

A. General metabolic functions

The metabolic functions of glutathione reductase are largely synonymous with those of the product GSH. Glutathione is the most abundant thiol-disulfide pair in cells by more than an order of magnitude and the GSH:GSSG ratio is generally 20:1 or larger (Jocelyn, 1972). The fact that GSSG normally can be detected in cell extracts is consistent with a nonequilibrium physiological state which is either limited by the enzyme level or by available NADPH. The thermodynamic equilibrium of the glutathione reductase reaction (5) is very high at pH 7.4 due to the differences in oxidation-reduction potential (Williams, 1976) (see section V).

Two recent symposia on glutathione (Flohé *et al.*, 1974; Arias and Jakoby, 1976) has covered many aspects of the biochemistry of this compound. GSH is utilized in several types of reactions which yield GSSG.

GSH is the substrate for enzymes such as GSH peroxidase (E.C. 1.11.1.9) and for a group of enzymes which utilize the monothiols of GSH for the reduction of disulfide bonds (for a review, see Freedman, 1979). Examples of enzymes of this group are GSH-insulin (or protein) transhydrogenase (E.C. 1.8.4.2), GSH-cystine-transhydrogenase (E.C. 1.8.4.4) and GSH-protein disulfide isomerase (E.C. 5.3.4.1). Several of the enzymes in this group, some of which may be identical, are microsomal and are glycoproteins in mammalian cells (see Freedman, 1979).

Another function of GSH, which may be enzymatic or non-enzymatic, is the oxido-reduction of protein thiols and the formation of mixed disulfides. A particular function of this type of reaction, is the control of enzyme activity. Examples of such metabolic control was first reported for fructose-1,6-phosphatase (E.C. 3.1.3.11) of liver by Pontremoli & Horecker (1970) and glycogen synthetase D (E.C. 2.4.1.11) (Ernest & Kim, 1973). Other examples are brain adenylate cyclase (Baba *et al.*, 1978), liver phosphorylase phosphatase (Takashi *et al.*, 1978) and *E. coli* ribonucleotide reductase (Holmgren, 1979b).

In determining possible essential functions for glutathione (and glutathione reductase) one problem has been the very high levels in almost all cells (0.5 - 10 mM). One way of studying the functions of GSH has been to use oxidizing agents such as diamide $[(CH_3)_2\text{-NCON = NCON-}(CH_3)_2]$ introduced by Kosower (see Kosower & Kosower, 1976). This drug, which oxidizes GSH to GSSG has extensive effects on cell functions and inhibits protein, RNA and DNA synthesis. A problem in interpreting the results from the use of diamide is that the cells treated are virtually emptied of all their NADPH through reduction of the GSSG by glutathione reductase.

A function for GSH in protecting red blood cell membranes against oxidative damage has been suggested from analysis of some hemolytic disorders. In some of these cases a defective glutathione reductase function in red cells has been identified (Beutler & Srivastava, 1970; Worthington & Rosemeyer, 1974). The conditions are characterized by a low GSH:GSSG ratio,which could be secondary to lack of NADPH caused by low glucose-6-phosphate dehydrogenase activity. Alternatively, glutathione reductase could have been lacking FAD because of riboflavin deficiency.

Mutants of *E. coli* in the biosynthesis of GSH have been identified independently by two groups (Apontoweil & Berends, 1975; Fuchs & Warner, 1975). The viable mutants are either defective in the glutathione synthetase gene gshB and accumulate γ-Glu-Cys or are defective in the γ-glutamyl-cysteine synthetase gene gshA. Neither mutant is absolutely tight because some enzyme activity is observed *in vitro*. However, the levels of GSH are below one percent of normal. It was recently shown, using GSH^- mutants, that the oxygen effect in radiation damage is linked to the GSH content of cells (Morse & Dahl, 1978). The increased survival of normal cells to irradiation in nitrogen atmosphere is absent in GSH^- cells, which have the higher sensitivity to irradiation both in oxygen and nitrogen atmosphere. Further analysis of this system should help to establish the function for GSH in radioprotection which has been postulated (see Modig & Révész, 1967). So far no mutants in the gene for glutathione reductase in *E. coli* have been characterized.

B. The glutaredoxin system

A novel function for GSH in the biosynthesis of deoxyribonucleotides was recently discovered in a mutant of *E. coli* (tsnC 7004) lacking thioredoxin (Holmgren, 1976). In the presence of a protein called glutaredoxin, GSH will function as hydrogen donor for *E. coli* ribonucleoside diphosphate reductase. This essential enzyme, composed of two subunits called protein B1 and protein B2, synthesizes deoxyribonucleotides (dNDP´s) from the corresponding ribonucleotides (rNDP´s) (for a recent review, see Thelander & Reichard, 1979). The function of the glutaredoxin system is shown in the following reactions:

$$2\ GSH + rNDP \xrightarrow[\text{glutaredoxin}]{\text{ribonucleotide reductase}} GSSG + dNDP + H_2O \qquad (6)$$

$$GSSG + NADPH + H^+ \xrightarrow[\text{reductase}]{\text{glutathione}} 2\ GSH + NADP^+ \qquad (7)$$

$$\text{Net reaction: } NADPH + H^+ + rNDP \longrightarrow NADP^+ + dNDP + H_2O$$

Reaction 6 above, is totally dependent on the small protein glutaredoxin and this formed the assay for the isolation of glutaredoxin in homogenous form from *E. coli* B cells (Holmgren, 1979a, b). Glutaredoxin from *E. coli* has the following properties: it is a small (12,000 molecular weight) acidic protein containing around 89 amino acid residues including two half-cystine forming a single disulfide bridge. This cystine disulfide bridge is readily reduced to a dithiol by GSH, NADPH and glutathione reductase. It is not reduced by NADPH and *E. coli* thioredoxin reductase, nor is it reduced by *E. coli* glutathione reductase in the absence of GSH. Tryptic peptide maps of glutaredoxin and thioredoxin from *E. coli* demonstrated that the two proteins are structurally unrelated and thus the products of two separate genes. The homogenous *E. coli* glutaredoxin showed, apart from the specific function with ribonucleotide reductase, also inherent activity as a powerful GSH-disulfide-transhydrogenase (Holmgren, 1979a). It was found that E. coli was a rich source of such transhydrogenase activity and that glutaredoxin represented a small fraction of the total activity of these enzymes which were separated in several different molecular species. The function of the other enzymes in E. coli are unknown.

Glutaredoxin showed an apparent K_m-value of 1.3 x 10^{-7} M for ribonucleotide reductase in the presence of 4 mM GSH. The molecular activity of glutaredoxin in CDP reduction was about 10-fold higher than that of thioredoxin and similar to the corresponding values for ribonucleotide reductase subunits (110 - 150 min^{-1}). The apparent V_{max}-values for ribonucleotide reductase with the thioredoxin or glutaredoxin systems were similar. An apparent K_m-value of GSH of 0.40 mM was determined in the presence of excess NADPH, glutathione reductase and CDP. A dithiol form of glutaredoxin,isolated after chemical reduction with dithiothreitol, reduced a stoichiometric amount of CDP to dCDP in the presence of ribonucleotide reductase. In the absence of NADPH and glutathione reductase, the velocity of the reduction of CDP to dCDP by GSH was highly influenced by the ratio of GSH to GSSG and even small amounts of GSSG inhibited strongly. Thus, inhibition by GSSG is a possible physiological control mechanism for deoxyribonucleotide synthesis, which also gives glutathione reductase a pivotal role in this important reaction.

Further experiments are required to establish if the thioredoxin or glutaredoxin system is the physiological hydrogen donor system for deoxyribonucleotide synthesis. Possibly both systems are active in the cell but have different roles (see also section IV).

The protein called T4 thioredoxin, which is induced in E. coli after phage T4 infection (Berglund & Sjöberg, 1970) has also glutaredoxin activity (Holmgren, 1978). It catalyzes GSH-dependent ribonucleotide reduction with a T4 induced ribonucleotide reductase. The T4 protein has properties of both thioredoxin and glutaredoxin since T4 thioredoxin can be either reduced by NADPH and thioredoxin reductase or by NADPH, GSH and glutathione reductase. The structure of T4 thioredoxin will be discussed in section IV.

Evidence for a mammalian glutaredoxin was recently obtained using calf thymus (Luthman et al., 1979). The protein in calf thymus had the following characteristics: (i) its molecular weight determined by gel chromatography was about 12,000; (ii) it was active with ribonucleotide reductase from thymus in the presence of GSH, NADPH and glutathione reductase but had no activity with NADPH and thioredoxin reductase and (iii) it was immunologically different from thioredoxin since it did not bind to antithioredoxin immunoadsorbents. Furthermore, species cross-reactivity of the E. coli and calf thymus glutaredoxin and thioredoxin systems with the corresponding ribonucleotide reductase enzymes showed essentially no species specificity for the homologous thioredoxin but a high species specificity for the homologous glutaredoxin.

C. Structure of glutathione reductase

Glutathione reductase from human erythrocytes was crystallized (Schulz et al., 1975) and a three-dimensional structure of the enzyme at 3Å resolution has been elucidated by X-ray crystallography (Schulz et al., 1978). The polypeptide chain has been traced and binding positions of FAD, NADP and GSSG have been determined. The latter two compounds were positioned after soaking crystals in 10 mM NADP and 10 mM GSSG (Schulz et al., 1978). As yet no complete amino acid sequence is known for glutathione reductase which has about 450 residues per subunit. The structure of the active center disulfide located in the "glupo-fragment" has been determined (Krohne-Ehrich et al., 1977); see section VA.

The enzyme subunit (50,000 in molecular weight) consists of three structural domains (Zappe et al., 1977) corresponding to binding of FAD and NADP and a somewhat flat interface-domain where GSSG is bound. A schematic figure is shown in Fig. 5.

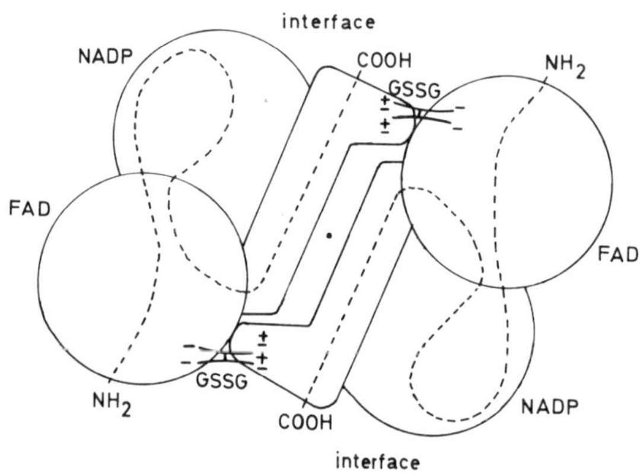

Fig. 5. Sketch of the dimeric enzyme glutathione reductase. The FAD-, NADP- and interface-domains are outlined. The general course of the polypeptide chain is given as a dashed line. FAD and NADP bind to their respective domains in an extended conformation. Except for the nicotinamide moiety, NADP can be described as binding at the protein surface. In contrast, FAD binds at the surface of the FAD-domain but at the boundary to the NADP-domain, so that only the adenine moiety extends to the protein surface. Oxidized glutathione (GSSG) binds between subunits. There is a close contact between the flavin ring of one subunit and the interface-domain of the other. (From Schulz et al. (1978).)

The NADP- and interface domains are formed by continuous sections of the peptide chain whereas the FAD domain consists of two noncontinuous parts of the chain. FAD and $NADP^+$ bind to their respective domains in an extended conformation reminiscent of NAD bound to dehydrogenases (Rossmann et al., 1975). GSSG, on the other hand, is bound between the FAD-domain and the interface domain of the other subunit, such that both tripeptide halves are stretched out between subunits. Thus, the active center is located between subunits.

Glutathione reductase binds the two dinucleotides, FAD and NADP in the same subunit. A super-secondary structure known as the dinucleotide fold or "Rossmann fold" (Rossmann et al., 1975) has been observed in several dinucleotide binding proteins (Schulz, 1977). This consists of three adjacent strands of β-pleated sheets connected through helices. Both the FAD and NADP-domain structures resemble the Rossmann folds both with respect to chain fold and binding mode of the cofactors. This may indicate a distant evolutionary relationship to other dehydrogenases. Furthermore, the FAD and NADP-domains resemble each other. The chain folds of residues 6 - 139 (first chain segment of the FAD-domain) and residues 168 - 272 (COOH-terminal part of NADP-domain) are topologically equivalent and closely similar in geometry. The similarities seem to be highly significant and indicate an evolutionary relationship between the two domains. A primordial gene duplication event accompanied or followed by gene splicing is very likely as an explanation of the origin of these structures (Schulz et al., 1978).

The structure of the active center of glutathione reductase as seen from the three-dimensional structure of the molecule (Schulz et al., 1978) is shown in Fig. 6. The FAD-residue and the redox-active disulfide formed by Cys-46 and Cys-41 (Williams, 1976) are in close contact. The sulfur of Cys-46 touches the isoalloxazine ring of FAD close to position 4a. The sulfur of Cys-41 is in van der Vaals contact with the disulfide of the bound GSSG. The structure clearly shows that a hydride ion cannot be transferred directly from NADPH to either of the two cysteines since the isoalloxazine ring forms an insurmountable barrier. The position of NADPH bound at the active center has not been observed. From model building it seems likely that NADPH makes contact with the isoalloxazine ring either at C9 or N1. The side chain of a lysine residue (Lys-49) (see section V), curls around the isoalloxazine ring (Fig. 6).

D. Catalytic mechanism

Mannervik and coworkers have recently used chemical modification to identify catalytically essential amino acid residues in glutathione reductase apart from the oxidation-reduction active disulfide (Mannervik et al., 1979). The use of 2,4,6-trinitrobenzenesulfonate (TNBS) selectively blocks the reduction of GSSG by glutathione reductase, but leaves unchanged the transhydrogenase activity of the NADPH binding site (Carlberg & Mannervik, 1979). Low concentrations of reagent were effective: 0.05 μM TNBS gave 60% inhibition in the GSSG reduction assay at pH 7. The modification was instantaneous and was reversi-

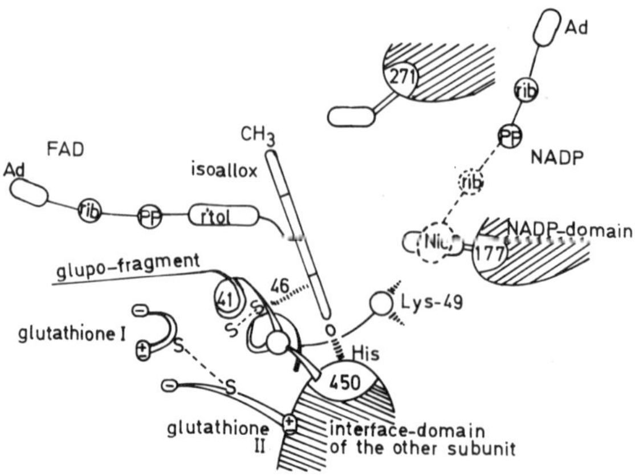

Fig. 6. Sketch of the active center of glutathione reductase. Flavin is hydrogen-bonded to the backbone of the other interface-domain at residue 450. The glupo-fragment containing the active disulfide is shown. The proximal Cys-46 is adjacent to the 4a-carbon of flavin. The side chain of Lys-49 curls around the isoalloxazine. Two large side chains of the NADP-domain are inserted. Nicotinamide may bind at position, where it would stack between these side chains. Inserted is the alternative binding position where the nicotinamide ring stacks on one of the large side chains and touches N1 of FAD. The observed densities prohibit stacking of the nicotinamide ring with the isoalloxazine and a contact between nicotinamide and N5 of FAD. GS_I and GS_{II} are stretched between subunits. The large side chain of residue 450 from the other subunit protrudes into the space enclosed by the isoalloxazine, the active disulfide of the protein, and the disulfide bridge of GSSG. This side chain is presumed to be a His, that takes part in catalysis. (From Schulz et al. (1978).)

ble by dialysis or dilution. No evidence of reduction of the TNBS was observed. The most likely effect of TNBS is that it interacts specifically with the dithiol of the active site and breaks the interaction between a thiolate anion and the flavin. Possibly TNBS forms an unstable covalent bond with the "distal" sulfhydryl group (Cys-41 in the human enzyme) (Schulz et al., 1978) of the active site dithiol in the 2-electron reduced enzyme.

An essential histidine residue could also be identified in glutathione reductase by use of ethoxyformic anhydride (Boggaram & Mannervik, 1978). This histidine residue is apparently involved in the reduction of GSSG by the dithiol of the 2-electron reduced enzyme since the pyridine nucleotide transhydrogenase activity of the enzyme was unaffected. It seems likely that this residue is His-450 in the human enzyme.

The catalytic mechanism for glutathione reductase has recently been discussed by Schulz _et al._ (1978); Mannervik _et al._ (1979); and Williams _et al._ (1978). As with lipoamide dehydrogenase (section II) the catalytic mechanism may be divided in two halves. The first involves the oxidized enzyme, E, NADPH and a proton:

$$E + NADPH + H^+ \longrightarrow EH_2 + NADP^+$$

The second involves the oxidation of EH_2 by GSSG:

$$EH_2 + GSSG \longrightarrow E + 2\ GSH$$

The three-dimensional structure of the enzyme (Fig. 6) shows that the isoalloxazine ring of the FAD separates the two binding sites of a subunit so that NADPH delivers reducing equivalents on one side and GSSG accepts them on the other side of the ring (Schulz _et al._, 1978). The formation of EH_2 may involve the formation of either a charge transfer complex between Cys-46 and the flavin ring or a sigma bond between C4a and Cys-46 (see Fig. 7). The internal, probably charged, Lys-49 may enhance the acceptor capacity of the isoalloxazine and promote the dissociation of $NADP^+$.

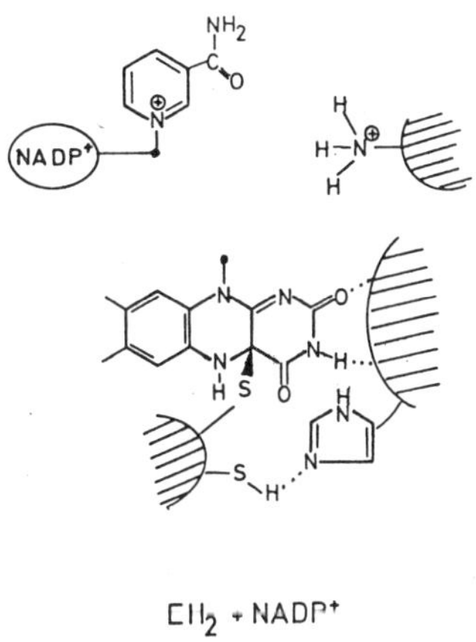

Fig. 7. Catalytic mechanism for glutathione reductase. (After Mannervik _et al._, 1979.)

The second part of the catalytic mechanism may involve the reaction of GSSG with Cys-41 (see Fig. 8). A base/thiol ion pair involving the thiolate of Cys-41 and the imidazolium side chain of His-450 from the other subunit is postulated. The positively charged His-450 attracts electrons from sulfur II in GSSG and polarizes the disulfide bond. This makes Cys-41 more nucleophilic and facilitates an attack of Cys-41 on sulfur I of GSSG. In a concentrated action His-450 is deprotonated, one molecule of GSH is released and a mixed disulfide is formed (Fig. 8). In the absence of a positive charge of His-450, the

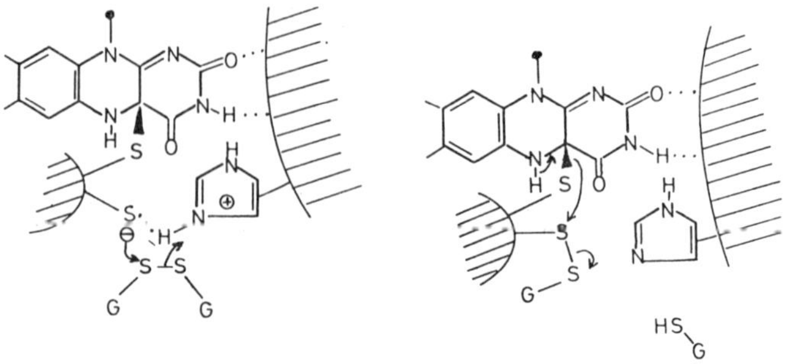

Fig 8. Catalytic mechanism for glutathione reductase. (After Mannervik et al., 1979.)

charge transfer complex becomes unstable and the thiolate ion Cys-46 is now available for a nucleophilic attack on the mixed disulfide and the second GSH is released (Fig. 8).

IV. THIOREDOXIN REDUCTASE

Thioredoxin reductase catalyzes the reversible reduction of thioredoxin-S_2 by NADPH:

$$\text{Thioredoxin-}S_2 + \text{NADPH} + H^+ \rightleftharpoons \text{thioredoxin-(SH)}_2 + \text{NADP}^+ \quad (8)$$

Thioredoxin-S_2 is a protein with a molecular weight of 12,000 containing a single oxidation reduction disulfide (Laurent et al., 1964; Holmgren, 1968). Thioredoxin reductase was originally identified as part of a hydrogen donor system for ribonucleotide reductase from *E. coli* (Moore et al., 1964; Thelander, 1968). The same enzyme was earlier studied in yeast (Black et al., 1960; Asahi et al., 1961) and has been purified from that source (Speranza et al., 1973; Porqué et al., 1970a). Thioredoxin reductase from mammalian cells has recently been obtained in highly purified form from Novikoff hepatoma (Chen et al., 1977), calf thymus (Holmgren, 1977) and rat liver (Luthman & Holmgren, 1979 unpublished experiments). As with glutathione reductase, affinity chromatography on 2',5'-ADP-Sepharose (Holmgren, 1977) has simplified the isolation of pure enzyme.

A. Structure of thioredoxin

The disulfide substrate for thioredoxin reductase is thioredoxin-S_2. This small protein has been isolated in pure form from E. coli (Laurent et al., 1964), yeast (Porqué et al., 1970), Novikoff hepatoma (Herrmann & Moore, 1973), calf liver (Engström et al., 1974) and rat liver (Luthman & Holmgren, 1979 unpublished results). In addition, a T4 phage induced thioredoxin has been purified to homogeneity (Berglund & Sjöberg,1970). Thioredoxins are species-specific proteins and thioredoxin from E. coli and phage T4 has been the subject of extensive structural studies in recent years. Thus, their complete amino acid sequences (Holmgren, 1968; Sjöberg & Holmgren, 1972) and three-dimensional structures to 2.8Å resolution (Holmgren et al., 1975; Söderberg et al., 1978) have been determined.

Thioredoxin from E. coli has an active site disulfide (Holmgren, 1968) with the following structure:

```
                        32         35
-Trp-Ala-Glu-Trp-Cys-Gly-Pro-Cys-Lys-Met-
                 └ S──────── S ┘
```

The three-dimensional structure of thioredoxin-S_2 at 2.8Å resolution (Holmgren et al., 1975) shows that this disulfide is located in a unique protrusion of the molecule (Fig. 9).

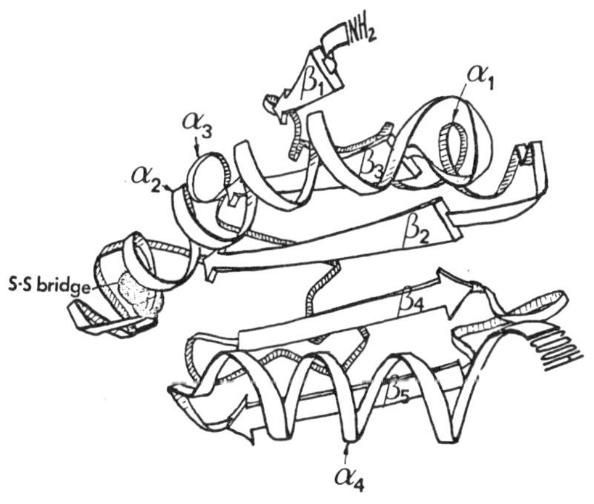

Fig. 9. Schematic drawing of E. coli thioredoxin-S_2 by Dr. Bo Furugren. (From Holmgren et al., 1975.)

The central part of the thioredoxin molecule is built of a core of three parallel and two antiparallel strands of pleated sheet surrounded by four α-helices. The region of the molecule building up the active center with two parallel strands joined by a protrusion and a helix is structurally very similar to corresponding functionally important regions in the nucleotide binding domains in the dehydrogenases (Rossmann et al., 1975). The only charged amino acid residue close to the disulfide bridge is

Lys-36 which may be involved in a thiolate/ion pair of catalytic significance in thioredoxin-$(SH)_2$. A localized conformational change occurs in thioredoxin-S_2 upon reduction to thioredoxin-$(SH)_2$. This results in a large increase in the quantum yield of the tryptophan fluorescence (Holmgren, 1972; Holmgren, 1973).

Thioredoxin from yeast has been isolated in two forms, I and II (Porqué *et al.*, 1970). The amino acid sequence of the active center disulfide of yeast thioredoxin is identical to that of *E. coli* thioredoxin-S_2 (Hall *et al.*, 1971) whereas other parts of the molecules have different, but probably homologous primary structures. Also the active center disulfide of calf liver thioredoxin is identical to that of *E. coli* thioredoxin (Holmgren, A., 1979, unpublished results). Thus, so far, all three thioredoxins from *E. coli*, yeast and mammalian cells have the following common active center sequence:

```
-Trp-Cys-Gly-Pro-Cys-Lys
     └ S ───── S ┘
```

Thioredoxins from mammalian cells contain two or more structural cysteine residues apart from the active center disulfide (Holmgren, 1977; Hermann & Moore, 1973). Oxidation of these extra thiols *in vitro* may lead to inactivation of thioredoxin through formation of aggregated protein. This makes studies with the mammalian thioredoxin-S_2 technically more complicated than use of the *E. coli* thioredoxin-S_2.

Phage T4 thioredoxin-S_2 is reduced by *E. coli* thioredoxin reductase and NADPH. The K_m-values for the *E. coli* and T4 thioredoxins (apparent, measured at 120 µM NADPH) are 10 and 6 µM, respectively (Berglund, 1969). T4 thioredoxin has an amino acid sequence which is completely different from the sequence of *E. coli* thioredoxin (Sjöberg & Holmgren, 1972); even the two residues between the half-cystines are different in T4 thioredoxin:

```
-Cys-Val-Tyr-Cys-
 └ S ───── S ┘
```

Yet, the three-dimensional structures of *E. coli* and T4 thioredoxin show extensive similarities in their folding (Söderberg *et al.*, 1978). As previously described (see section III), T4 thioredoxin has glutaredoxin activity, suggesting that it is a functional mixture in properties of thioredoxin and glutaredoxin.

Recently, it was found that thioredoxin isolated rapidly from *E. coli* log phase cells has phosphate bound in a phosphothiol linkage (Pigiet & Conley, 1978). As much as 96% of thioredoxin has been suggested to be present with phosphate predominantly located at Cys-32 (Conley & Pigiet, 1978). The functional significance of these results is not yet clear but they suggest functions for *E. coli* thioredoxin in phosphotransferase reactions.

B. Metabolic functions of thioredoxin and thioredoxin reductase

Thioredoxin reductase and thioredoxin were originally identified as a cyclic electron transport system required for the enzymatic formation of deoxyribonucleotides (Moore *et al.*, 1964; Thelander & Reichard, 1979) (see Fig. 10).

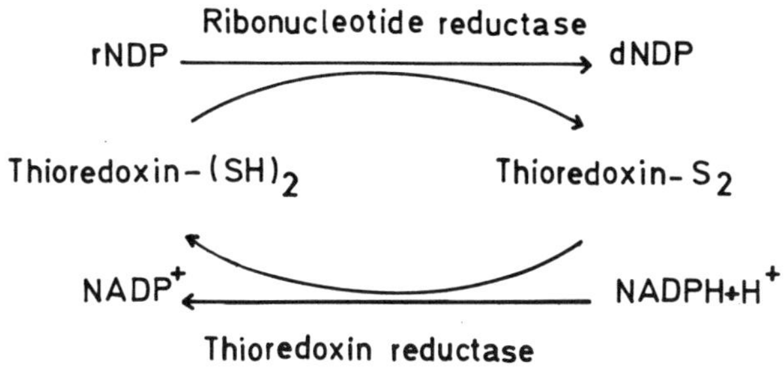

Fig. 10. Function of thioredoxin reductase in ribonucleotide reduction. rNDP = ribonucleotides: dNDP= deoxyribonucleotides.

Ribonucleotide reductase from *E. coli* is an essential, allosterically regulated enzyme, the function of which is to supply the deoxyribonucleotides for DNA replication (Thelander & Reichard, 1979). The *E. coli* enzyme consists of two nonidentical subunits, called protein B1 and B2, which both are absolutely required for activity. The B1 subunit is a dimer of molecular weight 160,000 and has binding sites for the substrates, which are each of the four ribonucleoside diphosphates. B1 also binds positive or negative allosteric effectors. These are deoxyribonucleoside triphosphates and ATP and they can regulate both the reaction rate and the substrate specificity of ribonucleotide reductase. Protein B2 is a dimer of molecular weight 78,000 and has non-heme iron and an essential free radical located on a tyrosine residue.

Studies of the *E. coli* ribonucleotide reductase have shown that thioredoxin-$(SH)_2$ reacts with ribonucleotide reductase with a ping-pong mechanism and that no stable ternary complex is formed *in vitro* (Thelander & Reichard, 1979). The following scheme describes the flow of electrons through thioredoxin reductase *via* a system of redox-active disulfides (Thelander, 1974):

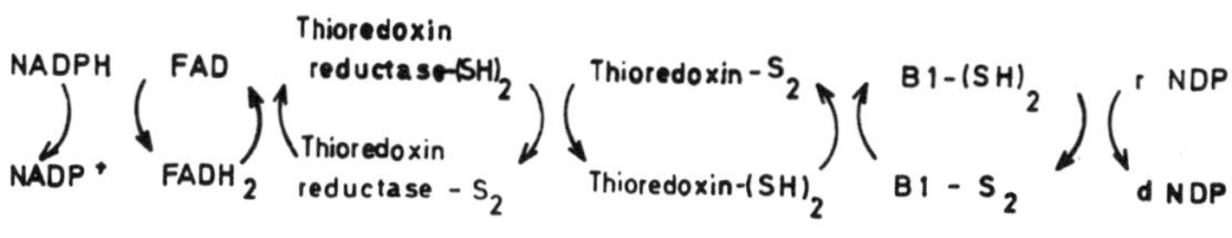

The simple picture of ribonucleotide reduction described so far has recently been upset by two discoveries: (1) prototrophic mutants of *E. coli* lacking either thioredoxin (Mark *et al.*, 1976; Holmgren, 1976; Holmgren *et al.*, 1978) or thioredoxin reductase (Fuchs, 1977) show no decreased ability to reduce ribonucleotides and (2) most thioredoxin in wild type *E. coli* log phase cells occurs as phosphothioredoxin, containing a phosphothiol linkage (Pigiet & Conley, 1978; Conley & Pigiet, 1978). These results clearly demonstrate that thioredoxin is not an obligatory intermediate in ribonucleotide reduction. Another hydrogen donor system for ribonucleotide reductase consisting of GSH, NADPH, glutathione reductase and a small protein called glutaredoxin was recently characterized from *E. coli* (Holmgren, 1979a, b; see section IIIB). Further studies are required to definitely establish the physiological roles of the thioredoxin and glutaredoxin systems.

The thioredoxin system has been implicated in enzymatic reduction of methionine sulfoxide and sulfate (see Table I). It is also an efficient NADPH-disulfide reducing system since thioredoxin-$(SH)_2$ is a powerful protein disulfide reductase.

Table I. Systems for which thioredoxin reductase and thioredoxin is the possible electron donor.

System	References
Ribonucleotide reductase (in bacteria, yeast, mammalian cells and plants)	Review by Thelander & Reichard, 1979
Methionine sulfoxide reductase (in yeast)	Porqué *et al.*, 1970
Active sulfate reductase (in yeast)	Porqué *et al.*, 1970
Reduction of protein disulfides (bacteria, yeast, mammalian cells)	Holmgren & Morgan, 1976
Degradation of insulin (mammalian cells)	Holmgren, 1977, unpublished

Recently functions for thioredoxin independent of thioredoxin reductase have been described; these are listed in Table II. The only genetically proven essential function of *E. coli* thioredoxin is to be subunit of a phage T7 DNA polymerase (Mark & Richardson, 1976). Plants seem to have one form of thioredoxin, which is utilized for specific regulation of photosynthetic enzymes such as fructose-bisphosphatase (Wolosiuk & Buchanan, 1977). The reduction of the chloroplast thioredoxin is carried out by a light-attenuated novel enzyme called ferredoxin-thioredoxin-reductase. Thioredoxin has been implicated as carrier of active sulfate (Tsang & Schiff, 1975; Schmidt &

Table II. System in which an independent function of thioredoxin is implicated.

System	References
Subunit of phage T7 DNA polymerase	Mark & Richardson, 1976
Regulation of photosynthetic chloroplast enzymes (plants)	Wolosiuk & Buchanan, 1977 Holmgren *et al.*, 1977
Carrier of active sulfate in sulfate reduction (*E. coli*, green algae)	Tsang & Schiff, 1976 Schmidt & Christen, 1978
General dithiol-disulfide oxidoreductase activity	Holmgren, unpublished 1979

Christen, 1978). Thioredoxin also catalyzes general dithiol-disulfide-oxidoreductions between *e.g.* dihydrolipoate and protein disulfides (Holmgren, 1979 unpublished). This may be of particular importance for "thiol redox control" of enzyme activities.

The gene for thioredoxin maps close to the origin of replication at 84 min. on the *E. coli* linkage map (Mark *et al.*, 1977). Preliminary mapping of thioredoxin reductase in *E. coli* shows a different location for this enzyme in the *leu* region (Fuchs, 1977).

In mammalian cells, the thioredoxin system is present in cells with little or no DNA synthesis consistent with functions besides ribonucleotide reduction (Holmgren & Luthman, 1978). Mammalian liver is a rich source of thioredoxin reductase.

C. Properties of thioredoxin reductase

Native thioredoxin reductase from *E. coli* and yeast have molecular weights of around 70,000 (Williams, 1976). Both enzymes are dimers and have two molecules of FAD. Thioredoxin reductases from mammalian cells also contain two FAD but have higher molecular weight; 116,000 reported for the enzyme from Novikoff rat hepatoma (Chen *et al.*, 1977) and about 110,000 from rat liver (Luthman & Holmgren, 1979 unpublished results). The mammalian enzyme is also quite susceptible to oxidative inactivation by thiol-disulfide exchange; this also applies to mammalian thioredoxin (see section IVA).

Thioredoxin reductases are highly specific for NADPH. The *E. coli* enzyme shows a very low K_m-value for this nucleotide (1.2 μM) at 25° and pH 7.6 (Willaims, 1976). NADP is a competitive inhibitor of the enzyme with a K_i of 15 μM. Plots of 1/V against 1/S give parallel lines consistent with a binary complex mechanism (Williams, 1976).

The *E. coli* thioredoxin reductase shows a very pronounced specificity for the disulfide of *E. coli* or phage T4 thioredoxin-S_2 when compared with GSSG, lip-S_2 and other disulfide containing substances. The K_m-value for *E. coli* thioredoxin is 2.8 μM (Williams, 1976). The turnover number of the enzyme/FAD (about 2,000 min^{-1}) is lower than for glutathione reductase and lipoamide dehydrogenase. This may relate to the much larger size of thioredoxin as the substrate for thioredoxin reductase.

The mammalian thioredoxin reductases seem to cross-react fully with other mammalian thioredoxins (Luthman & Holmgren, 1979 unpublished), despite species specificities in the structure of thioredoxin (Holmgren & Luthman, 1978). Furthermore, calf thymus thioredoxin reductase reduces *E. coli* thioredoxin-S_2 quite efficiently (Holmgren, 1977). The lower substrate specificity is also reflected in a direct reduction of DTNB (5,5'-dithiobis(2-nitrobenzoic acid)) with NADPH, without thioredoxin as intermediate electron carrier (Holmgren, 1977). The DTNB reduction can be used as a convenient assay during purification.

D. Catalytic mechanism

Thioredoxin reductase is not yet as well characterized as the more studied enzymes, glutathione reductase and lipoamide dehydrogenase. Almost all information published so far comes from the *E. coli* enzyme which will be exclusively discussed here. Since the mammalian enzymes have a considerably higher molecular weight they might turn out to be different.

The sequence of a peptide containing the active center disulfide in *E. coli* thioredoxin reductase has been determined (Thelander, 1970; Ronchi & Williams, 1972) (see Table III) and shows a cystine loop with only two residues. This structure is different from the corresponding structures in glutathione reductase and lipoamide dehydrogenase (see section V); it does however resemble the structure of thioredoxin, which also has two residues between the half-cystines.

The flavin spectrum in *E. coli* thioredoxin reductase has a maximum at 456 nm (11.3 $mM^{-1}\ cm^{-1}$). The spectral changes of thioredoxin reductase upon anerobic titration with NADPH show that 2 equivalents (4-electrons) are required to reduce the enzyme (Williams, 1976). This results in almost complete bleaching of the 456 nm absorbance and it gives no spectral evidence for a stable two-electron intermediate complex between a thiolate and the FAD (EH_2), which is present in lipoamide dehydrogenase and glutathione reductase. In the reduced form of the enzyme, the electrons are likely to be shared unequally between the disulfide active in the oxidation-reduction, and the FAD, the latter having the greater share.

Studies of the reaction between NADPH and thioredoxin reductase with a rapid reaction spectrophotometer have been carried out by Massey & Williams and coworkers (see Williams, 1976).

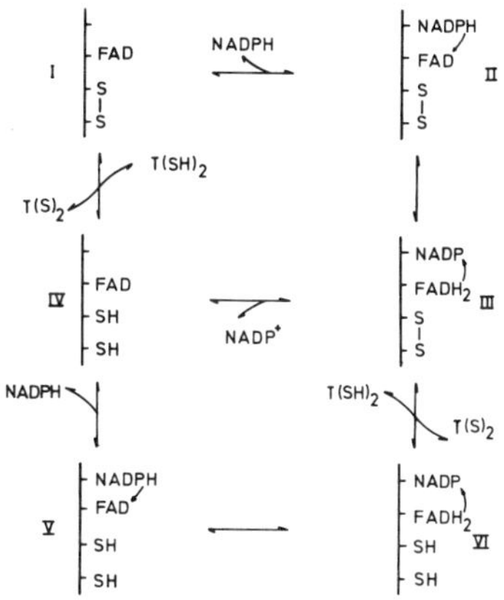

Fig. 11. Reaction mechanism for E. coli thioredoxin reductase. T = thioredoxin. Only one subunit is shown. (After Williams, 1976.)

The results have been used to formulate a reaction mechanism (Fig. 11).

If thioredoxin is present in excess the reaction proceeds clockwise in the upper square (I → II → III → IV → I) and the species IV is assumed to have a very transient appearance. If NADPH is present in excess, catalysis is assumed to proceed from I → II → III → IV → V → VI → III, etc. or counter clockwise around the lower square. The oxidized enzyme I reacts very rapidly to bind NADPH to give form II, which has an enhanced absorption at long wave lengths (600 - 900 nm) but little change at 450 nm. The decay of form II to give III and VI upon full reduction results in the almost complete bleaching of the enzyme at 450 nm. The rate-limiting step in the reaction is not identified.

The thioredoxin reductase reaction is readily reversibly and the enzyme may be reduced by thioredoxin-$(SH)_2$. The equilibrium constant of $[NADP+] \times [T\text{-}(SH)_2] / [T\text{-}S_2] \times [NADPH]$ falls by a factor of 10 for each unit rise in pH. At pH 7, a value of 48 for the equilibrium constant for E. coli thioredoxin-S_2 has been estimated (Moore et al., 1964).

V. COMPARISON OF THE ENZYME SYSTEMS

A. Structure of enzymes

Amino acid sequence analysis of peptides containing the active center disulfides of lipoamide dehydrogenase, glutathione reductase and thioredoxin reductase have been determined and are shown in Table III. All the bridges form small loops and reveal some remarkable homologies. The bridge sizes are

Table III. Amino acid sequences around catalytically active disulfide bridges.

Lipoamide dehydrogenase	
Pig heart[a]	-Asx-Glx-Thr-Leu-Gly-Gly-Thr-Cys-Leu-Asn-Val-Gly-Cys-Ile-Pro-Ser-Lys-
E. coli[b]	-Tyr-Asn-Thr-Leu-Gly-Gly-Val-Cys-Leu-Asn-Val-Gly-Cys-Ile-Pro-Ser-Lys-
Glutathione reductase	
Human erythrocytes[c]	-Leu-Gly-Gly-Thr-Cys-Val-Asn-Val-Gly-Cys-Val-Pro-----Lys-
Yeast[d]	-Leu-Gly-Gly-Thr-Cys-Val-Asn-Val-Gly-Cys-Val-Pro-----Lys-
Thioredoxin reductase	
E. coli[e]	-Ala-Cys-Ala-Thr-Cys-Asp-Gly-Phe-
Thioredoxin	
E. coli[f]	-Val-Asp-Phe-Trp-Ala-Glu-Trp-Cys-Gly-Pro-Cys-Lys-Met-Ile-Ala-Pro-Ile-

a. Brown & Perham, 1974.
b. Brown & Perham, 1972; Burleigh & Williams, 1972.
c. Krohne-Ehrich *et al.*, 1977.
d. Jones & Williams, 1975.
e. Thelander, 1970; Ronchi & Williams, 1972.
f. Holmgren, 1968.

thus identical for the lipoamide dehydrogenases and glutathione reductases from a phylogenetically wide range of eucaryotes and procaryotes. Furthermore, the amino acid sequence homology is very clear. In contrast, no homology in either the bridge size or the amino acid sequence is evident for the thioredoxin reductase of *E. coli*. The observed structures suggest that lipoamide dehydrogenase and glutathione reductase arose by divergence from a common ancestral gene through a gene duplication event (Perham *et al.*, 1979). Perhaps the origin of thioredoxin reductase is different and it may have evolved towards a similar catalytic mechanism.

B. Interaction between pyridine-nucleotide-disulfide systems

The three pyridine nucleotide-disulfide oxidoreductases are themselves remarkably specific for their respective disulfide substrate; the degree of cross-reaction is normally less than one percent (Williams, 1976). Is there any interaction between the substrates? These reactions are theoretically possible due to the fairly close redox potentials. Thus, the oxidation reduction potentials at pH 7 and 25° of the GSSG/GSH, lip-S_2/lip$(SH)_2$ and thioredoxin-S_2/thioredoxin-$(SH)_2$ couples are -0.25, -0.29 and -0.26 V, respectively (Gorin *et al.*, 1975; Clark, 1960 and Holmgren, 1968). A rapid reduction of thioredoxin-S_2 by lip-$(SH)_2$ has been observed *in vitro* (Holmgren, 1979 unpublished). In contrast GSH will only reduce thioredoxin-S_2 very slowly. Generally, GSH is a poor disulfide reductant mainly for entropic reasons, since two molecules have to come together to reduce a disulfide. Thioredoxin-$(SH)_2$ is a powerful protein disulfide reductase (Holmgren, unpublished 1979) but reacts rather slowly with GSSG and lip-S_2 *in vitro*.

The regulation of the intracellular oxidation states of the substrates and its significance is relatively unknown. It is clear from measurements of GSH (5.31 μmol/g wet weight) and GSSG (0.27 μmol/g wet weight in fed rat livers (Vina *et al.*, that the glutathione reductase reaction is far from thermodynamic equilibrium *in vivo*. This can be calculated from the experimentally observed ratio [free $NADP^+$]/[free NADPH] of 10^{-2} at pH 7.0 (Veech *et al.*, 1969). The discrepancy between *in vivo* measured values and *in vitro* calculated (10 nM *versus* 0.27 mM), is four orders of magnitude. This may be explained by a relatively low maximal capacity of the glutathione reductase enzyme and its high K_m-value for GSSG.

VI. PERSPECTIVES AND SUMMARY

Lipoamide dehydrogenase and glutathione reductase have impressive similarities in structure and catalytic mechanism. This is dependent on reversible electron transfer from pyridine nucleotides to FAD and a redox active disulfide. The three-dimensional structure of human glutathione reductase confirms earlier predictions from spectral and kinetic measurements of the close contact in the active center between the redox active cystine disulfide and the flavin ring. A catalytic mechanism has been formulated based on the protein structure, chemical modification and kinetic studies.

The structure of thioredoxin reductase is similar to the other two enzymes and it may have evolved towards a common mechanism. The structure of thioredoxin-S_2 shows a unique protrusion which contains the oxidation reduction disulfide. Further analysis of the thioredoxin structure and the pyridine-nucleotide-disulfide oxidoreductases will probably reveal common properties of functional disulfides. The range of metabolic functions of the enzyme systems is an area of further research. These metabolic functions of the pyridine-nucleotide-disulfide oxidoreductases range from energy metabolism, enzyme regulation and membrane protective functions to biosynthesis of deoxyribonucleotides and DNA. The function of the thioredoxin and glutaredoxin systems in the T4 virus infected *E. coli* cell is an example of the complex interplay of these systems.

VII. ACKNOWLEDGEMENTS

I am grateful to Dr. Richard Perham and Dr. Bengt Mannervik for kindly providing reprints and preprints in advance of publication. I wish to thank Mrs. Delphi Post for typing the manuscript. Research in the author´s laboratory and the preparation of this review was supported by the Swedish Medical Research Council Projects 13X-3529 and 13P-4292 and the Swedish Cancer Society, Project 961.

VIII. REFERENCES

Apontoweil, P. & Berends, W. (1975) Biochim. Biophys. Acta 399, 1-9.

Arias, J.M. & Jakoby, W.B. (1976) eds. Glutathione: Metabolism and function. Kroc Foundation Series, vol. 6, pp. 1-382, Raven Press, New York.

Asahi, T., Bandurski, R.S. & Wilson, L.G. (1961) J. Biol. Chem. 236, 1830-1835.

Asnis, R.E. (1955) J. Biol. Chem. 213, 77-85.

Baba, A., Lee, E., Matsuda, T., Kihara, T. & Iwata, H. (1978) Biochem. Biophys. Res. Commun. 85, 1204-1210.

Berglund, O. (1969) J. Biol. Chem. 244, 6306-6308.

Berglund, O. & Sjöberg, B.-M. (1970) J. Biol. Chem. 245, 6030-

6035.
Beutler, E. & Srivastava, S.K. (1970) Nature (Lond.) 226, 759-760.
Black, S., Harte, E.M., Hudson, B. & Wartofsky, L. (1960) J. Biol. Chem. 235, 2910-2916.
Boggaram, V. & Mannervik, B. (1978) Biochem. Biophys. Res. Commun. 83, 558-564.
Brodelius, P., Larsson, P.O. & Mosbach, K. (1974) Eur. J. Biochem. 47, 81-89.
Brown, J.P. & Perham, R.N. (1972) FEBS Lett. 26, 221-224.
Brown, J.P. & Perham, R.N. (1973) Eur. J. Biochem. 39, 69-73.
Brown, J.P. & Perham, R.N. (1974) Biochem. J. 137, 505-512.
Burleigh, B.D. Jr. & Williams, C.H. Jr. (1972) J. Biol. Chem. 247, 2077-2082.
Carlberg, I. & Mannervik, B. (1975) J. Biol. Chem. 250, 5475-5480.
Carlberg, I. & Mannervik, B. (1979) FEBS Letters 98, 263-266.
Chen, C.-C., Borns McCall, B.L. & Moore, E.C. (1977) Prep. Biochemistry 7, 165-177.
Clark, W.M. (1960) Oxidation Reduction Potentials of Organic Systems, Williams and Wilkins, Baltimore.
Collins, J.H. & Reed, L.J. (1977) Proc. Natl. Acad. Sci. USA 74, 4223-4227.
Conley, R.R. & Pigiet, V. (1978) J. Biol. Chem. 253, 5568-5572.
Danson, M.J., Hooper, E.A. & Perham, R.N. (1978a) Biochem. J. 175, 193-198.
Danson, M.J., Fersht, A.R. & Perham, R.N. (1978b) Proc. Natl. Acad. Sci. USA 75, 5386-5390.
Engström, N.-E., Holmgren, A., Larsson, A. & Söderhäll, S. (1974) J. Biol. Chem. 249, 205-210.
Ernest, M.J. & Kim, K.H. (1973) J. Biol. Chem. 248, 1550-1555.
Flohé, L., Benohr, H.C., Sies, H., Waller, H.D. & Wendel, A. eds. (1974) Glutathione "Proceedings of the 16th Conference of the German Society of Biological Chemistry", pp. 1-316, Thieme, Stuttgart.
Freedman, R.B. (1979) FEBS Letters 97, 201-210.
Frey, P.A., Ikeda, B.H., Govino, G.R., Speckhard, D.C. & Wong, S. S. (1978) J. Biol. Chem. 253, 7234-7241.
Fuchs, J.A. & Warner, H.R. (1975) J. Bacteriol. 124, 140-148.
Fuchs, J. (1977) J. Bacteriol. 129, 967-972.
Gorin, G., Esfandi, A. & Guthrie Jr., G.B. (1975) Arch. Biochem. Biophys. 168, 450-454.
Guest, J.R. & Creaghan, I.T. (1973) J. Gen. Microbiol. 75, 197-210.
Guest, J.R. & Creaghan, I.T. (1974) J. Gen. Microbiol. 81, 237-245.
Hall, D.E., Baldesten, A., Holmgren, A. & Reichard, P. (1971) Eur. J. Biochem. 23, 328-335.
Henning, U., Dennert, G., Hertel, R. & Shipp, W.S. (1966) Cold Spring Harbor Symp. Quant. Biol. 31, 227-234.
Herbert, A.A. & Guest, J.R. (1969) Mol. Gen. Genet. 105, 182-190.
Herrmann, E.C. & Moore, E.C. (1973) J. Biol. Chem. 248, 1219-1223.
Holmgren, A. (1968) Eur. J. Biochem. 6, 475-484.

Holmgren, A. (1972) J. Biol. Chem. 247, 1992-1998.
Holmgren, A. (1973) J. Biol. Chem. 248, 4106-4144.
Holmgren, A., Söderberg, B.-O., Eklund, H. & Brändén, C.-I. (1975) Proc. Natl. Acad. Sci. USA 72, 2305-2309.
Holmgren, A. (1976) Proc. Natl. Acad. Sci. USA 73, 2275-2279.
Holmgren, A. & Morgan, F.J. (1976) Eur. J. Biochem. 70, 377-383.
Holmgren, A., Buchanan, B.B. & Wolosiuk, R.A. (1977) FEBS Letters 82, 351-354.
Holmgren, A. (1977) J. Biol. Chem. 252, 4600-4606.
Holmgren, A., Ohlsson, I. & Grankvist, M.-L. (1978) J. Biol. Chem. 253, 430-436.
Holmgren, A. (1978) J. Biol. Chem. 253, 7424-7430.
Holmgren, A. & Luthman, M. (1978) Biochemistry 17, 4071-4077.
Holmgren, A. (1979a) J. Biol. Chem. 254, 3664-3671.
Holmgren, A. (1979b) J. Biol. Chem. 254, 3672-3678.
Husain, S.S. & Lowe, G. (1968) Biochem. J. 108, 855-859.
Jocelyn, P.C. (1972) In: Biochemistry of the SH-group, Academic Press, New York.
Jones, E.T. & Williams, C.H. Jr. (1975) J. Biol. Chem. 250, 3779-3784.
Koike, M. & Hayakawa, T. (1970) Methods in Enzymology 18, Part A, 298-307.
Kosower, E.M. (1966) In: Flavins and Flavoproteins (Slater, E.C. ed.) pp. 1-14, Elsevier Publishing Co., Amsterdam.
Kosower, N.S. & Kosower, E.M. (1976) In: Glutathione: Metabolism and Function (Arias, J.M. & Jakoby, W.B. eds.) pp. 159-174, Kroc Foundation, Series 6.
Krohne-Erich, G., Schirmer, R.H. & Untucht-Grau, R. (1977) Eur. J. Biochem. 80, 65-71.
Langley, D. & Guest, J.R. (1977) J. Gen. Microbiol. 99, 263-276.
Laurent, T.C., Moore, E.C. & Reichard, P. (1964) J. Biol. Chem. 239, 3436-3445.
Luthman, M., Eriksson, S., Holmgren, A. & Thelander, L. (1979) Proc. Natl. Acad. Sci. USA, 76, 2158-2162.
Mannervik, B., Jacobsson, K. & Boggaram, V. (1976) FEBS Letters 66, 221-224.
Mannervik, B., Boggaram, V., Carlberg, I. & Larson, K. (1979) In: Flavins and Flavoproteins (Yagi, I. & Yamano, T. eds.), Jap. Sci. Soc. Press, Tokyo, in press.
Mark, D.F. & Richardson, C.C. (1976) Proc. Natl. Acad. Sci. USA 73, 780-784.
Mark, D.F., Chase, J.W. & Richardson, C.C. (1977) Molec. Gen. Genet. 155, 145-152.
Massey, V. (1960) Biochim. Biophys. Acta 37, 314-322.
Massey, V. & Veeger, C. (1961) Biochim. Biophys. Acta 48, 33-47.
Matthews, R.G. & Williams, C.H.Jr. (1976) J. Biol. Chem. 251, 3956-3964.
Matthews, R.G., Wilkinson, K.D., Ballou, D.P. & Williams, C.H.Jr. (1976) In: Flavins and Flavoproteins (Singer, T.P., ed.) pp. 464-472, Elsevier Scientific Publishing Co., Amsterdam.
Matthews, R.G., Ballou, D.P., Thorpe, C. & Williams, C.H.Jr. (1977) J. Biol. Chem. 252, 3199-3207.
Meister, A. (1975) In: Metabolic Pathways, 3rd ed. (Greenberg, D.M., ed.) Vol. VII, pp. 101-188, Academic Press, New York.

Modig, H.G. & Révész, L. (1967) Int. J. Radiat. Biol. 13, 469-477.
Moore, E.C., Reichard, P. & Thelander, L. (1964) J. Biol. Chem. 239, 3445-3452.
Moore, E.C., Cardemil, E. & Massey, V. (1978) J. Biol. Chem. 253, 6413-6422.
Morse, M.L. & Dahl, R.H. (1978) Nature 271, 660-662.
Perham, R.N. & Thomas, J.O. (1971) FEBS Letters 15, 8-12.
Perham, R.N. (1975) Philos. Trans. R. Soc. London Ser. B 272, 123-136.
Perham, R.N., Harrison, R.A. & Brown, J.P. (1978) Biochem. Soc. Trans. 6, 47-50.
Perham, R.N., Harrison, R.A., Coggins, J.R. & Brown, J.P. (1979) In: Proceedings of the 6th International Symposium on Flavins and Flavoproteins (Yagi, K. & Yamano, T. eds.) Japan Scientific Press, Tokyo, in press.
Pettit, F.H. & Reed, L.J. (1-67) Proc. Natl. Acad. Sci. USA 58, 1126-1130.
Pigiet, V. & Conley, R.R. (1977) J. Biol. Chem. 252, 6367-6372.
Pigiet, V. & Conley, R.R. (1978) J. Biol. Chem. 253, 1910-1920.
Polgar, L. (1973) Eur. J. Biochem. 33, 104-109.
Pontremoli, S. & Horecker, B.L. (1970) Current Topics in Cellular Regulation 2, 173-199.
Porqué, P.G., Baldesten, A. & Reichard, P. (1970) J. Biol. Chem. 245, 2363-2370.
Reed, L.J. & Oliver, R.M. (1968) Brookhaven Symp. Biol. 21, 397-412.
Reed, L.J. (1974) Acc. Chem. Res. 1, 40-46.
Reed, L.J., Pettit, F.H., Eley, M.H., Hamilton, L., Collins, J.H. & Oliver, R.M. (1975) Proc. Natl. Acad. Sci. USA 72, 3068-3072.
Ronchi, S. & Williams, C.H.Jr. (1972) J. Biol. Chem. 247, 2083-2086.
Rossmann, M.G., Liljas, A., Brändén, C.-I. & Banaszak, L.J. (1975) In: The Enzymes (Boyer, P.D., ed.) vol. 11, pp. 61-102, Academic Press, New York.
Schmidt, A. & Christen, U. (1978) Planta 140, 239-244.
Schulz, G.E., Zappe, H., Worthington, D.J. & Rosemeyer, M.A. (1975) FEBS Letters 54, 86-88.
Schulz, G.E. (1977) Angew. Chem. 16, 23-32.
Schulz, G.E., Schirmer, R.H., Sachsenheimer, W. & Pai, E.F. (1978) Nature 273, 120-124.
Sjöberg, B.-M. & Holmgren, A. (1972) J. Biol. Chem. 247, 8063-8068.
Sperenze, M.L., Ronchi, S. & Minchiotti, L. (1973) Biochim. Biophys. Acta 327, 274-281.
Söderberg, B.O., Sjöberg, B.-M., Sonnerstam, U. & Brändén, C.-I. (1978) Proc. Natl. Acad. Sci. USA 75, 5827-5830.
Takashi, S., Tokutake, S. & Usami, M. (1978) J. Biol. Chem. 254, 7376-7382.
Thelander, L. (1968) Eur. J. Biochem. 4, 407-422.
Thelander, L. (1970) J. Biol. Chem. 245, 6026-6029.
Thelander, L. (1974) J. Biol. Chem. 249, 4858-4862.
Thelander, L. & Reichard, P. (1979) Ann. Rev. Biochem. 48, 133-

158.
Thorpe, C. & Williams, C.H.Jr. (1976a) J. Biol. Chem. 251, 3553-3557.
Thorpe, C. & Williams, C.H.Jr. (1976b) J. Biol. Chem. 251, 7726-7728.
Tsang, M. L.-S. & Schiff, J.A. (1976) J. Bacteriol. 125, 923-933.
Veech, R.L., Eggleston, L.V. & Krebs, H.A. (1969) Biochem. J. 115, 609-619.
Viña, J., Hems, R. & Krebs, H.A. (1978) Biochem. J. 170, 627-630.
Voetberg, H.V.M. & Weeger, C. (1973) Eur. J. Biochem. 33, 285-291.
Williams, C.H.Jr. & Arscott, L.D. (1971) Methods in Enzymology 17, Part B, 503-509.
Williams, C.H.Jr. (1976) The Enzymes 3rd Edition 13, 89-173.
Williams, C.H.Jr., Thorpe, C. & Arscott, L.D. (1978) In: Mechanisms of Oxidizing Enzymes (Singer, T.P. & Ondarza, R.N., eds.) pp. 3-16, Elsevier/North Holland Inc., New York.
Wolosiuk, R.A. & Buchanan, B.B. (1977) Nature 266, 565-567.
Worthington, D.J. & Rosemeyer, M.A. (1974) Eur. J. Biochem. 48, 167-177.
Worthington, D.J. & Rosemeyer, M.A. (1975) Eur. J. Biochem. 60, 459-466.
Worthington, D.J. & Rosemeyer, M.A. (1976) Eur. J. Biochem. 67, 231-238.
Zappe, H.A., Krohne- Ehrich, G. & Schulz, G.E. (1977) J. Mol. Biol. 113, 141-152.

DEHYDROGENASE ACTIVITIES OF FATTY ACID SYNTHESIZING ENZYME SYSTEMS

Sarvagya S. Katiyar and John W. Porter

CONTENTS

DEHYDROGENASE ACTIVITIES OF FATTY ACID SYNTHESIZING ENZYME SYSTEMS

Sarvagya S. Katiyar and John W. Porter

1. INTRODUCTION

1.1 Some Properties of Fatty Acid Synthesizing Enzyme Systems

The dehydrogenases of fatty acid synthetases form an integral part of the enzyme systems that synthesize long-chain fatty acids. These fatty acid synthesizing enzyme systems occur as soluble entities in the cytoplasm of plant, animal and microbial cells. Irrespective of the source of the enzyme system, the synthesis of fatty acids involves the same chemical mechanism and the following stoichiometry (Wakil & Ganguly, 1959; Brady *et al.*, 1960; Wakil, 1961).

$$\text{Acetyl-CoA} + 7\ \text{Malonyl-CoA} + 14\ \text{NADPH} + 14\ H^+ \longrightarrow$$

$$1\ \text{palmitic acid} + 14\ NADP^+ + 8\ \text{CoA} + 7\ CO_2 + 6\ H_2O \qquad ...1$$

1.2 Nonaggregated Fatty Acid Synthesizing Enzyme Systems

From an organizational standpoint fatty acid synthesizing systems can be classified into two fundamentally different types. To the first type belong systems which are nonaggregated and lack a higher level of structural organization. The second class consists of multienzyme complexes that are tightly aggregated. Nonaggregated enzyme systems occur in many bacteria (Lennarz *et al.*, 1962; Alberts *et al.*, 1963, 1964) and plants (Overath & Stumpf, 1964; Brooks & Stumpf, 1966; Simoni *et al.*, 1967) and their component enzymes exist as individual proteins in ruptured cell extracts. Their enzymes are separable and several have been partially or completely purified by standard methods of protein purification (Volpe & Vagelos, 1973; Katiyar & Porter, 1977). The *E. coli* fatty acid synthesizing system consists of seven individual enzymes, including a pantetheine-containing protein (acyl carrier protein). Acyl carrier protein (ACP) is a low molecular weight protein (mol. wt. of approximately 10,000) that functions as a carrier of acyl groups in the reaction sequences of fatty acid synthesis catalyzed by the nonaggregated fatty acid synthesizing systems (Goldman & Vagelos, 1962; Lennarz *et al.*, 1962; Wakil *et al.*, 1964; Majerus *et al.*, 1964; Majerus & Vagelos, 1967; Prescott & Vagelos, 1972). Another distinguishing feature of nonaggregated enzyme systems is the release of free intermediates into the solution. The rate of component reactions is therefore dependent on the rate of collision between the substrate and the participating enzyme at each step.

1.3 Aggregated Fatty Acid Synthetases-Multienzyme Complexes

By contrast, the enzyme systems of avian liver and specialized glands (Hsu *et al.*, 1965; Hsu & Yun, 1970; Buckner & Kolattukudy, 1976), mammalian liver and mammary gland (Burton *et al.*, 1968; Smith & Abraham, 1970; Carey & Dils, 1970; Knudson, 1972; Roncari, 1974a,b) and microbial organisms [yeast (Lynen, 1961) and *Mycobacterium smegmatis* (Bloch, 1977)] are classified as multienzyme complexes. These systems have been isolated as single physical entities and purified to homogeneity. Lynen (1961, 1967a,b) assumed in early studies that all the individual components of yeast fatty acid synthetase exist as single polypeptide chains and are assembled as such in the complex. The polypeptide chains were presumed to be held together by exceptionally strong noncovalent forces. Similarly, the finding of eight protein bands in phenol/acetic acid/urea gels and five different N-terminal amino acids led Yang *et al.* (1967) to propose that the pigeon liver fatty acid synthetase consists of eight monofunctional components. Until recently this has been the accepted model of the macromolecular structural organization of the catalytic entities of fatty acid synthetases of animal tissues. There were precedents from other sources supporting this conclusion. For example, the pyruvate dehydrogenase system is readily dissociable into three monofunctional enzymes which reassociate to again give active enzyme complex (Reed & Cox, 1966). However, it now appears that fatty acid synthetase complexes are not composed of individual proteins, inasmuch as they have persistently resisted all attempts at dissociation into individual peptides and subsequent fractionation of these components.

A different model for the fatty acid synthetase multienzyme complexes has been pioneered by Schweizer and coworkers (Schweizer & Bolling, 1970; Schweizer *et al.*, 1971, 1973; Kuhn *et al.*, 1972; Tauro *et al.*, 1974; Knobling *et al.*, 1975). In the early '70s this laboratory showed *via* a genetic approach that the component enzymes of the yeast fatty acid synthetase are coded by two unlinked gene loci designated as "fas 1" and "fas 2." This result suggested that the subunits of yeast fatty acid synthetase are composed of only two nonidentical multifunctional polypeptide chains. This result of the Schweizer group was in excellent agreement with the earlier date of Porter's group (Butterworth *et al.*, 1967; Kumar *et al.*, 1970a,b, 1972a,b; Muesing *et al.*, 1975) who demonstrated, and then extensively studied, the reversible dissociation of fatty acid synthetase into two subunits of almost equal size. Later, the dissociation of fatty acid synthetase complexes from different sources was studied by other investigators (Lynen *et al.*, 1968; Sumper *et al.*, 1969a; Smith & Abraham, 1971; Yun & Hsu, 1972; Vance *et al.*, 1973; Maitra & Kumar, 1974; Stoops *et al.*, 1975). As a result of these numerous studies it now appears that the dissociation of the fatty acid synthetase into two subunits in solutions of low ionic strength is an intrinsic property of all animal synthetases. In contrast, the yeast enzyme shows a similar dissociation behavior, but only in solutions of high ionic strength. Currently it is generally accepted, therefore, that the two subunits of the animal fatty acid synthetase are two closely-related multifunctional polypeptide chains.

Multienzyme complexes have an excellent structural organization for catalytic efficiency. They are efficient, in contrast to the nonaggregated enzymes, because the intermediates remain covalently bound to enzyme and are not released into the solution. Thus, there is no accumulation of free intermediates, as there is in nonaggregated enzyme systems.

A study of the increasing functional structuring of the fatty acid synthetase enzyme system from the nonaggregated type of *E. coli* to the highly organized multienzyme complex has been considered by some to be important from an evolutionary point of view. It was thought that in the course of evolution, nonaggregated systems might have been modified in stages as a result of increasing functional and organizational structuring of the interior of the cell. According to this hypothesis, the transition from a nonaggregated system to tightly bound multienzyme complexes might have occurred through a transitional fatty acid synthetase which had a loose organization that was intermediate between the two extremes. This consideration led Bloch (1977) to investigate the enzyme system of *Mycobacterium smegmatis*. However, extensive investigations with this system failed to show the existence of any transitional fatty acid synthetase complex. In fact, this enzyme system apparently has a structural organization similar to that of multienzyme complexes.

Fatty acid synthetase multienzyme complexes from various sources have molecular weights in the range of 4×10^5 to 2.5×10^6. Regardless of the source and size, and irrespective of the architecture, the enzyme system, whether nonaggregated or aggregated, catalyzes a series of highly organized chemical reactions (Fig. 1). Mechanisms of action for the yeast and pigeon liver enzyme complexes have been proposed by Lynen *et al.*, (1967a,1968) and Phillips *et al.*, (1970a), respectively. More recently, Katiyar *et al.* (1975) have used steady state kinetic

Site 1 — **Acetyl transacylase** ——> Acyl transfer

Acetyl-CoA + E-4′-PP ——> Acetyl-4′-PP-E + CoA ...2

Site 2 — **Malonyl transacylase** ——> Malonyl transfer

Malonyl-CoA + E-4′-PP ——> Malonyl-4′-PP-E + CoA ...3

Site 3 — **Acetyl-4′-PP-E + ES_{cys}-H** ——> Acetyl-S_{cys}-E + 4′-PP-E

or

Acyl-4′-PP-E + ES_{cys}-H ——> Acyl-S_{cys}-E + 4′-PP-E ...4

Site 3′ — **β-Ketoacyl synthetase** ——> Condensation-decarboxylation

Acetyl- (or acyl-) S_{cys}-E + malonyl-4′-PP-E ——> Acetoacetyl- (or β-ketoacyl-) 4′-PP-E + CO_2 + ES_{cys}- H ...5

Site 4 — **β-Ketoacyl reductase** ——> Reduction 1

β-Ketoacyl-4′-PP-E + NADPH + H^+ ——> D(-)-β-Hydroxyacyl-4′-PP-E + $NADP^+$ ...6

Site 5 — **β-Hydroxyacyl dehydrase** ——> Dehydration

D(-)-β-Hydroxyacyl-4′-PP-E ——> *trans*-α,β-Unsaturated acyl-4′-PP-E + H_2O ...7

Site 6 — **Enoyl reductase** ——> Reduction 2

trans-α,β-Unsaturated acyl-4′-PP-E + NADPH + H^+ ——> Acyl-4′-PP-E + $NADP^+$ ...8

Site 7 — **Palmitoyl deacylase** ——> Palmitoyl transfer

Palmitoyl-4′-PP-E ——> Palmitic acid + 4′-PP-E ...9

Fig. 1. Reaction sequence of fatty acid biosynthesis.

Katiyar & Porter (1977).

analysis in the further study of the mechanisms of the reactions catalyzed by the pigeon liver enzyme. Both dehydrogenase reactions are a part of this reaction sequence. Since the enzymes of the *E. coli* (nonaggregated) system are separable, the properties of the individual dehydrogenases of this system have been studied. However, the separation and purification of these multifunctional components has not yet been accomplished for the multienzyme complex. Thus, in understanding the dehydrogenases of fatty acid synthetase we need to discuss the overall mechanism of action of the fatty acid synthetase complex. Moreover, further understanding of these dehydrogenases could be obtained if their location in the complex could be ascertained. We will, therefore, discuss in some detail the present state of knowledge of the structural and functional organization of synthetase complexes.

In the fatty acid synthetase system the dehydrogenases that use NADPH are commonly designated as 'reductases.' There are two reductases: β-ketoacyl reductase and enoylacyl reductase. In this chapter we will be using the term 'reductase' interchangeably with dehydrogenase.

2. SUBSTRATES, COENZYMES AND PRODUCTS

The substrates and coenzyme requirements of various enzyme systems for fatty acid synthesis and the products formed are summarized in Table I.

2.1 Acetyl-CoA and Malonyl-CoA

The substrates for the nonaggregated enzyme systems are acetyl and malonyl thioesters of acyl carrier protein. In contrast, aggregated multienzyme systems require acetyl- and malonyl-CoA. K_m values for acetyl-CoA and malonyl-CoA for the pigeon liver complex (Katiyar *et al.*, 1975) are 1.4 and 30 μM, respectively. In general, various fatty acid synthetase complexes have K_m values for acetyl-CoA in the range of 20 to 50 μM and for malonyl-CoA around 40 μM. The enzyme system of *Mycobacterium smegmatis* shows a unique dependence of the acetyl-CoA requirement on the time point of measurement (Bloch, 1977). In the initial 5 to 10 s of incubation the apparent K_m for acetyl-CoA is 6 μM, and then it abruptly rises to an extraordinarily high value of 400 μM.

2.2 Pyridine Nucleotides

In the reaction sequence (Fig. 1) catalyzed by the fatty acid synthesizing enzyme system two of the catalytic activities (reactions 6 and 8) are dehydrogenases. These dehydrogenases of fatty acid synthetase complexes from avian and mammalian

TABLE I

Substrates and Coenzymes for the Fatty Acid Synthesizing Enzyme Systems from Various Sources

Enzyme	Substrates	Coenzyme	Products
E. coli[a]	Acetyl-ACP & malonyl-ACP	NADPH, NADH	Vaccenic (75%) and stearic (25%) acids
Plants[b]	Acetyl-ACP & malonyl-ACP	NADPH, NADH	Palmitic (20%) and stearic (80%) acids
Yeast[c]	Acetyl-CoA & malonyl-CoA	NADPH, FMN	Palmitoyl and stearyl-CoA
Mycobacterium smegmatis[d]	Acetyl-CoA & malonyl-CoA	NADPH, NADH FMN (?)	(C_{16} + C_{18}) and (C_{24} + C_{26}) acyl-CoA
Euglena gracilis[e]	Acetyl-CoA & malonyl-CoA	NADPH, NADH	Palmitic acid
Mammary gland[f]	Acetyl-CoA or butyryl-CoA & malonyl-CoA	NADPH	Palmitic acid
Pigeon liver[g]	Acetyl-CoA & malonyl-CoA	NADPH	Palmitic acid
Chicken liver[h]	Acetyl-CoA & malonyl-CoA	NADPH	Palmitic acid
Rat liver[i]	Acetyl-CoA & malonyl-CoA	NADPH	Palmitic acid
Brevibacterium ammoniagenes[j]	Acetyl-CoA & malonyl-CoA	NADPH	Palmitic, stearic and oleic acids

[a]Vagelos *et al.* (1966); Weeks & Wakil (1968).
[b]Brooks & Stumpf (1966); Simoni *et al.* (1967).
[c]Lynen (1961,1967a,b).
[d]Bloch (1977)
[e]White *et al.*(1971); Delo *et al.* (1971).
[f]Smith & Abraham (1970); Carey & Dils (1970); Lin & Kumar (1971, 1972).
[g]Hsu *et al.* (1965).
[h]Hsu & Yun (1970).
[i]Burton *et al.* (1968).
[j]Kawaguchi & Okuda (1977).

sources carry out NADPH-specific reductions. NADPH used in the synthesis of fatty acids effects two separate reductions, namely the conversion of β-ketoacyl enzyme to hydroxyacyl enzyme (reaction 6) and the conversion of α,β-enoylacyl enzyme to acyl enzyme. Each of these reactions is repeated six times in the synthesis of one molecule of palmitic acid. Nicotinamide coenzyme requirements for various fatty acid synthetase systems also differ. In the nonaggregated enzyme system of *E. coli* (Vagelos *et al.*, 1966; Weeks & Wakil, 1968) and plants (Brooks & Stumpf, 1966) both NADPH and NADH are needed, whereas in the fatty acid synthetase systems of animal tissues NADPH is a far more effective electron donor than NADH for both β-ketoacyl reduction and α,β-enoylacyl reduction.

Fatty acid synthetase of *M. smegmatis* shows maximal activity when both NADPH and NADH are present (White *et al.*, 1971). When only NADPH is supplied, the reaction is much slower and the K_m value is abnormally high (600 μM). This indicates that NADPH is the more specific electron donor for one of the reductive steps and NADH for the other. It has been demonstrated that NADH is preferentially utilized at the enoylacyl reductase step, the second reduction catalyzed by the synthetase. Yet another multienzyme complex, which utilizes both NADPH and NADH for fatty acid synthetase activity, has recently been purified from *Brevibacterium ammoniagenes* (Kawaguchi, 1977). The K_m values for NADPH and NADH for this enzyme have been found to be 18 and 11 μM, respectively.

2.3 Flavin Mononucleotide (FMN)

The presence of FMN in the fatty acid synthesizing enzyme complex has been reported for the synthetases of yeast and *M. Smegmatis* (Ilton *et al.*, 1971). In the yeast enzyme FMN serves as a hydrogen carrier between NADPH and the double bond in the enoylacyl reduction step and the flavin prosthetic group is tightly bound to the enoylacyl reductase site (Lynen, 1961). Because of the flavin content, the hexagonal crystals of yeast fatty acid synthetase show a bright yellowish-green fluorescence in the fluorescence microscope. Previous to 1970 the requirement for FMN was unique to the yeast system, since no other fatty acid synthetase was known to depend on this coenzyme for activity. According to recent reports, though, the enzyme system of *M. smegmatis* is also suspected of possessing FMN. It has been reported (Bloch, 1977) that the purified fatty acid synthetase of *M. smegmatis* shows absorption peaks at 370 and 460 nm, which are consistent with the presence of flavin. However, the earlier observation with the partially purified *M. smegmatis* synthetase, showing a 2- to 3-fold rise in reaction rate in the presence of FMN (Ilton *et al.*,1971) is no longer observed with the homogeneous enzyme.

2.4 Products

The products of the fatty acid synthesizing systems and their amounts vary appreciably for enzymes obtained from different sources. In general, palmitate and stearate, or their CoA derivatives are the major products. Yeast and *M. smegmatis* enzymes produce fatty acyl-CoA thioesters, whereas bacterial, avian and mammalian enzyme systems form nonesterified long-chain fatty acids. The products formed by pigeon (Hsu *et al.*, 1965), chicken (Hsu & Yun, 1970), rat (Burton *et al.*, 1960) and human (Roncari, 1974a) fatty acid synthetases are primarily palmitic acid. On the other hand, fatty acid synthetases from lactating bovine (Kinsella *et al.*, 1975), rat (Smith & Abraham, 1970), and rabbit (Carey & Dils, 1970) mammary glands synthesize palmitate and smaller amounts of C_4 to C_{14} acids. The pattern of fatty acids formed *in vitro* by bovine mammary tissue enzyme is altered by varying the malonyl-CoA to acetyl-CoA ratio and by increasing the enzyme concentration in the assay (Kinsella *et al.*, 1975). Higher concentrations of enzyme (0.5 mg/ml) generate greater amounts of short and medium chain fatty acids. Likewise, the relative percentages of fatty acids of various chain lengths formed by rat and rabbit mammary gland enzymes (Carey & Dils, 1970; Hansen *et al.*, 1970) can be altered by changing the concentrations of acetyl-CoA and malonyl-CoA. A more important factor, though, in the production of short-chain fatty acids by the mammary gland enzyme is a deacylase with a specificity for short-chain fatty acids. This deacylase is not a constituent of the mammary gland fatty acid synthetase complex, but it functions in association with this complex to produce short-chain fatty acids.

Yeast enzyme can also be manipulated to form shorter- or longer-chain fatty acids by changing the ratio of acetyl-CoA to malonyl-CoA concentrations (Sumper *et al.*, 1969b). However, the major product is always stearate, except under very unusual conditions. *M. smegmatis* enzyme is unique in the sense that it shows a bimodal product pattern over a wide range of assay conditions (Flick & Bloch, 1974). In one of the modes palmitoyl- and stearyl-CoA are the predominant components, whereas C_{24} and C_{26} acyl-CoAs are the major components in the other.

Another unique aggregated enzyme system, in terms of products, has recently been isolated and purified from *Brevibacterium ammoniagenes*. This enzyme system has been reported to synthesize monounsaturated, as well as saturated, fatty acids from the substrates acetyl- and malonyl-CoA. The major products of this enzyme are palmitic, stearic and oleic acids.

2.5 Inhibitors

Inhibition of fatty acid synthesis is effected by $NADP^+$, free coenzyme A and palmitoyl-CoA (Katiyar *et al.*, 1975).

Palmitoyl-CoA inhibition occurs at the critical micellar concentration of this compound (2 to 4 μM), and it is dependent on pH and ionic strength (Dorsey & Porter, 1968; Zahler *et al.*, 1968; Bloch, 1977). The physiological significance of the inhibitory effects of palmitoyl-CoA on the fatty acid synthetase is questionable in view of the detergent effects of this compound. However, palmitoyl-CoA appears to play an important role in the regulation of the fatty acid synthetase of *M. smegmatis* (Bloch, 1977).

Katiyar *et al.* (1975) have reported detailed kinetic investigations on the product inhibition of the pigeon liver enzyme. Free coenzyme A was found to be a competitive inhibitor of acetyl-CoA and malonyl-CoA and an uncompetitive inhibitor against NADPH. In contrast, the inhibition by $NADP^+$ of the rate of fatty acid synthesis produced a competitive pattern with NADPH as the varied substrate and an uncompetitive pattern with acetyl-CoA and malonyl-CoA as varied substrates.

The antibiotic cerulenin has been reported to be a potent inhibitor of fatty acid synthesizing enzyme systems from various microorganisms and from rat liver (Kawaguchi *et al.*, 1978). For example, yeast synthetase activity is completely inhibited at a cerulenin concentration of 10 μg/ml. Kawaguchi *et al.* (1978) have also investigated the effect of cerulenin on the fatty acid synthetase from *Cephalosporium caerulens* (a cerulenin-producing microorganism) and they have found that this synthetase is much less sensitive to cerulenin than the yeast enzyme.

3. REACTION MECHANISM OF FATTY ACID SYNTHESIS

3.1 Chemical Insight Into Individual Reactions

The synthesis of fatty acids by the enzyme systems from animal (Phillips *et al.*, 1970b), microbial (Lynen, 1967a; Bloch, 1977), including *E. coli* (Alberts *et al.*, 1963,1964; Volpe & Vagelos, 1973), and plant (Overath & Stumpf, 1964; Simoni *et al.*, 1967) sources involves the same chemical mechanism. This schematic reaction sequence for fatty acid synthesis in multienzyme complexes is given in Fig. 1.

The sequence of reactions of Fig. 1 occurs in an organized way in all multienzyme complexes. The intermediates are covalently bound to the 4′-phosphopantetheine group (Lynen, 1961; Brodie *et al.*, 1964), which is part of the acyl carrier protein (Qureshi *et al.*, 1974). This covalent chemistry of fatty acid synthesis is a remarkably invariant feature of all fatty acid synthesizing enzyme systems.

The biosynthesis of fatty acids is initiated by the transfer of acetyl and malonyl groups to the enzyme. The acetyl and malonyl transacylation is accomplished by three key functional groups on the enzyme, *viz*, one —OH group and two different types of –SH groups. The binding sites containing these groups have been designated as follows (Porter *et al., 1971*):

(a) 'loading site' (the –OH group of a serine),
(b) 'transfer site' (the –SH group of 4′-phosphopantetheine on the acyl carrier protein), and
(c) 'acyl storage site' (the –SH group of a cysteine on the condensing enzyme).

The loading site has a preferential affinity for the acetyl group which first binds to the serine residue, then is transferred to the 4′-phosphopantetheine –SH of the acyl carrier protein, and finally to the cysteine –SH of the condensing enzyme (Porter *et al.*, 1971). The malonyl moiety is similarly transferred from malonyl-CoA through the loading site to the 4′-phosphopantetheine –SH. These acetyl and malonyl groups, activated in the form of their esters, condense to yield a β-ketoacyl thioester in a concerted process which involves a simultaneous decarboxylation of the malonyl carboxyl group. The β-carbonyl oxygen of the β-ketoacyl thioester is then converted to a methylene group in three subsequent reactions, *viz,* β-ketoacyl reduction (in the presence of NADPH), β-hydroxylacyl dehydration and an α,β-enoylacyl reduction. This sequence results in the formation of butyryl thioester, which is then transferred to the cysteine –SH of the condensing enzyme. Six repetitions of the above cycle yield palmitate. The reactions of fatty acid synthesis are terminated by the hydrolysis of the thioester bond between the palmitoyl moiety and 4′-phosphopantetheine. In multienzyme complexes the acyl chains are transferred either to water in the pigeon or rat liver enzyme systems (Porter *et al.*, 1971) or to coenzyme A in the yeast enzyme (Lynen, 1967a).

The mechanism of synthesis of fatty acids by the pigeon liver multienzyme complex, described above, has been summarized by Phillips *et al.* (1970a) in Fig. 2. One complete cycle of the reactions has been shown here. A repetition of this cycle six or seven times would yield palmitic or stearic acid.

In the enzyme systems from *E. coli* or plants which do not possess the higher level of organization of the multienzyme complexes, each intermediate is released into the solution. Thus, like any reaction in solution, the rates of the component reactions will be dependent on collisions between the substrate and the individual enzyme component at each step. It seems probable, though, that in nonaggregated systems some kind of organization may exist, as suggested by the localized occurrence of most of the *E. coli* acyl carrier protein in the bacterial membrane.

3.2 Kinetic Mechanism of Fatty Acid Synthesis

Katiyar *et al.* (1975) have employed steady state kinetic analysis in the further study of the reactions catalyzed by the pigeon liver enzyme complex. This has given greater insight

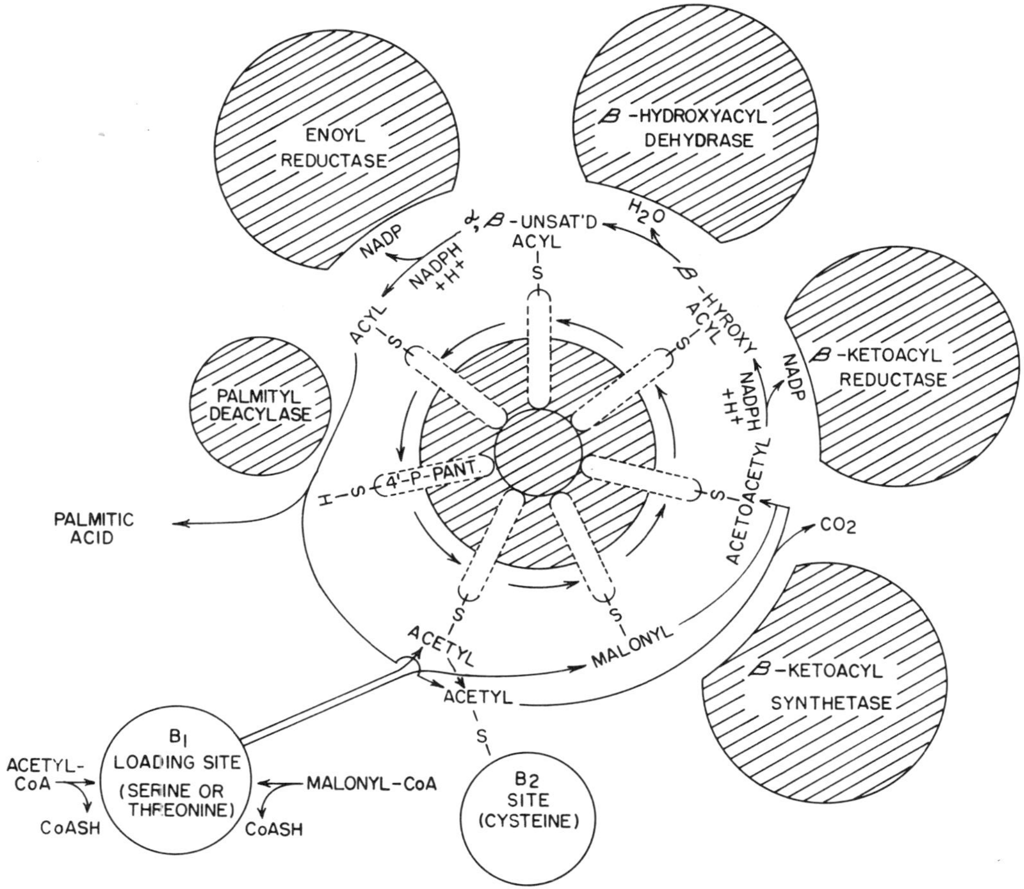

Fig. 2. A proposed mechanism of fatty acid synthesis.

Phillips et al. (1970a)

into various kinetic events leading to the formation of fatty acids. A novel seven-site ping-pong kinetic mechanism has been proposed on the basis of initial velocity and product inhibition data. The model that explains various kinetic events of the pigeon liver enzyme-catalyzed reactions is as follows.

Acetyl-CoA combines with the enzyme in the vicinity of the transacylase site (Site 1, Fig. 1), where free CoA is released and an acetyl group is bound covalently to the enzyme (Phillips *et al.*, 1970a,b; Joshi *et al.*, 1970; Katiyar & Porter, 1975). Then the enzyme-bound 4′-phosphopantetheine (4′PP) binds to the transacylase site. A transfer of the acetyl group occurs, and the acetyl-4′-phosphopantetheine enzyme then leaves the transacylase site. This overall process is a ping-pong event.

Site 1.

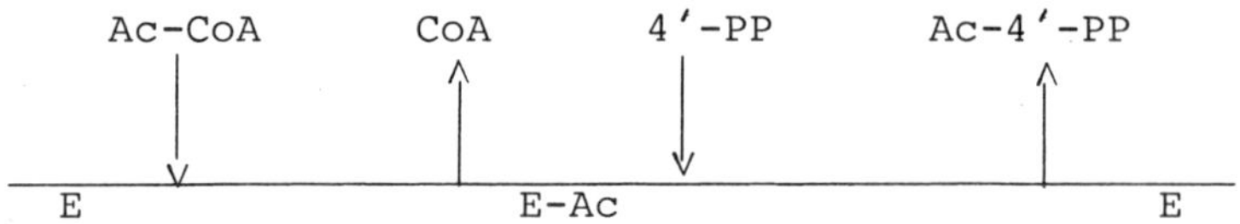

Similarly, the reaction of malonyl-CoA and 4′-phosphopantetheine at the other transacylase site (Site 2, Fig. 1) is also a ping-pong event.

Site 2.

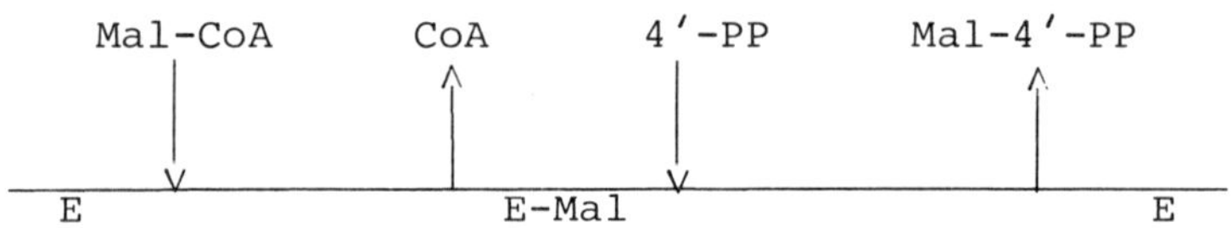

Acetyl 4′-phosphopantetheine, formed in the Site 1 reaction, then transfers its acetyl moiety to the cysteine –SH (Site 3, Fig. 1) and the liberated 4′-phosphopantetheine goes to Site 2 to bind a malonyl group. Malonyl 4′-phosphopantetheine thus formed then combines at the decarboxylation-condensation site (Site 3′, Fig. 1), which is in the vicinity of acetyl-cysteine. The products of this reaction dissociate from their respective enzyme sites as shown below. The reactions at 3 and 3′ are also ping-pong in nature.

Site 3 and 3′:

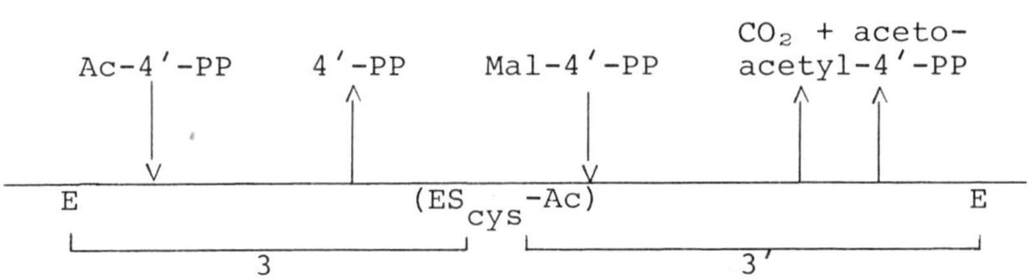

β-Ketoacyl-4′-phosphopantetheine, formed at Site 3, and NADPH combine at the ketoreductase site (or dehydrogenase 1) with the subsequent formation of β-hydroxyacyl-4′-phosphopantetheine as the product of the reduction. The binding of NADPH to the enzyme has been shown to be independent of the covalent binding of either acetyl or malonyl groups (Dugan & Porter, 1970,1971). The reaction at this dehydrogenase site may thus be random sequential or ordered, with NADPH adding first. A ping-pong reaction at this site is ruled out on the basis of NADPH binding data (Dugan & Porter, 1970,1971). No direct hydride transfer from NADPH to enzyme occurs, and the fluorescence of NADPH increases with increased NADPH binding. The stereospecific transfer of hydrogen from the B side of NADPH to the substrate is thus accomplished without the involvement of a direct hydride transfer from NADPH to enzyme (Dugan *et al.*, 1970). β-Hydroxyacyl 4′-phosphopantetheine is then transported to catalytic Site 5, where it is dehydrated. The product of dehydration, α,β-unsaturated acyl-4′-phosphopantetheine and NADPH then combine at the enoylacyl reductase site (dehydrogenase 2). At this site there is direct transfer of hydrogen from the A side of NADPH to the substrate (Dugan *et al.*, 1970). Here also, the reaction may be random sequential or ordered, with NADPH adding first. The reactions occurring at Sites 2 to 6 repeat six more times to give palmitoyl-4′-phosphopantetheine, which is finally deacylated at Site 7. In shorthand notation the complete sequence of reactions for the synthesis of fatty acids could be written as $1(323'456)_7 7$.

A pictorial model showing different catalytic sites and the kinetic events leading to the formation of various intermediates and products described above is shown in Fig. 3.

In the above scheme of kinetic events both the loading site (serine-OH) and the transfer site (pantetheine-SH) are used by both of the substrates. It is to be expected, therefore, that a competition between the two substrates could occur. We have found that, indeed, acetyl-CoA and malonyl-CoA, at certain concentrations, inhibit the rate of fatty acid synthesis by the pigeon liver enzyme (Katiyar & Porter, 1974). This double substrate inhibition is of a competitive type (Katiyar *et al.*, 1975). Katiyar *et al.* (1975) have also derived the steady state equation for the abbreviated model shown in Fig. 3. This equation successfully explains the double competitive substrate inhibition shown by acetyl-CoA and malonyl-CoA and the formation of fatty acids from malonyl-CoA and NADPH alone (Katiyar *et al.*, 1974). Another new insight given by the steady state rate equation is that the decarboxylation of the free malonyl carboxyl group at Site 3′ is presumably a slower reaction than the other site reactions (Kumar *et al.*, 1972b).

The two dehydrogenases of the fatty acid synthetase complex become nonfunctional in the absence of NADPH. When fatty acid synthetase is incubated with acetyl-CoA and malonyl-CoA the formation of triacetic acid lactone (Nixon *et al.*, 1968; Yalpani

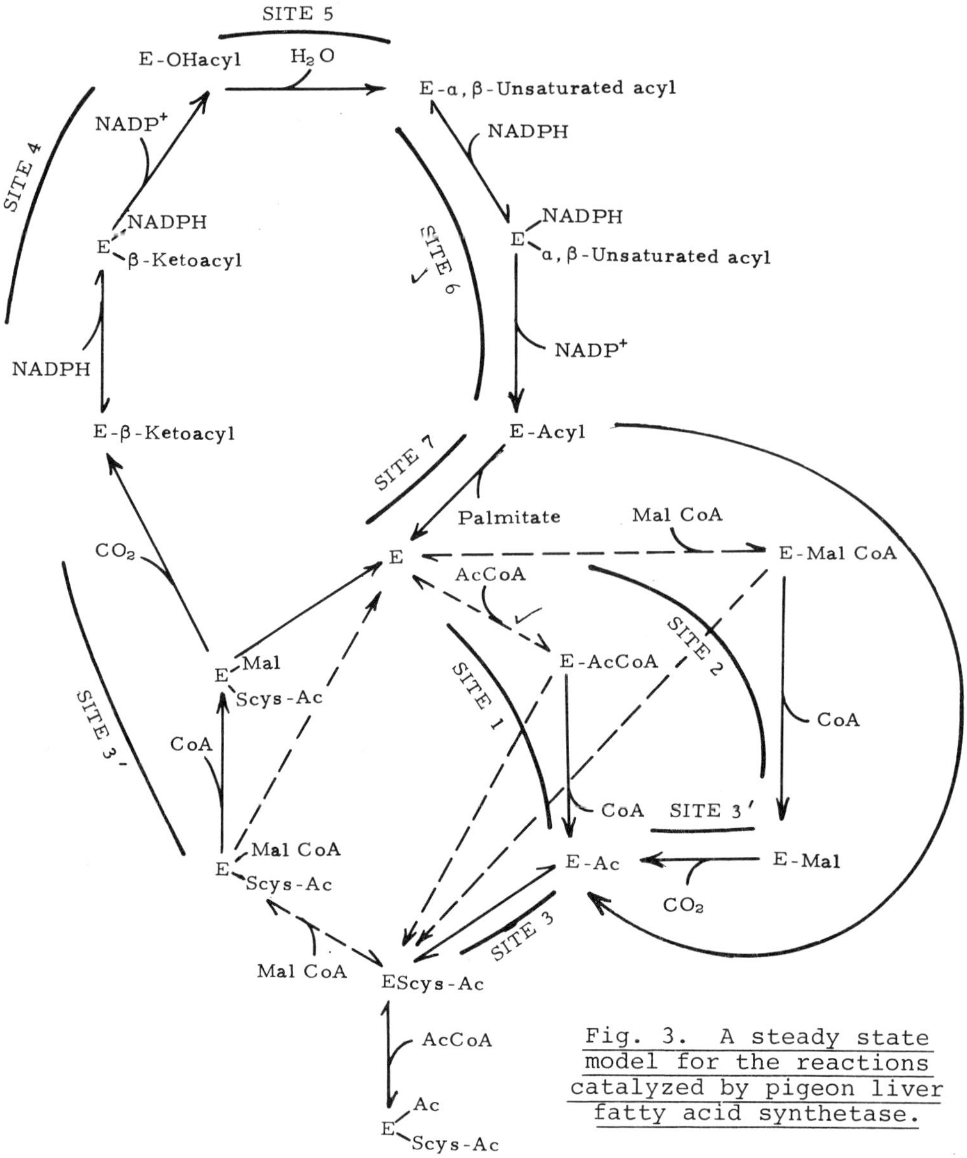

Fig. 3. A steady state model for the reactions catalyzed by pigeon liver fatty acid synthetase.

Katiyar & Porter (1977).

Steps involving substrates are shown as reversible, whereas those involving products are shown as irreversible processes, since the forward reaction resulting in the formation of palmitate is shown in the absence of products. Various catalytic sites with which individual reactions are associated are also shown.

et al., 1969), a derailment product of fatty acid biosynthesis occurs. The kinetic mechanism discussed above is also capable of explaining the synthesis of triacetic acid lactone.

4. STRUCTURAL AND FUNCTIONAL ORGANIZATION OF FATTY ACID SYNTHETASE MULTIENZYME COMPLEXES

The two dehydrogenases of the fatty acid synthetase systems from various sources are a part of the highly organized fatty acid synthetase multienzyme complexes. Even in the non-aggregated type of enzyme systems of *E. coli* and plants, the two hydrogen-transferring components take part in a highly integrated sequence of reactions. Their location in the fatty acid synthetase complex in relation to the other component enzyme activities is, therefore, an important structural issue. In order to gain an insight into the location of the dehydrogenases, we will need to discuss the available information on the structural and functional organizations of fatty acid synthetase multienzyme complexes.

4.1 Dissociation and Reassociation of the Multienzyme Complexes

4.1.1 Reversible Dissociation

The earliest probes to unravel the architecture of the subunits of fatty acid synthetase multienzyme complexes (14S) were initiated in Porter's laboratory (Butterworth *et al.*, 1967) where the reversible dissociation of the pigeon liver fatty acid synthetase to half-molecular weight subunits (9S) was accomplished. Dissociation of the pigeon liver complex was carried out by reaction of the complex with carboxymethyl disulfide (Butterworth *et al.*, 1967). Inactivation and dissociation of the complex were attributed to the loss of essential –SH groups of the enzyme, either through auto-oxidation or by disulfide exchange with the disulfide formed from 2-mercaptoethanol. In subsequent studies (Kumar *et al.*,1970a,b, 1972a,b) it was found that oxidation of –SH groups of the enzyme is not a prerequisite for its dissociation into half-molecular weight subunits. In fact, the complete dissociation of the pigeon liver enzyme into two subunits of nearly equal size, without recourse to chemical modification of the –SH groups of the enzyme, can be effected by dialysis against low ionic strength buffer in the presence of dithiothreitol or 2-mercaptoethanol.

The factors affecting the dissociation, reassociation and reactivation of the half-molecular weight non-identical subunits of the pigeon liver complex have now been well studied (Kumar *et al.*, 1970a,b, 1972a,b; Muesing *et al.*, 1975). Low ionic strength Tris-glycine or phosphate buffer (10 to 50 mM), low temperature (0 to 5°C), and mildly alkaline pH (8.35 to 9.5) favor the dissociation of synthetase complexes to different extents depending on the source. Later studies have reported on the reversible dissociation of fatty acid synthetases from lactating rat mammary gland (Smith & Abraham, 1970,1971), bovine mammary gland (Maitra & Kumar, 1974), Chicken (Yun & Hsu, 1972; Stoops *et al.*, 1975) and rat liver (Stoops *et al.*, 1975) and *Mycobacterium smegmatis* (Vance *et al.*, 1973) into inactive subunits. Yeast synthetase, in contrast, retains its native structure even in distilled water, and is dissociated only by alternate freezing and thawing in the presence of high salt concentrations, *viz*, 1 to 2 M NaCl or LiCl (Sumper *et al.*, 1969a). The molecular weights of the subunits, depending on the source of the enzyme, are in the range of 175,000 to 250,000.

The dissociation and loss of activity of the enzyme at low ionic strength can be prevented by the addition of 10 to 50 μM NADPH (Kumar & Porter, 1972). At 0.2 M phosphate buffer, pH 7.0, where the pigeon liver enzyme complex is very stable, two NADPH binding sites are present on the enzyme. However, the number of NADPH binding sites increases to three as the ionic strength of the buffer is lowered (Dugan & Porter, 1970). Obviously, this increased NADPH binding site is not needed for fatty acid synthesis, since the complex becomes enzymatically inactive at lower ionic strength. The exact function of this additional exposed NADPH binding site at low ionic strength is therefore not clear. It may be suggested, though, that the freshly exposed site is in some way involved in the stabilization of the complex at low ionic strength. In addition, the effectiveness of NADPH, even at low concentrations, in preventing the dissociation of the complex, points to the existence of very specific interactions between this nucleotide and the two subunits.

4.1.2 Reassociation and Reactivation

The half-molecular weight subunits obtained from the pigeon liver enzyme complex can be reassociated and reactivated in the presence of high ionic strength buffer (0.2 M phosphate), room temperature and dithiothreitol. The extent and rate of reactivation of the complex is dependent on ionic strength, NADPH concentration, pH and temperature (Muesing *et al.*, 1975). This reversible dissociation-reassociation phenomena can be explained by the scheme shown on the next page (Muesing *et al.*, 1975).

In this scheme active complex, in the presence of dithiothreitol, is converted to reduced but inactive subunits through an inactive intermediate complex. These subunits can then be reassociated to active enzyme. However, during prolonged treatment at 0° the subunits may be converted irreversibly to inactive subunits that cannot be reasociated. Reactions similar to the above occur when –SH groups of the active enzyme complex are oxidized through disulfide bond formation. This enzyme is then reversibly dissociated to oxidized subunits.

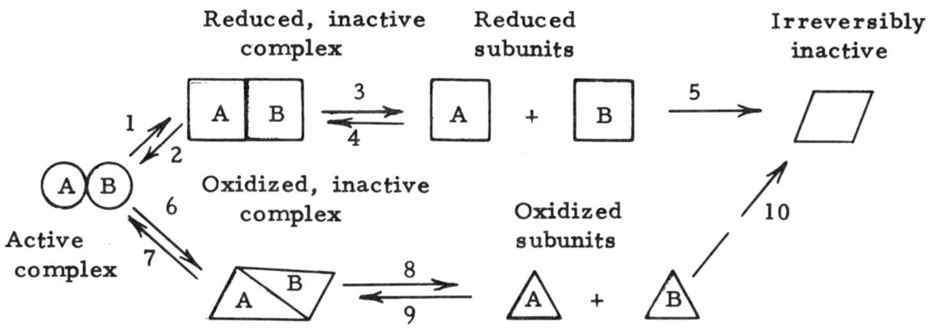

Reassociation of the oxidized or reduced subunits to enzymatically-active complex occurs by a reversal of the dissociation process. Reduced subunits are reassociated in the presence of 0.1 to 0.2 M phosphate buffer, 20 to 30°C, and a pH of 7.0, whereas oxidized subunits also require the presence of dithiothreitol for reassociation to active enzyme complex. Inactive complex formation precedes the restoration of enzyme activity on reassociation. In contrast to the avian liver system, reactivation of the yeast enzyme occurs on decreasing the ionic strength.

4.1.3 Activities of Dehydrogenases and Other Component Enzymes of the Half-Molecular Weight Subunits

The reversible dissociation of the fatty acid synthetase complex to subunits is associated with the complete loss of activity of the enzyme for the overall synthesis of fatty acids. However, the loss of overall synthetic activity does not result in the complete loss of more than one of the component activities shown in Fig. 1. The half-molecular weight subunits of pigeon liver enzyme obtained by treatment with carboxymethyl disulfide show a reduced capacity to bind acetyl or malonyl groups, and they are unable to catalyze the condensation-CO_2 exchange reaction. Similarly, the two subunits obtained at low ionic strength in the presence of dithiothreitol catalyze all the partial reactions of fatty acid synthesis except the condensation-CO_2 exchange reaction (Kumar *et al.*, 1970b). Acetyl- and malonyl-CoA pantetheine transacylase activities are, respectively, 30% and 50% lower in the dissociated enzyme than in the complex. Other partial activities of fatty acid synthesis are reduced, but all are still present in the dissociated subunits.

The K_m values for the complex and the subunits obtained

with model substrates are quite similar for the reduction of S-acetoacetyl-N-acetyl cysteamine, dehydration of β-hydroxybutyryl-N-acetyl cysteamine and deacylation of palmitoyl-CoA. V_{max} values for each of the partial reactions for the dissociated enzyme are within 20 to 30% of the values obtained with the undissociated complex. Presumably, therefore, the binding sites of the component enzymes for these substrates maintain their conformational integrity in the dissociated enzyme.

The significant differences between the undissociated and dissociated enzyme is the absence of the condensation-CO_2 exchange reaction. This loss of condensation activity is not due to loss of covalent binding of acetyl or malonyl groups to the serine -OH site or transfer to the 4′-phosphopantetheine –SH site of the enzyme. It is found, though, that the covalent binding of acetyl groups to the cysteine –SH site, which is essential for the condensation reaction, is eliminated in the dissociated subunits. Thus, the loss of condensing activity in the inactive complex or subunits can be attributed to alterations of the interactions between the 4′-phosphopantetheine and cysteine –SH sites, such that the transfer of the acetyl group from 4′-phosphopantetheine to the cysteine –SH site is prevented.

In contrast to the pigeon liver enzyme, the subunits of the yeast enzyme obtained by freezing and thawing of the complex in 1 M NaCl or LiCl, lose more catalytic activities. The subunits do not show condensing enzyme or either dehydrogenase activities. However, acetyl and malonyl transferase activities and activity for the dehydrase are present to the full extent, as measured with model substrates (Sumper *et al.*, 1969a, Lynen, 1973). Yeast enzyme appears to be unique in the loss of both β-ketoacyl reductase and enoylacyl reductase activities on dissociation into subunits. The loss of enoylacyl reductase activity in the subunits of the bovine mammary gland enzyme has also been reported. However, Stoops *et al.* (1977) have questioned the interpretation of these data. It appears from the above studies that the active conformations for both reductases require specific protein-protein interactions which come into play when the two polypeptide chains are held together in the native yeast, and possibly mammary gland enzyme systems. Dissociation of these complexes eliminates these protein-protein interactions and renders the β-ketoacyl and enoylacyl reductases non-functional.

The dissociation of avian and mammalian liver (Kumar *et al.*, 1970a,b, 1972a,b; Meusing *et al.*, 1975) and mammary gland (Smith & Abraham, 1970, 1971) fatty acid synthetase complexes into two subunits of almost equal size is now universally accepted. A further confirmation of the earlier findings has been reported by Stoops *et al.*, 1975, 1977, 1978a,b) who have shown that the subunits of rat, chicken and yeast fatty acid synthetases consist of two polypeptides of nearly the same size.

Determined efforts of various laboratories to dissociate fatty acid synthetases from various sources into monofunctional enzymatic components as the next step in the study of the structural organization of fatty acid synthetase complexes have failed, even though earlier work indicated that such components existed (Lynen, 1967b; Yang *et al.*, 1967). As a result of these failures to dissociate yeast and animal tissue fatty acid synthetase complexes into monofunctional enzymatic entities under conditions in which protease activity could not be a factor, plus the results of genetic studies by Schweizer *et al.* (1970,1973), a new proposal of the structural organization of fatty acid synthetase complexes has arisen. According to this proposal the yeast fatty acid synthetase complex consists of two multifunctional polypeptide chains. Further support for this model has come from Lynen's laboratory (Lynen, 1973). Support for this model has also come from recent studies by Nepokroeff & Porter (1978) who have shown that the molecular weight of the mRNA for rat liver fatty acid synthetase is in the range of 2.0 to 3.0 $\times 10^6$ daltons. This molecular weight is consistent with a mRNA coding for the half-molecular weight species of rat liver fatty acid synthetase.

The new model for the structural organization of the yeast fatty acid synthetase proposed by Schweizer and associates is based on the results of a comprehensive genetic analysis of yeast fatty acid synthetase auxotrophs (Schweizer & Bolling, 1970; Schweizer *et al.*, 1971,1973; Kuhn *et al.*, 1972; Tauro *et al.*, 1974; Knobling *et al.*, 1975). It has been shown that the component enzymes of the yeast fatty acid synthetases are coded by two unlinked gene loci designated as 'fas 1' and 'fas 2.' These gene loci are polycistronic and have been defined by Schweizer as complex gene loci that each code for a multifunctional protein. Each of these gene clusters codes for individual catalytic activities or enzymes, giving rise to two protein chains which are multifunctional. Collectively, the enzyme activities of the two polypeptide chains account for all of the reactions catalyzed by the complex. The two gene clusters have been assigned the following specific functions on the basis of complementation and tetrad analysis, as well as the biochemical characterization of various complementation groups (Schweizer *et al.*, 1973).

fas 1	fas 2
Acetyl transferase	4′-Phosphopantetheine binding region
Malonyl (palmitoyl) transacylase	β-Ketoacyl synthetase
Dehydratase	β-Ketoacyl reductase
Enoyl reductase	

The two multifunctional subunits of yeast fatty acid synthetase are nonidentical (see Addendum). Only one of the polypeptide chains contains the 4′-phosphopantetheine binding or acyl carrier region (Stoops *et al.*, 1978a). Native yeast fatty acid synthetase (molecular weight 2.44 $\times 10^6$) therefore consists of six subunits of each type, thereby implying the presence of six residues of 4′-phosphopantetheine per mole of the intact synthetase.

Experimentally, four to six residues per mole have been found (Dietlin & Schweizer, 1975), which is in good agreement with the $\alpha_6\beta_6$ structure proposed for the enzyme. Unfortunately, it has not been possible to isolate a free subunit from a noncomplementing mutant for further *in vitro* studies. It has been suggested, therefore, that the unassociated subunits are rapidly degraded *in vivo* or that their synthesis is under reciprocal positive control by the fas 1 and fas 2 loci† (Dietlin & Schweizer, 1975; Knobling & Schweizer, 1975).

Animal synthetases, like yeast, consist of only two subunits, and they appear to be nonidentical. The enzyme systems from chicken liver (Yun & Hsu, 1972), bovine mammary gland (Maitra & Kumar, 1974), pigeon liver (Lornitzo *et al.*, 1974, 1975) and rat liver (Qureshi *et al.*, 1976a) have been reported to consist of two nonidentical subunits. The subunits of the pigeon liver enzyme complex have been separated by affinity chromatography on ε-aminocaproyl pantetheine-sepharose (Lornitzo *et al.*, 1974, 1975). The subunits I (designated 'reductase') and II (designated 'transacylase') catalyze different partial reactions of fatty acid synthesis, and when reassociated, 75% of the original fatty acid synthetase activity is regained. The separated subunit I (reductase) contains the β-ketoacyl and enoylacyl reductase activities and the acyl carrier protein. The separated subunit II (transacylase) contains the serine —OH and cysteine —SH sites, and the acetyl- and malonyl-CoA:pantetheine transacylase activities. Both of the subunits contain activities for β-hydroxybutyryl thioester dehydration and palmitoyl-CoA deacylation. The existence of two deacylase sites on the pigeon liver enzyme has been independently confirmed by Kumar (1975). It has been shown that the palmitoyl-CoA deacylase activity of the native enzyme is selectively inhibited by phenylmethane sulfonyl-fluoride. Binding experiments demonstrated that two moles of [^{14}C]phenylmethane sulfonyl-fluoride are bound covalently per mole of the enzyme. Thus, the presence of deacylase activity on each of the subunits is consistent with this finding. The reductase subunit has been reported to contain eight sulfhydryl groups per mole, whereas the transacylase has 60.

It is of interest to compare the partial activities of the separated pigeon liver subunits with the partial activities associated with yeast fas 1 and fas 2, Table II. Though the correspondence shown by yeast fas 1 and liver subunit II, and yeast fas 2 and liver subunit I is good, there are some significant differences. First, both of the reductases are found on subunit I of pigeon liver enzyme, whereas enoylacyl reductase is found on yeast fas 1. Secondly, dehydrase activity is present in both of the liver subunits and only one of the yeast subunits. Thirdly, condensing activity is not detected on any of the liver subunits,

---

†Schweizer *et al.* (1979). In press.

but it is on one of the yeast subunits. The loss of condensing activity in the dissociated complex and in the separated subunits of avian enzyme has been attributed to a lack of interaction between 4′-phosphopantetheine-SH and cysteine-SH binding sites. As a result, it has been suggested that the acetyl group is not transferred from the 4′-phosphopantetheine-SH to the cysteine-SH site.

TABLE II

Component Activities of the Subunits of Yeast[a] and Pigeon Liver Fatty Acid Synthetases[b]

Activity	Yeast fas 1[a]	Pigeon liver subunit II[b]	Yeast fas 2[a]	Pigeon liver subunit I[b]
Acetyl-CoA transacylase	+	+		
Malonyl-CoA transacylase	+	+		
Palmitoyl-CoA transacylase	+	-		-
β-Ketoacyl thioester synthetase			+	
β-Ketoacyl thioester reductase			+	+
β-Hydroxyacyl thioester dehydrase	+	+		+
Enoyl thioester reductase	+			+
Palmitoyl-CoA deacylase	-	+	-	+
4′-phosphopantetheine site			+	+

\- Indicates absence of the activity in intact synthetase.
\+ Indicates presence of the activity in the subunit.

[a] Schweizer & Bolling (1970); Schweizer (1971, 1973); Kuhn *et al.* (1972); Tauro *et al.* (1974); Knobling *et al.* (1975).
[b] Lornitzo *et al.* (1975).

Presumably, a specific protein-protein interaction between the two subunits is essential for condensing enzyme activity in avian liver fatty acid synthetase complex. Apparently, however, such interactions are not required for the functionality of the condensing enzyme of yeast fatty acid synthetase. It would appear from these results, therefore, that there is a basic differecne in the architecture of the multifunctional polypeptide chains in the fatty acid synthetases from yeast and pigeon liver. The reported separation of the subunits of the fatty acid synthetases obtained from rat, human and chicken liver gives further support to the nonidentical nature of the subunits of these complexes. Additional support for the nonidentity of the subunits comes from the work on chicken liver (Yun & Hsu, 1972; Bratcher & Hsu, 1975) and bovine mammary gland enzymes (Maitra & Kumar, 1974). The nonidentity of the subunits of these enzymes would be unquestionable if the separation of these subunits were achieved by a second independent method and their partial reactions redetermined. However, difficulties in the reproducibility of the affinity chromatographic method restricts us from making an all-out assertion about the nonidentity of the subunits at this time (Lornitzo *et al.*, 1974, 1975).

Another possible unambiguous proof to establish the nonidentity of subunits is the number of 4′-phosphopantetheine prosthetic groups of the complex. Earlier analytical data showed the presence of only one 4′-phosphopantetheine residue per mole in all animal fatty acid synthetase systems studied (Volpe & Vagelos, 1976). This result was consistent with the nonidentity of subunits. However, evidence for two such residues per mole in rat and chicken liver enzymes has been reported by Stoops et al. (1975) and in goose uropygial gland fatty acid synthetase by Buckner & Kolattukudy (1976). The implication of this finding is the existence of two identical subunits, each containing a prosthetic group and all the catalytic activities for fatty acid synthesis. However, the contention that the two subunits are identical, on the basis of two prosthetic groups per mole of enzyme, is premature and not convincing. Thus, the issue of the identity of the two multifunctional polypeptide chains has become a matter of controversy, which can only be clarified by separating the halves by another independent method or by proving by more rigorous methods that the halves are identical. [See also p. 231.]

The preponderance of evidence presented above and the failure to separate the individual enzyme activities from mammalian and avian liver and yeast strongly suggests that the synthetases of yeast and animal tissues are composed of multifunctional polypeptide chains. It appears, therefore, that the earlier model of Lynen (1961, 1967a,b) in which the yeast enzyme consists of seven individual enzymes and of Yang *et al.* (1967) who proposed that pigeon liver enzyme consists of 7 to 8 proteins must be tabled until there is further conclusive evidence for or against the separation of individual proteins under conditions where no proteolysis occurs. In the meantime, the multiprotein concept has received its strongest support from the isolation of ACPs from yeast (Willecke *et al.*, 1969; Schweizer *et al.*, 1970) and pigeon, rat, human (Qureshi *et al.*, 1974, 1976b), dog (Roncari, 1974b) and chicken (Bratcher & Hsu, 1975) liver enzymes. Bratcher & Hsu (1975) also reported the separation of active enzyme components from the fatty acid synthetase of chicken liver by preparative disc gel electrophoresis. In contrast, Stoops *et al.* (1977) found that chicken liver fatty acid synthetase preparations contain a protease which they could not remove by various chromatographic procedures. They suggested, therefore, that proteolysis of the chicken liver complex has led to erroneous interpretations of the observed gel electrophoretic pattern of this enzyme (Bratcher & Hsu, 1975). Recently, Smith *et al.* (1976) and Lin & Smith (1978) have been able to release a functional thioesterase component of rat mammary gland fatty acid synthetase by limited trypsinization without affecting the other activities of the complex. This specific proteolytic cleavage is consistent with the earlier suggestion that the observation of multiple polypeptide components of fatty acid synthetases is a consequence of proteolytic action.

5. β-KETOACYL REDUCTASE AND α,β-ENOYLACYL REDUCTASE ACTIVITIES OF THE MULTIENZYME COMPLEX

Fatty acid synthetase enzyme complexes are more complicated than most dehydrogenases. These enzyme systems, besides carrying out several reactions in addition to hydrogenation, also carry out two different types of hydrogenation reactions. These reductive reactions are:

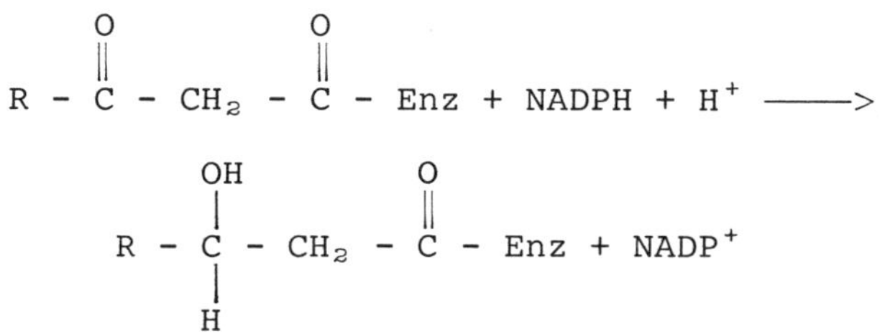

β-Ketoacyl Reduction

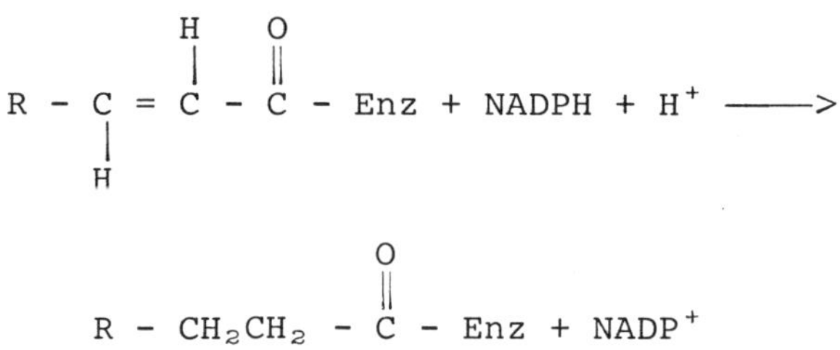

α,β-Enoylacyl Reduction

In the course of synthesis of one mole of palmitic acid, these reactions are repeated seven times. NADPH in both of these reactions donates a hydride ion to the acyl chain. In order to understand the role of NADPH in these reactions, we will discuss the mechanism of binding of NADPH to the fatty acid synthetase and the stereochemistry of the transfer of hydrogen from NADPH to the acyl chain.

The pH optima for the reactions of ketoacyl and enoylacyl reductases of rat mammary gland fatty acid synthetase are 6.8 and 7.2, respectively (Agradi & Smith, 1976). Both reductases are relatively unaffected by treatment of the native fatty acid synthetase with proteases such as trypsin or chymotrypsin (Smith *et al.*, 1976; Lin & Smith, 1978). Apparently both of the activities are associated with a tightly-folded core section of the polyfunctional peptide. Similar conclusions in respect to the location of the dehydrogenases in the pigeon liver fatty acid synthetase have been reached from immunological cross-reactivity of catalytic reactions (see section 6).†

Kumar *et al.*‡ have carried out extensive studies on the dehydrogenase reactions of fatty acid synthetase from lactating bovine mammary gland. Ketoacyl reductase activity of bovine mammary gland fatty acid synthetase is specific for NADPH (K_m = 5 μM) and this enzyme can use acetoacetyl-CoA as the substrate. Also, it appears that acetoacetyl-CoA is a better substrate (K_m = 5 μM) than acetoacetyl pantetheine (K_m = 1 mM) and s-acetoacetyl-N-acetylcysteamine (K_m > 10 mM) for this reaction. Acetoacetyl-CoA reductase is sensitive to –SH reagents and it is inhibited by phenylmethyl sulfonyl fluoride. Methylmalonyl-CoA is a competitive inhibitor of this reaction. Enoylacyl reductase of bovine mammary gland fatty acid synthetase can also use crotonyl-CoA as the substrate.

5.1 Binding of NADPH to Avian and Mammalian Liver Fatty Acid Synthetases

The nature and number of binding sites of NADPH to the pigeon and rat liver enzyme has been studied extensively in the laboratory of Porter (Dugan & Porter, 1970,1971). These sites were examined by the techniques of fluorescence enhancement, fluorescence polarization and equilibrium dialysis. Pigeon and rat liver enzyme complexes increase the fluorescence yield of NADPH by 4.6- and 3-fold, respectively, at different enzyme and buffer concentrations. Fluorescence yield appears to be independent of buffer concentration, amount of NADPH bound to enzyme and the nature of buffer components. The number of NADPH binding sites on each synthetase is independent of the enzyme concentration. The number of sites and dissociation constants for binding of NADPH to fatty acid synthetase are given in Table III.

Pigeon liver enzyme exhibits only two binding sites for NADPH in 0.2 M phosphate, pH 7.0, with a K_d of 8 μM (Dugan &

† Katiyar *et al.* Unpublished results.
‡ Kumar, Soma, Personal communication.

Porter, 1970). At 0.1 M or lower phosphate concentration, another site appears to yield NADPH binding sites which have equivalent but decreased dissociation constants of 3.5 μM. Since the third binding site is exposed only at low ionic strength, two types of NADPH binding sites are indicated. The equivalence of these binding sites at 0.1 M buffer concentration can be rationalized by assuming that the differences in binding under these conditions are too small to be detected by present methods. The presence of two NADPH binding sites in 0.2 M phosphate, where fatty acid synthesis occurs at the maximal rate, is consistent with the conclusion that a different binding site is required for each of the two types of reductions (β-ketoacyl and α,β-enoylacyl enzyme). The function of the third site appears to be related to the prevention of dissociation of the complex at low ionic strength.

TABLE III

Number of Binding Sites and Dissociation Constants for Binding of NADPH to Pigeon and Rat Liver Fatty Acid Synthetases as a Function of Buffer and Buffer Concentration[a]

Enzyme	Buffer	Number of binding sites	Dissociation constants K_{diss} (μM)
Pigeon liver	0.2 M phosphate	2	8
	0.1 M phosphate	3	3.5
	0.1 M phosphate	3	3
Rat liver	0.2 M phosphate	2	1 (K_{d_1})
		2	50 (K_{d_2})
	0.1 M phosphate	2	<1 (K_{d_1})
		2	~1 (K_{d_2})

[a] Dugan & Porter, 1970.

NADPH binding studies carried out on the separated subunits of the pigeon liver enzyme have confirmed the above findings (Lornitzo *et al.*, 1975). Subunit I was found to possess two binding sites. This conclusion was based on a calculation from the maximum fluorescence enhancement which gave a value of 1.6 moles of NADPH bound per mole of protein. This agrees very well with the location of β-ketoacyl and α,β-enoylacyl thioester activities on Subunit I. However, the affinity of NADPH for one of these sites on the separated subunit I is low. It has been suggested that the weaker binding site is associated with the α, β-enoylacyl thioester reductase activity. The assignment of the weaker binding site to this enzyme resulted from the effect of NADPH on the rates of both dehydrogenase activities in Subunit I and in the pigeon liver fatty acid synthetase complex. The results of this comparative study are shown in Fig. 4. The slopes for double reciprocal plots of velocity *versus* NADPH concentration are the same for β-ketoacyl thioester reductase in the complex and in the Subunit I (Fig. 4A). In contrast, the slope for α, β-enoylacyl thioester reductase activity decreases at low concentrations of NADPH with the intact complex, whereas with Subunit I it remains constant.

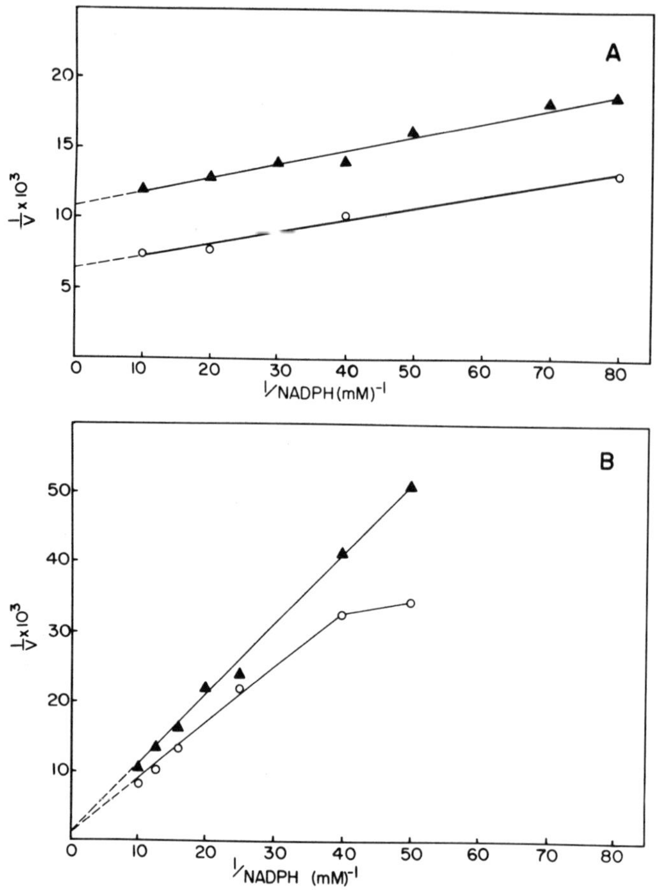

Fig. 4. Double reciprocal plots of 1/NADPH *versus* 1/*V* (where *V* is expressed as specific activity).

Lornitzo *et al.* (1975).

A, β-ketoacyl thioester reductase and B, crotonyl thioester reductase (crotonyl thioester level in B is 7 mM). Standard fatty acid synthetase (o——o); Subunit I (▲——▲).

Subunit II, in contrast to Subunit I, showed no binding of NADPH. However, an equimolar mixture of Subunit I and Subunit II bound 3.2 moles of NADPH per mole of protein as compared to 3.1 moles of NADPH per mole of native fatty acid synthetase complex. The third binding site thus does not appear until Subunit II, which by itself does not bind NADPH, is present. As discussed elsewhere, the lowering of ionic strength results in the dissociation of the fatty acid synthetase complex into two subunits. It also results in the appearance of the third NADPH binding site. Since NADPH prevents the dissociation of the fatty acid synthetase complex into subunits, it is strongly suggested that one of the NADPH binding sites is in a region between the two subunits which is drastically affected by the conformational changes that lead to the dissociation of the complex. Furthermore, reactivation studies (Muesing *et al.*, 1975) with the dissociated subunits indicate that each of the three binding sites are intimately involved in the reactivation of pigeon liver fatty acid synthetase subunits by NADPH in low ionic strength buffer. The exact sequence of NADPH binding and the reassociation of the enzyme prior to its reactivation is unknown.

Rat liver enzyme has four NADPH binding sites (Dugan & Porter, 1970; Yu & Burton, 1972). Even at high ionic strength, one pair of binding sites has a much stronger affinity than the other pair. In 0.2 M phosphate buffer, pH 7.0, the first pair of these sites has a K_d of 1 μM, and the second pair has a K_d of 50 μM (Table III). As the ionic strength is lowered, the differences in the K_d values for the four sites become insignificant. This means that the binding affinity of the second pair of sites increases relative to the first pair, with a decrease in the ionic strength. This equivalence of K_ds at low ionic strength is reminiscent of the equivalent nature of three binding sites of the pigeon liver enzyme at lower ionic strength. The difference in values obtained for the number of NADPH binding sites for the pigeon and rat liver enzymes, is surprising. It may be pointed out here, however, that if the earlier determined molecular weight of 540,000 (Hsu *et al.*, 1965) is used instead of the later value of 450,000 (Yang *et al.*, 1967) the pigeon liver enzyme would also show four binding sites. Yu and Burton (1972) have studied the kinetics of NADPH binding at the two pairs of NADPH binding sites of the rat liver enzyme. They determined the rate constants for the binding reaction and for the dissociation of the enzyme:coenzyme complex using a stopped-flow technique for the stronger and weaker pairs of binding sites. The rate constants for the binding reactions were found to be approximately 1×10^7 M^{-1} sec^{-1} and 1.2×10^6 M^{-1} sec^{-1} for the reactions occurring at the stronger and weaker pairs of binding sites, respectively. Furthermore, these values remained nearly constant in the buffer concentration range of 0.12 to 0.20 M, pH 7.0. In contrast, the rate constants for the dissociation reactions of the enzyme-NADPH complex at the stronger and weaker pair of binding sites were found to vary with the buffer concentration in the range of 1 to 9 sec^{-1} and 10 to 30 sec^{-1}, respectively.

Hsu & Wagner (1970) reported four independent and non-interacting NADPH binding sites with K_d values of 0.29 μM per mole of chicken liver enzyme. In contrast, Srinivasan & Kumar (1976) found only three binding sites. These sites are apparently identical above 12°C with a K_d value of 0.6 μM. At 12°C these sites split into two subsets - a pair of sites with a K_d of 0.3 μM and the third binding site with a K_d of 1.1 μM. As the temperature was further lowered to 5°C, the site with weak affinity disappeared, leaving only two binding sites with a K_d of 0.5 μM. Modification of the chicken enzyme with phenylmethanesulfonyl fluoride to an extent that deacylase activity was completely lost, showed that two NADPH binding sites still survived. Also, both of the reductase activities, as measured with the model substrates S-acetoacetyl-N-acetyl cysteamine and crotonyl-CoA were retained on the modified enzyme. Thus, the data obtained with the pigeon and chicken liver enzymes demonstrate that out of three NADPH binding sites only two are needed for the functionality of both dehydrogenases of the complex.

Similarly, in the case of rat liver enzyme, it is reasonable to assume that the pair of sites with the stronger affinity are involved in the catalysis effected by the two reductases. Further, it is probable that the weaker pair of binding sites are not catalytically functional. Two lines of argument may be advanced to explain the presence of the weaker affinity NADPH binding sites. First, negative cooperativity may be responsible for the presence of binding sites with different binding affinities. Thus, the binding of NADPH at two sites would incude a conformational change in the enzyme, thereby decreasing the binding affinity at the other sites. Levitzki & Koshland (1969) have suggested that negative cooperativity may be an attribute of many, rather than a few, enzymes. Secondly, it is possible that conformational changes may take place in the enzyme during purification which result in the presence of weaker binding sites. Such a change is possible in the rat liver enzyme, which is inactivated on storage at buffer concentrations of less than 0.5 M.

There is a competition between salt ions and NADPH for binding to the enzyme (Dugan & Porter, 1970, 1971) which is indicative of an ionic binding between NADPH and protein. The electrostatic nature of this binding is supported by the results obtained on the effects of salts on the binding of NADPH to pigeon and rat liver enzymes, Figs. 5 and 6.

The binding of NADPH to enzyme is greatest at the most dilute salt concentrations, and is inhibited as the salt concentration increases. This inhibition has been found to be of a competitive type (Fig. 6). Cations like K^+ and Na^+ produced the same degree of inhibition, whereas anions of chloride, phosphate and acetate showed considerable differences in inhibition (Fig. 5). The strongest inhibition of NADPH binding was exhibited by chloride ion, which might be due to the production of an enzyme conformation that is unfavorable for the binding of NADPH. In fact, above 0.5 M chloride ion concentration, no binding of NADPH to enzyme occurs. In general, though, the binding affinity of NADPH to enzyme increases with decreasing ionic strength, regardless of the buffer system.

Similarly, analogues of NADPH, namely NADH, $NADP^+$, acetylpyridine-$NADP^+$ and deamino-$NADP^+$ also competitively inhibit the binding of NADPH to the synthetase (Dugan & Porter, 1971).

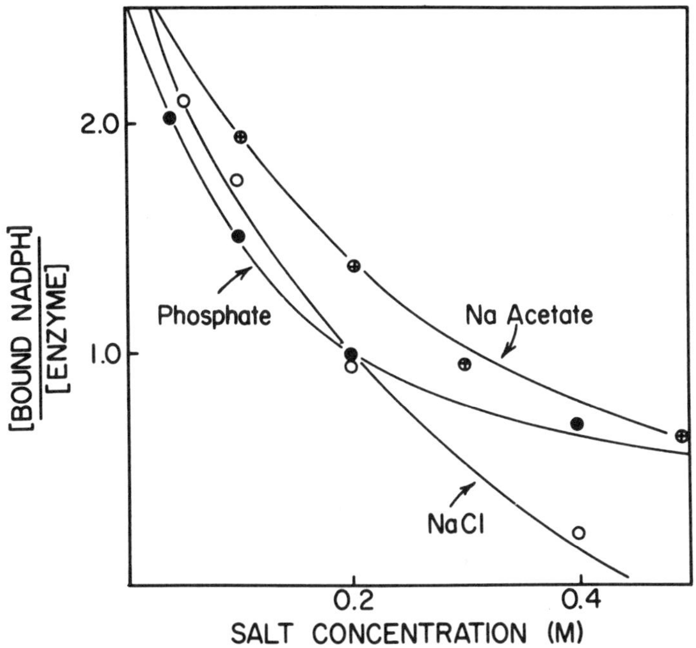

Fig. 5. Moles of bound NADPH per mole of pigeon liver fatty acid synthetase as a function of salt concentration.

Solutions of 1 ml containing NADPH, 10 μM; enzyme, 2 μM; and salts (as indicated on the figure) were assayed for the binding of NADPH by the enzyme by the method of fluorescence enhancement as described in the text. Similar results were achieved by fluorescence polarization.

A comparison of the dissociation constants of these analogues and NADPH (Table IV) is interesting, inasmuch as it gives some insight into a probable correlation between the structure of the coenzyme and the binding affinity.

It is evident that the binding affinity of NADH to the enzyme is much less than that of NADPH. Structurally, these two nucleotides differ only in the presence of a 2′-ribose phosphate group. It is possible that the absence of the 2′-phosphate moiety may be responsible for the inefficient utilization of NADH by the fatty acid synthetase of animal tissue. It should be remembered, however, that NADH is the preferred electron donor for the enoylacyl reduction step in the nonaggregated enzyme systems and the multienzyme complex of mycobacteria (Table I). Binding data for NADH for these complexes is not available. However, the data on the binding of NADH to rat and pigeon liver enzyme systems bring out several points of interest (Dugan & Porter, 1971),

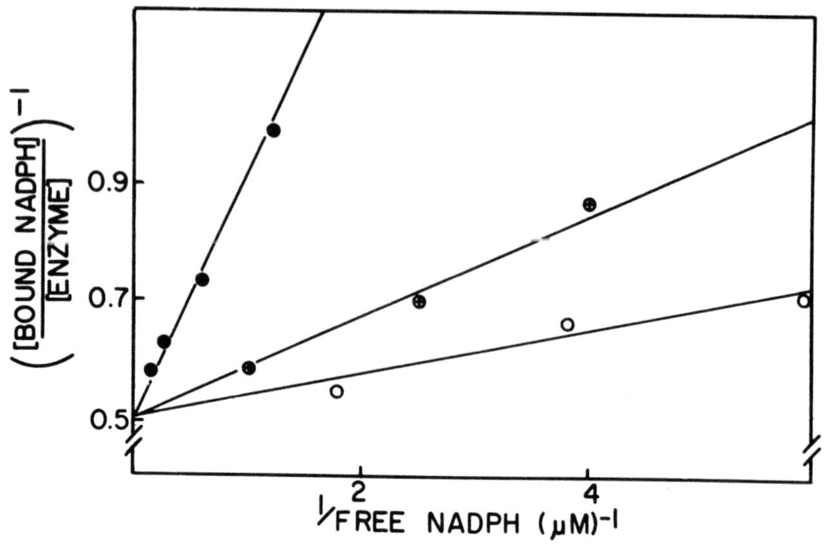

Fig. 6. A double reciprocal plot of moles of bound NADPH per mole of rat liver fatty acid synthetase *versus* free NADPH determined at different potassium phosphate buffer concentrations.

Solutions of 1 ml containing NADPH, through the range of 5 to 12 μM, and rat liver fatty acid synthetase, 4 μM, were assayed for the binding of NADPH by the enzyme. Assays were made in phosphate buffer, pH 6.8, by the method of fluorescence enhancement as described in the text. •——•, 0.2 M potassium phosphate, ⊗——⊗, 0.1 M potassium phosphate; o——o, 0.025 M potassium phosphate. (Dugan & Porter, 1971).

namely, that NADH can (a) function as a hydrogen donor in fatty acid synthesis, (b) quench the fluorescence of the fatty acid synthetase, (c) increase the polarization of fluorescence and (d) inhibit the binding of NADPH to enzyme. These facts and the data of Table IV suggest that NADH binds rather weakly to the enzyme. However, its nonutilization by the dehydrogenases of animal tissue fatty acid synthetases indicates that some differences exist in the binding alignment of the two nucleotides with the enzyme protein.

The weaker binding of $NADP^+$, in comparison to NADPH, is favorable for the synthesis of fatty acids. As discussed earlier, both dehydrogenases, and therefore the NADPH binding sites, are used repeatedly during the synthesis of palmitic acid. This necessitates the replacement of the product $NADP^+$ by the substrate NADPH. A weakly bound $NADP^+$ facilitates this displacement and the synthesis of fatty acids.

TABLE IV

Dissociation Constants for Binding of Coenzymes to Fatty Acid Synthetases[a]

Coenzyme (pyridine nucleotide)	Enzyme source	Phosphate buffer M	K_{diss} μM
NADPH	Pigeon	0.01	0.17
	Pigeon	0.20	8.0
	Rat	0.20	1.0
NADH	Pigeon	0.01	4.2
$NADP^+$	Pigeon	0.01	1.7
	Pigeon	0.20	175.
	Rat	0.20	27.
Deamino-$NADP^+$	Rat	0.20	31.
Acetylpyridine-$NADP^+$	Rat	0.20	40.

[a] Dugan & Porter, 1971.

As fatty acid synthesis is optimum at ionic strengths of 0.2 M or higher, whereas binding affinities and capacities are maximal at low ionic strengths, an attempt to assign individual binding sites to different dehydrogenases appears to be of no consequence. However, it seems reasonable to assume that only two binding sites of equal strength exist on the avian enzyme and that one pair of sites on the rat liver enzyme has a 50-fold higher binding affinity than the other pair. These binding data strongly suggest that the inhibition of fatty acid synthesis that occurs at low ionic strength is not due to any change in the functionality of the dehydrogenases.

Changes in enzyme conformation, as a consequence of the binding of other substrates, sometimes modifies the form of inhibition or activation of dehydrogenases (Colowick *et al.*, 1966). In fatty acid synthetases, a multisubstrate system, the binding of one of the substrates could influence the binding of the others. However, Dugan *et al.* (1970) found that neither of the substrates acetyl-CoA or malonyl-CoA nor their related compounds such as malonic acid, acetate, free coenzyme A or palmitic acid had any effect on the binding of NADPH to fatty acid synthetase. Only palmitoyl-CoA inhibited the NADPH binding (98.5%), but this inhibitory effect was presumably due to its detergent action. All of the evidence points to NADPH sites being separate from binding sites of other substrates and to the absence of any sequential order of binding of NADPH with respect to other substrates.

A comparison of the binding affinity of fatty acid synthetase for NADPH with other dehydrogenases shows that the synthetase dehydrogenases bind NADPH more strongly than most other dehydrogenases. For example, enzymes of carbohydrate metabolism have K_ds in the range of 130 to 0.31 μM (Colowick *et al.*, 1966), and glutamate dehydrogenase has a K_d near 2 μM (Bayley & Radda, 1966).

5.2 Stereochemical Studies of Hydrogen Incorporation Into Fatty Acids During Their Synthesis by the Synthetase Complex

During the biosynthesis of palmitic acid, hydrogen atoms are provided by the methyl group of acetyl-CoA, methylene group of malonyl-CoA, NADPH and water. The complete identification of the sources of hydrogen atoms in fatty acids synthesized by fatty acid synthetases of pigeon and rat liver, yeast and *Brevibacterium ammoniagenes* has been carried out by Dugan *et al.*, (1970) and Seyama *et al.* (1977a,b, 1978).

5.2.1 Stereospecificity of Hydrogen Incorporation from Nucleotides

Dugan *et al.*, (1970) convincingly established that the two reductive steps of fatty acid synthesis (β-ketoacyl enzyme and α,β-enoylacyl enzyme) by rat and pigeon liver enzyme complexes proceed stereospecifically. It was found that β-ketoacyl reduction proceeds by the transfer of hydrogen from the B side of position 4 of the pyridine ring of NADPH, whereas the enoylacyl reduction proceeds by transfer from the A side. The demonstration of this stereospecificity was achieved by studying the partial reactions with model substrates. When acetoacetyl-CoA or S-acetoacetyl-N-acetyl cysteamine was reduced to the β-hydroxybutyryl ester with pigeon or rat liver fatty acid synthetase in the presence of tritiated NADPH, tritium was incorporated into the product only from the B form of NADPH. In contrast, only the A form of tritium-labeled NADPH produced tritiated product when S-crotonyl-N-acetyl cysteamine was similarly reduced to the butyryl ester. The quantity of tritium incorporated into the product was nearly the same from the A and B forms of [^{3}H]NADPH. These results have been confirmed by Seyama *et al.* (1978), who have carried out an extensive gas chromatographic and mass spectrometric analysis of deuterium-labeled fatty acids. In these studies the model substrates S-acetoacetyl-N-acetyl cysteamine and S-crotonyl-N-acetyl cysteamine were reduced with deuterium-labeled NADPH and rat liver enzyme complex. The reaction products, namely β-hydroxybutyryl-N-acetyl cysteamine and S-butyryl-N-acetyl cysteamine, were analyzed by chemical ionization mass spectrometry. Chemical ionization mass spectra of the products are given in Figs 7 and 8. For the β-ketoacyl reductase, the quasimolar ion, $[MH]^+$ at m/e 350, remained unchanged when

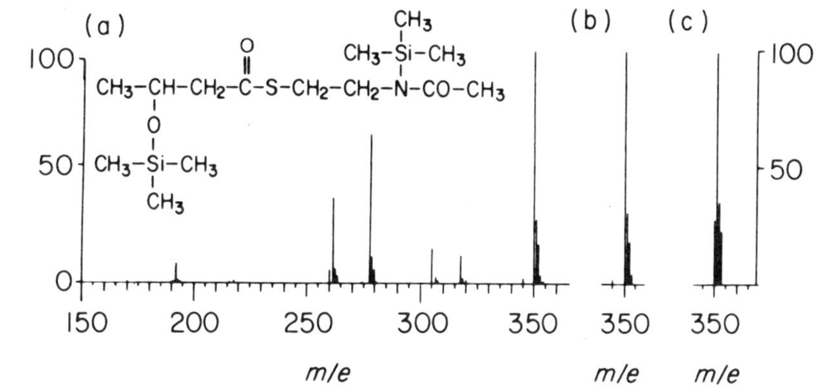

Fig. 7. Chemical ionization mass spectra of the trimethylsilyl derivative of S-β-hydroxybutyryl-N-acetyl cysteamine.

Seyama *et al.* (1978)

(a) Standard incubation mixture; (b) with [α-^{2}H]NADPH instead of NADPH; (c) with [4β-^{2}H]NADPH instead of NADPH.

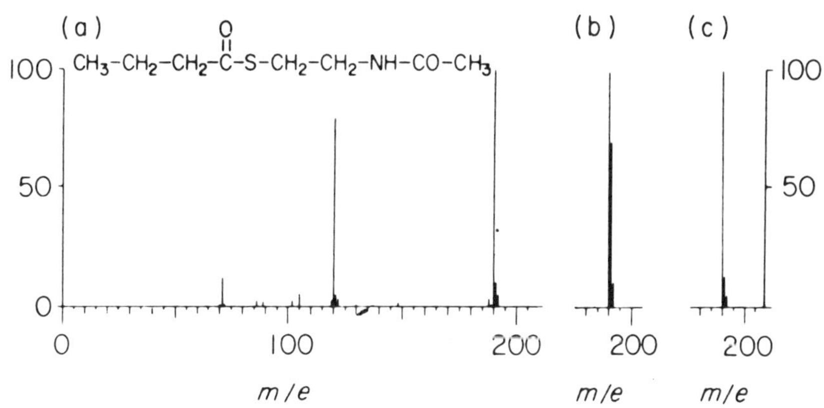

Fig. 8. Chemical ionization mass spectra of S-butyryl-N-acetyl cysteamine.

Seyama *et al.* (1978)

(a) Standard incubation mixture; (b) with [4α-^{2}H]NADPH instead of NADPH; (c) with [4β-^{2}H]NADPH instead of NADPH.

incubation was performed with [4α-^{2}H]NADPH (Fig. 7b), but it shifted to m/e 351 on incubation with [4β-^{2}H]NADPH (Fig. 7c). These results confirmed that β-ketoacyl reductase has B stereospecificity for NADPH. Similarly, in the mass spectrum of the product of enoylacyl reduction, some of the quasimolar ion $[MH]^+$ at m/e 190 of unlabeled S-butyryl-N-acetyl cysteamine (Fig. 8a) shifted to m/e 191 when (4α-^{2}H]NADPH was used for the reaction, but no change was observed when [4β-^{2}H]NADPH was incubated (Seyama *et al.*, 1978). These results demonstrate an A stereospecificity for NADPH in the enoylacyl reduction.

Similar experiments to elucidate the mechanism of hydrogen incorporation from NADPH into products were performed with the enzyme systems from yeast (Seyama *et al.*, 1977b) and *B. ammoniagenes* (Seyama *et al.*, 1977a). The stereospecificity of β-ketoacyl reductase for NADPH with both of these enzyme systems was identical (B stereospecificity) to that obtained with rat and pigeon liver enzymes. In contrast to the rat and pigeon liver enzymes, the enoylacyl reductase of yeast enzyme showed a B-stereospecificity for NADPH. The enzyme system of *B. ammoniagenes* which is known to require NADH for enoylacyl reductase activity, also used the 4β-hydrogen of the nicotinamide ring of NADH (*i.e.*, B stereospecificity) at the enoylacyl reductase step. The stereospecificities of various fatty acid synthetases for NADPH or NADH are summarized in Table V.

TABLE V

Stereospecificity of Dehydrogenase Catalyzed Reactions of Various Fatty Acid Synthetases for NADPH or NADH

Enzyme source	Dehydrogenase activity	
	β-Ketoacyl reductase	α,β-Unsaturated acyl reductase
Pigeon liver[a]	B	A
Rat liver[a,b]	B	A
Brevibacterium ammoniagenes[c]	B	B[e]
Yeast[d]	B	B

[a] Dugan *et al.* (1970)
[b] Seyama *et al.* (1978)
[c] Seyama *et al.* (1977a)
[d] Seyama *et al.* (1977b)
[e] Needs NADH for this reduction.

The stereospecificity of the α,β-unsaturated acyl reductase activity of fatty acid synthetase multienzyme complexes appears to be an exception to the following well-accepted generalizations of enzyme catalyzed reactions (Bentley, 1970). These generalizations state that (a) the stereospecificity of a particular reaction is fixed and is independent of the source of enzyme, and (b) the stereospecificity of a particular reaction is not affected by whether NAD^+ or $NADP^+$ is used as the coenzyme. As evident from Table V, the stereospecificity of the α,β-unsaturated acyl activity is dependent on the enzyme source as well as on the type of pyridine nucleotide.

5.2.2 Intramolecular Distribution of Deuterium Provided by Deuterium-Labeled NADPH in Fatty Acids

Further information brought out by the use of deuterium-labeled NADPH and gas chromatographic mass spectrometric analysis locates the hydrogen atoms provided by NADPH in the synthesis of palmitate. Differences in the rate of exchange of deuterium in fatty acid synthesis are also demonstrated by this technique. The mass spectrum of methyl stearate synthesized by the yeast fatty acid synthetase in the presence of [4β-^{2}H]NADPH, showed the presence of 14 deuterium atoms in the odd-numbered methylene carbon atoms (2 deuterium atoms per carbon atom) (Seyama *et al.*, 1977b). Seven deuterium atoms (1 per carbon atom) were incorporated from the 4β-position of deuterium-labeled NADPH (B form) at the β-ketoacyl reductase step and the other 7 (1 per carbon atom) from the same position of NADPH at the enoylacyl reductase step. Similar experiments with the enzyme systems from rat liver (Seyama *et al.*, 1978) and *B. ammoniagenes* (Seyama *et al.*, 1977a) have established that 14 hydrogen atoms transferred from NADPH or NADH in both reductive steps during fatty acid synthesis go to odd-numbered methylene carbon atoms (2 hydrogen atoms per carbon atom). Thus, fatty acid synthetases from *B. ammoniagenes* and rat liver, like the yeast enzyme, also incorporate 7 4β-hydrogens of NADPH on the odd-numbered carbon atoms (1 hydrogen per carbon) in the first reduction, but in the second reduction rat liver enzyme puts in 7 4α-hydrogens of NADPH (1 hydrogen per carbon) on the odd-numbered methylenes, whereas the *B. ammoniagenes* enzyme transfers 7 4β-hydrogens of NADH. No exchange between hydrogen of NADPH and water was observed at the enoylacyl reduction step with yeast and rat liver fatty acid synthetases, though there was some exchange of 4β-hydrogen of NADH with water at this step in the enzyme system of *B-ammoniagenes*.

5.2.3 Identification of Source of Other Hydrogen Atoms of Fatty Acids

The origin of the remaining hydrogen atoms of the fatty acids has now been identified (Seyama *et al.*, 1977a,b; 1978). Methyl hydrogen atoms of acetyl-CoA are quantitatively incorporated into palmitate (3 atoms per mole of palmitic acid) (Bressler & Wakil, 1961). Identification of the other hydrogen atoms was carried out by mass chromatography of the methyl esters of the fatty acids synthesized during incubation of substrates in deuterium oxide with the fatty acid synthetases of rat liver, yeast and *B. ammoniagenes*. It was found that in stearate (formed by the yeast enzyme) and palmitate (formed by the rat liver enzyme) on an average of 16 and 14 deuterium atoms were incorporated from deuterium oxide, respectively. These 14 deuterium atoms were located on the even-numbered methylene carbon atoms (2 deuterium atoms per carbon atom). Seven of these 14 deuterium atoms are incorporated as the result of a deuterium-hydrogen exchange between the methylene group of malonyl-CoA and deuterium oxide (Seyama *et al.*, 1977b, 1978). The exchange of methylene hydrogen atoms of malonyl-CoA with protons of water was suggested by ^{13}C nuclear magnetic resonance spectral analysis.

5.3 Kinetics of β-Ketoacyl Reduction

The kinetics of the overall reaction of fatty acid synthesis by the pigeon liver multienzyme complex was discussed earlier (section 3.2). Some aspects of the two dehydrogenase reactions in relation to the kinetic mechanism of the sequence of reactions involved in fatty acid synthesis were described. Since all of the intermediates are covalently bound and are not released into the solution, investigations of the kinetics of the individual dehydrogenase activities have been carried out with model substrates (Kumar *et al.*, 1970b). S-acetoacetyl-N-acetyl cysteamine has commonly been used as a model substrate for the first reductase of synthetase complexes. With the pigeon liver enzyme, maximum initial rates were obtained at a NADPH concentration of 0.1 mM for intact, as well as dissociated, enzyme. The apparent K_m for NADPH was 20 μM. Concentrations of NADPH greater than 0.2 mM exhibited substrate inhibition. Double reciprocal plots of initial rates *versus* initial substrate concentrations for the intact and dissociated enzyme are shown in Fig. 9. The $K_{m(app)}$ for S-acetoacetyl-N-acetyl cysteamine for the whole and dissociated enzyme was the same, namely 30 mM. On the other hand, the V_{max} for the dissociated enzyme (2.5 x 10^3 mmoles of NADPH oxidized per min per mg of protein) was about 30% lower than that for the intact complex (3.5 x 10^3 mmoles of NADPH oxidized per min per mg protein).

5.4 Kinetics of α,β-Enoylacyl Reduction

This dehydrogenase activity of fatty acid synthetase complexes requires NADPH or NADH (Table I) as the coenzyme. For the pigeon liver enzyme where the kinetics of this component reaction have been studied in detail by Kumar *et al.* 1970b), NADPH is the electron donor. The concentration of NADPH, which gives the highest initial rates for the formation of S-butyryl-N-acetyl cysteamine, is found to be 0.1 mM. The reaction is very slow, though, requiring the use of high concentrations of the complex

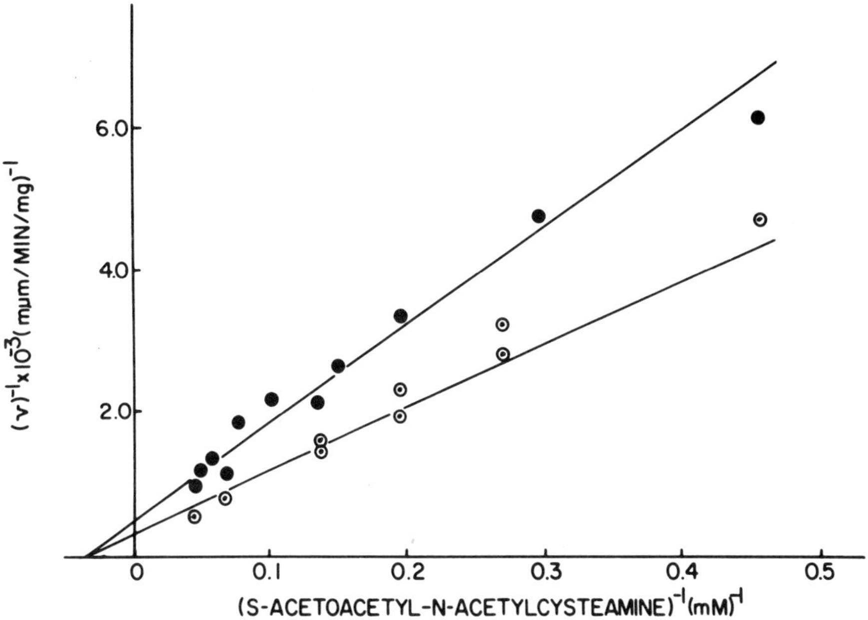

Fig. 9. Kinetics of the reduction of S-acetoacetyl-N-acetyl cysteamine in the presence of NADPH by undissociated (⊙) and dissociated (•) fatty acid synthetase.

Kumar *et al.* (1970b)

Measurements were made at 30°C in 0.18 M potassium phosphate, pH 6.8. The double reciprocal plots of the initial rates of reduction against the initial S-acetoacetyl-N-acetyl cysteamine contration at a constant level of NADPH (0.1 mM) are given. Protein concentrations were approximately 5 μg per ml and $K_{m(app)}$ values of 30 mM were calculated for both of the enzyme species.

(1.0 mg protein per ml). Double reciprocal plots for the undissociated fatty acid synthetase (Fig. 10) show that saturation kinetics were not observed under the experimental conditions, as is evident by the passage of double reciprocal plots through the origin. A comparison of the values of $V_{max}/K_{(app)}$, obtained from the slopes of the plots, shows that the complete enzyme is twice as active as the dissociated enzyme for this dehydrogenase.

Using model substrates, Kumar *et al.* (1970b) suggested that the component enzymes involved in the conversion of the acetoacetyl derivative to the butyryl derivative show decreasing activity. The rate of reduction of acetoacetyl-N-acetyl cysteamine is 100 times greater than the rate of dehydration of the β-hydroxy butyryl derivative, which in turn is five times greater than the rate of reduction of crotonyl-N-acetyl cysteamine.

However, it seems improbable that the differences in rates, if they exist, would be as great as reported above. There could be an alternative and more probable explanation for these rate differences. In contrast to the normal intermediates, the model substrates do not bind covalently to the enzyme, and poor binding results in poor catalysis by the component enzyme activities. This is presumably reflected by the high $K_{m(app)}$ values for N-acetyl cysteamine derivatives.

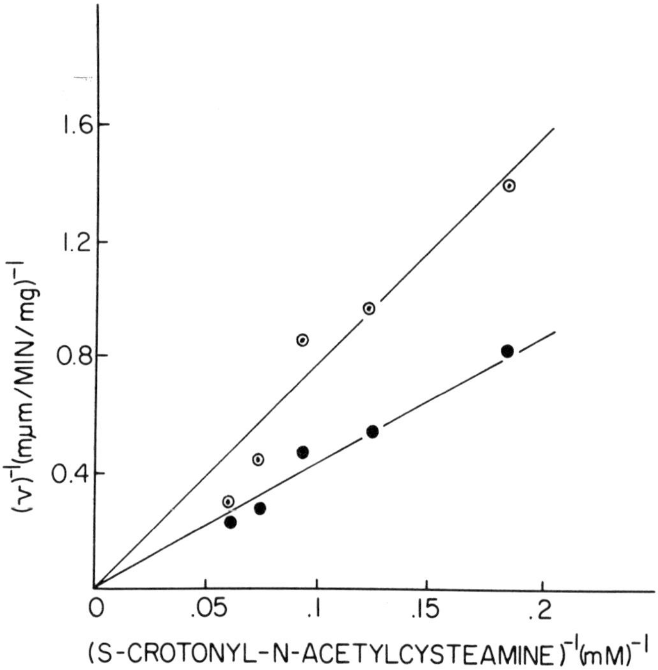

Fig. 10. Double reciprocal plots for the reduction of S-crotonyl-N-acetyl cysteamine by undissociated (•) and dissociated fatty acid synthetase (⊙) at 30°C in 0.18 M potassium phosphate, pH 6.8, and in the presence of a constant concentration of NADPH (0.1 mM).

Kumar *et al.* (1970b)

Substrate concentrations were varied between 5 and 17 mM and protein concentrations were approximately 1.0 mg per ml.

5.5 Bicyclic Diones as Model Substrates for β-Ketoacyl Reduction

Dutler *et al.* (1971a,b) have shown that certain non-physiological bicyclic 1,4-diones are reduced by the β-ketoacyl reductase component of the fatty acid synthetase complex of pig liver. This specificity of ketoreductase for the NADPH-dependent reduction of bicyclic diones has been utilized in the development of a convenient and rapid spectrophotometric assay for fatty acid synthetase activity (Ullman *et al.*, 1978). The four diones used in these studies have the following structures:

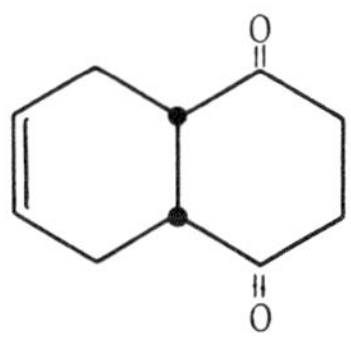

cis-1,2,3,4,4a,5,8,8a-Octahydro-1,4-dioxonaphthalene; abbreviated *cis*-octa

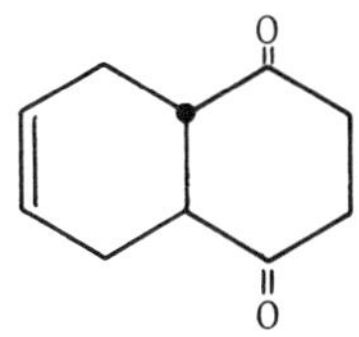

(±)*trans*-1,2,3,4,4a,5,8,8a-Octahydro-1,4-dioxanaphthalene; abbreviated *trans*-octa

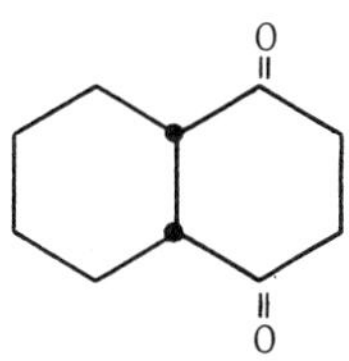

cis-Decalin-1,4-dione; abbreviated *cis*-per

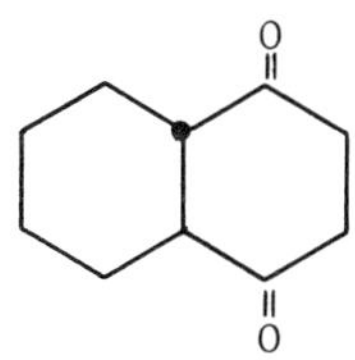

(±)*trans*-Decalin-1,4-dione; abbreviated *trans*-per

The K_m values for the four dione substrates *viz*, *cis*-octa, *trans*-octa, *cis*-per and *trans*-per have been found to be 3.8, 4.7, 1.2 and 1.1 mM, respectively (Ullman *et al.*, 1978). Similarly, partially purified fatty acid synthetases from monkey, turtle and chicken liver gave values of 2.0 to 2.5 mM for the K_m for *cis*-decalin-1,4-dione.

The substrate analog S-acetoacetyl-N-acetylcysteamine is the commonly used model substrate for assays for ketoreductase activity of fatty acid synthetases in most laboratories.

Therefore, a comparison of the kinetic data for ketoreductase activity of chicken liver fatty acid synthetase for the two model substrates is of interest, Table VI (Ullman *et al.*, 1978)

TABLE VI

K_m and V_{max} Values for Ketoreductase Activity on Model Substrates

Substrate	K_m mM	V_{max} (μmol/min/ml)
cis-Decalin-1,4-dione	2.5	2.2
S-Acetoacetyl-N-acetyl cysteamine	67.0	2.0

It appears from these results that *cis*-decalin-1,4-dione is a better substrate than S-acetoacetyl-N-acetyl cysteamine, since it has a lower K_m value and the same or higher V_{max} value. This finding suggests that the substrate binding site for ketoreductase activity is hydrophobic in nature and that the reductase has a significant lack of specificity.

5.6 Formation of Triacetic Acid Lactone

The fatty acid synthesizing enzyme systems from pigeon (Nixon *et al.*, 1968), yeast (Yalpani *et al.*, 1979) and *E. coli* Brock & Bloch, 1966) synthesize triacetic acid lactone from acetyl and malonyl-CoA in the absence of NADPH. In the absence of NADPH the ketoreductase (Site 4, Fig. 1) cannot function, and thus subsequent reactions leading to the synthesis of long-chain fatty acids do not occur. Triacetic acid lactone is formed from two molecules of malonyl-CoA and one of acetyl-CoA. The mechanism of synthesis of triacetic acid lactone is envisaged to occur through the formation of acetoacetyl-4′-phosphopantetheine-enzyme by the pathway shown in Fig. 1. In the absence of NADPH, acetoacetate is then transferred to the cysteine site and another condensation with malonyl-4′-phosphopantetheine-enzyme occurs. This product is then hydrolyzed and cyclizes to yield triacetic acid lactone.

6. IMMUNOLOGICAL REACTIVITIES OF THE SITES OF PARTIAL REACTIONS IN FATTY ACID SYNTHETASE MULTIENZYME COMPLEXES

The comparative immunological activities of the catalytic sites for the partial reactions of pigeon† and chicken liver fatty acid synthetases have given little insight into the location of each of the hydrogenases and other activities in these enzyme molecules (Kumar *et al.*, 1977). The loss of activities for some of the component reactions of pigeon liver fatty acid synthetase, as a function of antibody concentration, is shown in Table VII. It is of interest to note that on a molar basis the condensation-CO_2 exchange reaction is inhibited by six moles of purified antibody per mole of fatty acid synthetase, whereas larger molar excesses of antibody are required for the inactivation of other partial activities. Ketoreductase required the largest molar excess of antibody of any of the partial reactions tested, except for the palmitoyl-CoA deacylase. It appears from these results that the condensing site on the enzyme is more exposed to antibody than the other active sites. The implication of this finding is that the other catalytic sites are located inside the multi-enzyme complex and are therefore less affected by the binding of antibody molecules.

TABLE VII

Inhibition of Catalytic Activities of Pigeon Liver Fatty Acid Synthetase by Reaction with Monospecific Antibody†

Catalytic activity	Molar ratio[a] antibody/antigen
Fatty acid synthesis	6.6
Condensation reaction	6.0
Malonyl transacylation	12.0
Acetyl transacylation	18.0
Ketoreduction	30.0
Deacylation	∞

[a]Moles of fatty acid synthetase antibody required to inhibit the catalytic activity of one mole of fatty acid synthetase, as determined by extrapolation of the linear rate of inhibition obtained under conditions of antigen excess.

7. ISOLATION AND PURIFICATION OF DEHYDROGENASES OF NONAGGREGATED FATTY ACID SYNTHESIZING ENZYME SYSTEMS

†Katiyar *et al.* (1979). Unpublished results.

7.1 β-Ketoacyl-Acyl Carrier Protein (ACP) Reductase

This enzyme catalyzes the first reduction of fatty acid synthesis *viz* the NADPH-dependent reduction of the β-ketoacyl ACP to D(-)-β-hydroxyacyl-ACP (Fig. 1, Site 4). This reaction has an absolute requirement for NADPH. The substrates for the *E. coli* enzyme are ACP derivatives, but this enzyme can also reduce β-ketoacyl thioesters of CoA and pantetheine, though at slower rates (Alberts *et al.*, 1965; Prescott & Vagelos, 1972). Substrate inhibition of enzyme was obtained with acetoacetyl-ACP at concentrations higher than 0.05 M.

The β-ketoacyl reductase has a broad specificity for the chain length of the β-ketoacyl-ACP substrate (Toomey & Wakil, 1966) and is active with β-ketoacyl-ACP derivatives which are intermediates in the synthesis of both saturated and unsaturated fatty acids (Prescott & Vagelos, 1972). The activity of β-ketoacyl-ACP reductase is increased 20- to 40-fold by several mono and divalent cations (Schulz *et al.*, 1969). It has been suggested that this stimulation results from the formation of a complex between the cations and the protein moiety of ACP. Supposedly this complex facilitates the binding of substrate to the enzyme. In contrast, no activation of the enzyme by cations was seen when acetoacetyl-N-acetyl cysteamine was used as substrate.

7.2 α,β-Enoylacyl-Acyl Carrier Protein (ACP) Reductase

Preparations of enoylacyl-ACP reductase have been purified 250-fold from crude extracts of *E. coli* (Weeks & Wakil, 1968). The steps of purification from crude extracts of *E. coli* included DEAE-cellulose chromatography, Sephadex G-100 filtration and hydroxylapatite chromatography, followed by another Sephadex G-100 gel filtration. Enzyme preparations purified in this way contained two activities *viz* an NADPH-dependent and an NADH-dependent reductase. These activities were attributed to the presence of two distinct enzymes.

The NADPH-dependent enoylacyl-ACP reductase is unstable at pH values greater than 7.5 and it is inactivated above pH 8.0. It exhibits an absolute specificity for the acyl-ACP substrates, and it is more active on C_4-C_6 derivatives than on longer chain α,β-unsaturated acyl-ACP. This enzyme possesses an active thiol group, and it is inhibited by N-ethylmaleimide, iodoacetate and p-hydroxymercurobenzoate.

The other enzyme, the NADH-dependent enoylacyl-ACP reductase, is enzymatically-active over a wide range of pH values. It also possesses a functional thiol group, and it is readily inhibited by p-hydroxymercuribenzoate and iodoacetate. N-ethylmaleimide, on the other hand, stimulates the activity of this enzyme 2- to 3-fold. The substrate specificity of the NADH-dependent enoylacyl-ACP reductase is different than that of the

NADPH-dependent enzyme (Weeks & Wakil, 1968). This enzyme had a higher activity with decenoyl-ACP substrate than with crotonyl-ACP. Unlike the corresponding NADPH-dependent enzyme, it utilizes enoylacyl coenzyme A esters as substrates with K_m values of 2.5×10^{-3}M, compared to apparent K_m values of 4×10^{-5} M for the ACP derivatives. In terms of stability, the NADPH reductase activity was always the least stable of the two activities.

8. SUMMARY

Fatty acid synthetases catalyze the *de novo* formation of long-chain saturated fatty acids from acetyl- and malonyl-CoA. The overall reaction in this process can be resolved into seven different component (partial) reactions, five of which give rise to seven repetitive cycles. Each cycle elongates the fatty acyl chain by one unit of two carbon atoms. Two of the seven reactions of fatty acid synthesis are catalyzed by dehydrogenases. Both of these dehydrogenases use NADPH in most fatty acid synthesizing systems and these enzymes are designated as reductases.

The chemical mechanism of fatty acid synthesis appears to be the same in all biological systems studied so far. However, this uniformity does not hold with respect to the architectural organization of the component enzymes. These systems can be divided into two different groups. One group consists of enzyme systems that are tightly aggregated, while the other group includes nonaggregated systems and is composed of a series of independent and monofunctional enzyme entities. Tightly aggregated multienzyme fatty acid synthetase complexes of mammalian, avian and yeast dissociate reversibly into two subunits. These two subunits, in the case of the yeast enzyme, are nonidentical polyfunctional polypeptide chains which have all of the known partial enzyme activities on one or the other of the chains.

The kinetic and chemical mechanisms of the pigeon liver fatty acid synthetase have been established. A novel seven-site ping-pong kinetic mechanism for this reaction has been proposed on the basis of initial velocity and product inhibition data. The reactions occurring at the two dehydrogenase sites may be random sequential or ordered, with NADPH adding first. Since all the intermediates are covalently bound to the mammalian and avian liver enzyme complexes, and are not released into the solution, investigations of the kinetics of the individual dehydrogenase activities have been carried out with model substrates. Both of the dehydrogenases of fatty acid synthetase complexes from avian and mammalian sources carry out NADPH specific reductions. In contrast, the nonaggregated systems of *E. coli* and plants require NADPH and NADH as coenzymes.

The nature and number of binding sites of NADPH on the pigeon and rat liver enzymes has been established. Pigeon liver enzyme has only two binding sites in 0.2 M phosphate at pH 7.0, and these have a K_d of 8 μM. At lower ionic strengths a third site appears, and each of the three binding sites appears to be equivalent. In contrast, the rat liver enzyme has four NADPH binding sites. At high ionic strengths, one pair of these binding sites has a stronger affinity and the other a weaker affinity for NADPH. However, as the ionic strength is lowered, the four binding sites appear to become equivalent. A comparison of the binding affinities of fatty acid synthetase dehydrogenases with those of other dehydrogenases has shown that the synthetase dehydrogenases bind NADPH more strongly than do most other dehydrogenases.

The stereospecificity of hydrogen incorporation from nucleotides for the two reductive steps of fatty acid synthesis in rat, pigeon and yeast systems has been established. It was found that β-ketoacyl reduction proceeds by the transfer of hydrogen from the B side of position 4 of the pyridine ring of NADPH, whereas the enoylacyl reduction proceeds by the transfer of hydrogen from the A side for rat and pigeon liver enzyme. In contrast, the enoylacyl reductase of yeast shows a B stereospecificity for NADPH. In the synthesis of one molecule of palmitic acid, 14 hydrogen atoms are transferred from NADPH or NADH to the odd-numbered methylene carbon atoms (two hydrogen atoms per carbon atom).

The authors wish to acknowledge that the writing of this chapter was supported in part by Research Grant AM 01383 from the National Institutes of Arthritis, Metabolic and Digestive Diseases, National Institutes of Health, United States Public Health Service, and by the Medical Research Service of the Veterans Administration.

REFERENCES

Agradi, E. & Smith, S. (1976) *Int. J. Biochem.* 7, 467-472

Alberts, A.W., Goldman, P. & Vagelos, P.R. (1963) *J. Biol. Chem.* 238, 557-565

Alberts, A.W., Majerus, P.W., Talamo, B. & Vagelos, P.R. (1964) *Biochemistry* 3, 1563-1571

Alberts, A.W., Majerus, P.W. & Vagelos, P.R. (1965) *Biochemistry* 4, 2265-2274

Bayley, P.M. & Radda, G.K. (1966) *Biochem. J.* 98, 105-111

Bentley, R. (1970) *Molecular Asymmetry in Biology*, Vol. 2, pp. 1-89. Academic Press, New York

Bloch, K. (1977) *Adv. Enzymol.* 45, 1-84

Brady, R.O., Bradley, R.M. & Trams, E.G. (1960) *J. Biol. Chem.* 235, 3093-3098

Bratcher, S.C. & Hsu, R.Y. (1975) *Biochim. Biophys. Acta* 410, 229-235

Bressler, R. & Wakil, S.J. (1961) *J. Biol. Chem.* 236, 1643-1651

Brock, D.J. & Bloch, K. (1966) *Biochem. Biophys. Res. Commun.* 25, 473-480

Brodie, J.D., Wasson, G. & Porter, J.W. (1964) *J. Biol. Chem.* 239, 1346-1356
Brooks, J.L. & Stumpf, P.K. (1966) *Arch. Biochem. Biophys.* 116, 108-116
Buckner, J.S. & Kolattukudy, P.E. (1976) *Biochemistry* 15, 1948-1957
Burton, D.N., Haavik, A.G. & Porter, J.W. (1968) *Arch. Biochem. Biophys.* 126, 141-154
Butterworth, P.H.W., Yang, P.C., Bock, R.M. & Porter, J.W. (1967) *J. Biol. Chem.* 242, 3508-3516
Carey, E.M. & Dils, R. (1970) *Biochem. Biophys. Acta* 210, 371-387
Colowick, S.P., Van Eys, J. & Park, J.H. (1966) in *Comprehensive Biochemistry* (Florikin, M. & Stotz, E.H., eds.), Vol. 14, pp. 1-98. American Elsevier, New York
Delo, J., Ernst-Fonberg, M.L. & Bloch, K. (1971) *Arch. Biochem. Biophys.* 143, 392-400
Dietlin, G. & Schweizer, E. (1975) *Eur. J. Biochem.* 59, 177-184
Dorsey, J.A. & Porter, J.W. (1968) *J. Biol. Chem.* 243, 4667-4670
Dugan, R.E. & Porter, J.W. (1970) *J. Biol. Chem.* 245, 2051-2059
Dugan, R.E. & Porter, J.W. (1971) *J. Biol. Chem.* 246, 637-642
Dugan, R.E., Slakey, L.L. & Porter, J.W. (1970) *J. Biol. Chem.* 245, 6312-6316
Flick, P. & Bloch, K. (1974) *J. Biol. Chem.* 249, 1031-1036
Goldman, P. & Vagelos, P.R. (1962) *Biochem. Biophys. Res. Commun.* 7, 414-418
Hansen, H.J.M., Carey, E.M. & Dils, R. (1970) *Biochem. Biophys. Acta* 210, 400-410
Hsu, R.Y. & Wagner, B.J. (1970) *Biochemistry* 9, 245-251
Hsu, R.Y. & Yun, S. (1970) *Biochemistry* 9, 239-245
Hsu, R.Y., Wasson, G. & Porter, J.W. (1965) *J. Biol. Chem.* 240, 3736-3746
Ilton, M., Jevans, A.W., McCarthy, E.D., Vance, D., White, H.B., III, & Bloch, K. (1971) *Proc. Nat. Acad. Sci. USA* 68, 87-91
Joshi, V.C., Plate, C.A. & Wakil, S.J. (1970) *J. Biol. Chem.* 245, 2857-2867
Katiyar, S.S. & Porter, J.W. (1974) *Arch. Biochem. Biophys.* 163, 324-331
Katiyar, S.S. & Porter, J.W. (1975) *Arch. Biochem. Biophys.* 170, 220-227
Katiyar, S.S. & Porter, J.W. (1977) *Life Sciences* 20, 737-760
Katiyar, S.S., Briedis, A.V. & Porter, J.W. (1974) *Arch. Biochem. Biophys.* 162, 412-420
Katiyar, S.S., Cleland, W.W. & Porter, J.W. (1975) *J. Biol. Chem.* 250, 2709-2717
Kawaguchi, A. & Okuda, S. (1977) *Proc. Nat. Acad. Sci. USA* 74, 3180-3183
Kawaguchi, A., Tomada, H., Okuda, S., Awaya, J. & Omura, S. (1978) *Federation Proc.* 37, 1707. Abst.
Kinsella, J.E., Bruns, D. & Infante, J.P. (1975) *Lipids* 10(4), 227-237
Knobling, A. & Schweizer, E. (1975) *Eur. J. Biochem.* 59, 415-421

Knobling, A., Schiffman, D., Sickinger, H.D. & Schweizer, E. (1975) *Eur. J. Biochem.* 56, 359-367
Knudsen, J. (1972) *Biochem. Biophys. Acta* 280(3), 408-414
Kuhn, L., Castorph, H. & Schweizer, E. (1972) *Eur. J. Biochem.* 24, 492-497
Kumar, S. (1975) *J. Biol. Chem.* 250, 5150-5158
Kumar, S. & Porter, J.W. (1972) *J. Biol. Chem.* 246, 7780-7789
Kumar, S., Dorsey, J.K. & Porter, J.W. (1970a) *Biochem. Biophys. Res. Commun.* 40, 825-832
Kumar, S., Dorsey, J.A., Muesing, R.A. & Porter, J.W. (1970b) *J. Biol. Chem.* 245, 4732-4744
Kumar, S., Muesing, R.A. & Porter, J.W. (1972a) *J. Biol. Chem.* 247, 4749-4762
Kumar, S., Phillips, G.T. & Porter, J.W. (1972b) *Int. J. Biochem.* 3, 15-32
Kumar, S., Srinivasan, K.R. & Asato, Noritake (1977) *Biochim. Biophys. Acta* 489, 32-47
Lennarz, W.J., Light, R.J. & Bloch, K. (1962) *Proc. Nat. Acad. Sci. USA* 48, 840-846
Levitzki, A. & Koshland, D.E., Jr. (1969) *Proc. Nat. Acad. Sci. USA* 62, 1121-1128
Lin, C.Y. & Kumar, S. (1971) *J. Biol. Chem.* 246, 3284-3290
Lin, C.Y. & Kumar, S. (1972) *J. Biol. Chem.* 247, 604-610
Lin, C.Y. & Smith, S. (1978) *J. Biol. Chem.* 253, 1954-1962
Lornitzo, F.A., Qureshi, A.A. & Porter, J.W. (1974) *J. Biol. Chem.* 249, 1654-1656
Lornitzo, F.A., Qureshi, A.A. & Porter, J.W. (1975) *J. Biol. Chem.* 250, 4520-4529
Lynen, F. (1961) *Federation Proc.* 20, 941-951
Lynen, F. (1967a) *Biochem. J.* 102, 381-400
Lynen, F. (1967b) in *Progress in Biochemical Pharmacology* (Kritchevsky, D., Paoletti, R. & Steinberg, M.D., eds.), Vol. 3, pp. 1-31. S. Karger, Basel, New York
Lynen, F. (1973) in *Proc. Third International Specialized Symposium on Yeast*, Part II, pp. 7-29. Otaniemi, Helsinki
Lynen, F., Oesterhelt, D., Schweizer, E. & Willecke, K. (1968) in *Cellular Compartmentalization and Control of Fatty Acid Metabolism* (Gran, F.C., ed.), pp. 1-24. Academic Press, New York
Maitra, S.K. & Kumar, S. (1974) *J. Biol. Chem.* 249, 118-125
Majerus, P.W. & Vagelos, P.R. (1967) *Advan. Lipid Res.* 5, 1-33
Majerus, P.W., Alberts, A.W. & Vagelos, P.R. (1964) *Proc. Nat. Acad. Sci. USA* 51, 1231-1238
Muesing, R.A., Lornitzo, F.A., Kumar, S. & Porter, J.W. (1975) *J. Biol. Chem.* 250, 1814-1823
Nepokroeff, C.M. & Porter, J.W. (1978) *J. Biol. Chem.* 253, 2279-2283
Nixon, J.E., Putz, G.R. & Porter, J.W. (1968) *J. Biol. Chem.* 243, 5471-5478
Overath, P. & Stumpf, P.K. (1964) *J. Biol. Chem.* 239, 4103-4110
Phillips, G.T., Nixon, J.E., Dorsey, J.A., Butterworth, P.H.W., Chesterton, C.J. & Porter, J.W. (1970a) *Arch. Biochem. Biophys.* 138, 380-391

Phillips, G.T., Nixon, J.E., Abramovitz, A.S. & Porter, J.W. (1970b) *Arch. Biochem. Biophys.* 138, 357-371
Porter, J.W., Kumar, S. & Dugan, R.E. (1971) *Proc. Biochem. Pharmacol.* 6, 1-101
Prescott, D.J. & Vagelos, P.R. (1972) *Adv. Enzymol.* 36, 269-311
Qureshi, A.A., Lornitzo, F.A. & Porter, J.W. (1974) *Biochem. Biophys. Res. Commun.* 60, 158-165
Qureshi, A.A., Lornitzo, F.A., Jenik, R.A. & Porter, J.W. (1976a) *Arch. Biochem. Biophys.* 177, 364-378
Qureshi, A.A., Lornitzo, F.A., Hsu, R.Y. & Porter, J.W. (1976b) *Arch. Biochem. Biophys.* 177, 379-393
Reed, L.J. & Cox, D.J. (1966) *Ann. Rev. Biochem.* 35, 57-84
Roncari, D.A.K. (1974a) *Can. J. Biochem.* 52(3), 221-230
Roncari, D.A.K. (1974b) *J. Biol. Chem.* 249, 7035-7037
Schulz, H., Weeks, G., Toomey, R.E., Shapiro, M. & Wakil, S.J. (1969) *J. Biol. Chem.* 244, 6577-6583
Schweet, R.S., Katchman, B., Bock, R.M. & Jagannathan, V. (1951) *J. Biol. Chem.* 196, 563-567
Schweizer, E. & Bolling, H. (1970) *Proc. Nat. Acad. Sci. USA* 67, 660-666
Schweizer, E., Willecke, K., Winnewisser, W. & Lynen, F. (1970) *Vitam. Horm. (N.Y.)* 28, 329-343
Schweizer, E., Kuhn, L. & Castorph, H. (1971) *Hoppe-Seyler's Z. Physiol. Chem.* 242, 377-384
Schweizer, E., Kniep, B., Castorph, H. & Holzner, U. (1973) *Eur. J. Biochem.* 39, 353-362
Seyama, Y., Kasama, T., Yamakawa, T., Kawaguchi, A. & Okuda, S. (1977a) *J. Biochem.* 81, 1167-1173
Seyama, Y., Kasama, T., Yamakawa, T., Kawaguchi, A., Saito, K. & Okuda, S. (1977b) *J. Biochem.* 82, 1325-1329
Seyama, Y., Kawaguchi, A., Kasama, T., Sasaki, K., Arai, K., Okuda, S. & Yamakawa, T. (1978) *Biomedical Mass Spectrometry* 5, 357-361
Simoni, R.D., Criddle, R.S. & Stumpf, P.K. (1967) *J. Biol. Chem.* 242, 573-581
Smith, S. & Abraham, S. (1970) *J. Biol. Chem.* 245, 3209-3217
Smith, S. & Abraham, S. (1971) *J. Biol. Chem.* 246, 6428-6435
Smith, S., Agradi, E., Libertini, L. & Dileepan, K.N. (1976) *Proc. Nat. Acad. Sci. USA* 73, 1184-1188
Srinivasan, K.R. & Kumar, S. (1976) *J. Biol. Chem.* 251, 5352-5360
Stoops, J.K., Arslanian, M.J., Oh, Y.H., Aune, K.C., Vanaman, T.C. & Wakil, S.J. (1975) *Proc. Nat. Acad. Sci. USA* 72, 1940-1944
Stoops, J.K., Arslanian, M.J., Chalmers, J.H., Joshi, V.C. & Wakil, S.J. (1977) in *Bioorganic Chemistry* (Van Ramleen, ed.), Vol. 1, pp. 339-370. Academic Press, New York
Stoops, J.K., Awad, E.S., Arslanian, M.J., Gunsberg, S. & Wakil, S.J. (1978a) *J. Biol. Chem.* 253, 4464-4475
Stoops, J.K., Arslanian, M.J., Aune, K.C. & Wakil, S.J. (1978b) *Arch. Biochem. Biophys.* 188, 348-359
Sumper, M., Riepertinger, C. & Lynen, F. (1969a) *FEBS Lett.* 5, 45-49

Sumper, M., Oesterhelt, D., Riepertinger, C. & Lynen, F. (1969b) *Eur. J. Biochem.* 10, 377-387
Tauro, P., Holzner, U., Castorph, H., Hill, F. & Schweizer, E. (1974) *Mol. & Gen. Genet.* 129, 131-148
Toomey, R.E. & Wakil, S.J. (1966) *Biochim. Biophys. Acta* 116, 189-197
Ullman, A.H., Harding, J.W., Jr. & White, H.B., III (1978) *Anal. Biochem.* 84, 85-96
Vagelos, P.R., Majerus, P.W., Alberts, A.W., Larrabee, A.R. & Ailhaud, G.P. (1966) *Federation Proc.* 25, 1485-1494
Vance, D.E., Mitsuhashi, O. & Bloch, K. (1973) *J. Biol. Chem.* 248, 2303-2309
Van den Bosch, H., Williamson, J.R. & Vagelos, P.R. (1970) *Nature* 228, 338-341
Volpe, J.J. & Vagelos, P.R. (1973) *Ann. Rev. Biochem.* 42, 21-60
Volpe, J.J. & Vagelos, P.R. (1976) *Physiol. Rev.* 56, 339-417
Wakil, S.J. (1961) *J. Lipid Res.* 2, 1-24
Wakil, S.J. & Ganguly, J.J. (1959) *J. Am. Chem. Soc.* 81, 2597-2598
Wakil, S.J., Pugh, E.L. & Sauer, F. (1964) *Proc. Nat. Acad. Sci. USA* 52, 106-114
Weeks, G. & Wakil, S.J. (1968) *J. Biol. Chem.* 243, 1180-1189
White, H.B., Mitsuhashi, A. & Bloch, K. (1971) *J. Biol. Chem.* 246, 4751-4754
Willecke, K., Ritter, E. & Lynen, F. (1969) *Eur. J. Biochem.* 8, 503-509
Yalpani, M., Willecke, K. & Lynen, F. (1969) *Eur. J. Biochem.* 8, 495-502
Yang, P.C., Butterworth, P.H.W., Bock, R.M. & Porter, J.W. (1967) *J. Biol. Chem.* 242, 3501-3507
Yu, H.L. & Burton, D.N. (1972) *Arch. Biochem. Biophys.* 149, 498-504
Yun, S.L. & Hsu, R.Y. (1972) *J. Biol. Chem.* 247, 2689-2698
Zahler, W.L., Barden, R.E. & Cleland, W.W. (1968) *Biochim. Biophys. Acta* 165, 1-11

ADDENDUM

Subsequent to completion of this review, a paper by Wieland *et al.* [Eur. J. Biochem. 94, 189-197 (1979)] was published. In this paper it was shown that yeast fatty acid synthetase complex can be dissociated by acylation with methyl maleic anhydride. The dissociated subunits (α and β species) were then separated by analytical ultracentrifugation and by disc gel electrophoresis. The dissociated subunits were reassociated by mild acid hydrolysis of the acylated derivatives. It was also shown that the acylated subunits could be separated by sucrose density gradient centrifugation. Analyses of the separated β subunit showed it to contain malonyl, palmitoyl and acetyl transferase, enoyl reductase and dehydratase activities. These results corroborate earlier evidence on the nonidentity of the two subunits of yeast fatty acid synthetase. [Authors]

Much interest centres on the subunit composition of fatty acid synthetases, and after the typescript of this chapter had been received, further evidence was published concerning the fatty acid synthetases of mammals. Stoops *et al.* [J. Biol. Chem., 254 (1979), 7418-7426] described studies with rat liver and adipose tissue fatty acid synthetases, from which they concluded that in these cases the active enzyme consisted of two identical subunits, each of mol. wt. 220,000 to 240,000. No active form larger than this dimer was detected, nor was any fatty acid synthetase activity detected that could definitely be attributed to the monomeric form.

Smith and Stern [Arch. Biochem. Biophys., 197 (1979) 379-387] concluded from immunochemical studies that the fatty acid synthetase of lactating rat mammary gland consisted of two subunits, each of mol. wt. 240,000, and each containing one thioesterase I domain. From studies of the fragments obtained by limited proteolysis using trypsin, they further concluded (*ibid idem*) that the two subunits are most likely to be *identical*. Non-identity of the subunits would be compatible with their evidence only if the structures had certain features which they regarded as less probable (such as the thioesterase I domain at the C-terminal portion of one subunit, and at the N-terminal portion of the other subunit).

The 4'-phosphopantetheine moiety was located on a fragment of 95,000 mol. wt. obtained by limited trypsin digestion [Libertini & Smith, Arch. Biochem. Biophys. 192 (1979) 47-60]

[Editor]

HYDROXYMETHYLGLUTARYL-CoA REDUCTASE

Willis E. Brown and Victor W. Rodwell

CONTENTS

HYDROXYMETHYLGLUTARYL-CoA REDUCTASE (EC 1.1.1.34)

Willis E. Brown and Victor W. Rodwell

I. INTRODUCTION*

The reaction catalyzed by 3-hydroxy-3-methylglutaryl-coenzyme A (HMG-CoA) reductase [hydroxymethylglutaryl-CoA reductase (NADPH), EC 1.1.1.34] involves reductive deacylation of HMG-CoA to mevalonate and requires two moles of NADPH (Fig. 1).

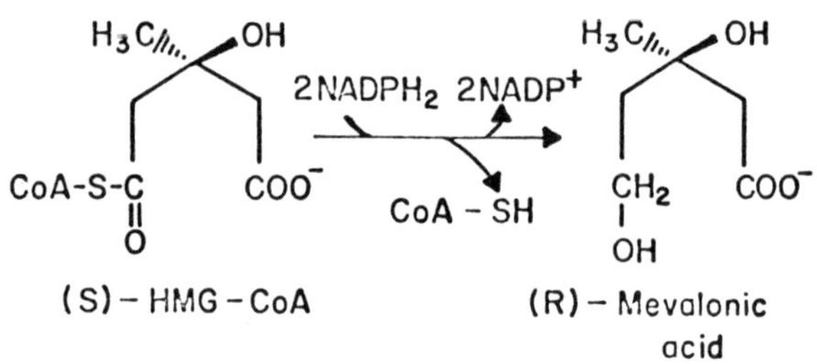

Fig. 1. The HMG-CoA reductase reaction.

Biochemical interest in HMG-CoA reductase arises both from the unusual nature of the reaction (two-step

* Literature review concluded March 1979.

reduction without formation of detectable free intermediates) and its physiological role as catalyst for a major rate-limiting reaction of polyisoprenoid biosynthesis. In particular, HMG-CoA reductase catalyzes the rate-limiting reaction of cholesterol biosynthesis in liver and other tissues. Medical interest in HMG-CoA reductase arises both from elevated levels of plasma LDL and VLDL cholesterol as a risk factor for coronary artery disease and the utility of HMG-CoA reductase as an indicator of uptake of cholesterol in cultured cells and other model systems.

At present, only limited evidence suggests that the activity of non-vertebrate HMG-CoA reductase is regulated. That of vertebrate tissues is, however, subject to stringent regulation by diverse factors such as concentration and composition of circulating plasma lipoproteins, quantity and composition of the diet, time of day, genetic heritage and various hormones. These factors appear to act primarily or exclusively via changes in the specific rate of synthesis of HMG-CoA reductase protein. At present, no convincing evidence documents a physiological role for the regulation of HMG-CoA reductase via control of its rate of degradation. Similarly, while the recently discovered modulation of HMG-CoA reductase activity *in vitro* by phosphorylation/dephosphorylation may prove significant for short-term regulation of reductase activity *in vivo*, convincing evidence is presently lacking.

This review will not discuss regulation of HMG-CoA reductase synthesis or degradation. For information on these matters the reader is referred to the following reviews: Sabine (1977), Rodwell *et al.* (1973,1976), Dempsey (1974), McNamara and Rodwell (1972) and Siperstein (1970).

II. DISTRIBUTION

Since isoprenoids and polyisoprenoids are ubiquitous in nature, HMG-CoA reductase probably occurs in all life forms and in all tissues that biosynthesize polyisoprenoids. As recently as 1975, however, evidence documenting the widespread distribution of HMG-CoA reductase was limited and in part inferential. Since then, improved assay techniques and increased interest in this enzyme have provided evidence for this assertion (Tables I-III).

TABLE I. Life Forms Known to Catalyze the HMG-CoA Reductase Reaction

Bacteria	Moulds	Insects
Yeast	Higher plants	Vertebrates

TABLE II. Livers Known to Catalyze the HMG-CoA Reductase Reaction

Chicken	Guinea Pig	Rabbit
Dog	Hamster	Rat
Fish	Human	Swine
Frog	Mouse	Turkey
Gerbil	Pigeon	

TABLE III. Mammalian Tissues Known to Catalyze the HMG-CoA Reductase Reaction[a]

Adrenal	HeLa	Liver
Aortic smooth muscle	Hepatoma	Lung
	HTC Cells	Lymphocyte
Brain	Intestine	Lymphoid
Glial	Kidney	Neuronal
Hair Root	Leukocyte	Skin

[a] Based on ability to incorporate acetate into digitonin-precipitable sterols, smooth, striated and cardiac muscle, adipose tissue, testis, ovary, spleen, monocytes and pancreas might be added to this list.

III. ASSAY

A. Choice of Cell Fraction for Analysis

The HMG-CoA reductase activity of cultured vertebrate cells, of yeast mitochondrial extracts or of certain bacterial extracts is readily assayed in crude homogenates. Crude homogenates of liver and certain other tissues are, however, difficult to assay directly due in part to the presence of enzymes that act on HMG-CoA, mevalonate or NADPH [e.g.: HMG-CoA deacylase (EC 3.1.2.5) and mitochondrial HMG-CoA lyase (EC 4.1.3.4)]. HMG-CoA reductase activity may be assayed in the postmitochondrial supernatant fraction of homogenates of tissues, including liver. This supernatant fraction is, however, customarily centrifuged (60-90 min at 100,000 x g) to obtain a crude microsomal membrane preparation which is then assayed.

B. Technique

The activity of yeast (Kawachi & Rudney 1970), certain bacterial (Fimognari & Rodwell 1965a) and partially purified rat liver HMG-CoA reductase (Kleinsek et al. 1977, Edwards et al. 1979a) may be determined by conventional spectrophotometric measurement of the rate of oxidation of NADPH at 340 nm. Note, however, that the enzyme units based on NADPH differ by a factor of two from those based on mevalonate due to the 2:1 stoichiometry of the reaction (Fig. 1). In this review, we express HMG-CoA reductase activity in pU (pmoles of mevalonate formed per min).

Standard spectral methods are generally inapplicable to vertebrate HMG-CoA reductase present in crude homogenates or microsomal membranes. One complication is that these membranes rapidly oxidize NADPH. Another is the low concentration of HMG-CoA reductase in most cells. This also limits application of colorimetric assay of liberated coenzyme A (Hulcher & Oleson 1973), although a coupled assay for liberated coenzyme A using α-ketoglutarate dehydrogenase appears suitable for assay of microsomal reductase (Baquir & Booth 1977).

The methods most widely used to assay vertebrate HMG-CoA reductase activity measure the incorporation of radioactivity from [^{14}C]HMG-CoA into mevalonate. An internal standard of [^{3}H]mevalonate commonly is included to permit correction for losses incurred during subsequent isolation of enzymically formed [^{14}C]mevalonate (Goldfarb & Pitot 1971). Incubation generally is for 10-60 min at 37°C at pH 7-8. Reaction mixtures are then acidified to inactivate HMG-CoA reductase and to convert mevalonate to the less polar δ-mevalonolactone. This is then either extracted into a relatively nonpolar solvent and counted directly (Ackerman et al. 1977) or isolated from such an extract or from the acidified reaction mixture by column chromatography (Huber et al. 1973), GLC (Hamprecht & Lynen 1970), electrophoresis (Berndt & Gaumert 1971) or TLC (Shapiro et al. 1974). The sensitivity of radioassay techniques is limited primarily by the specific activity of [^{14}C]HMG-CoA.

We employ direct TLC of aqueous reaction mixtures on activated silica gel G in benzene:acetone::1:1 (v/v) to separate mevalonolactone from more polar materials (Shapiro et al. 1974). Toluene may be substituted for benzene with little decrease in mobility or resolving power of the system (W. R. Bensch, personal communication). Incubation volumes of 40-150 µl are employed to conserve radioisotope and to facilitate application to TLC sheets.

A major improvement in the ease of assay of HMG-CoA reductase activity is a "mixed phase" assay that utilizes [^{3}H]HMG-CoA of >99% purity. Portions of acidified aqueous reaction mixtures are added directly to toluene-based scintillation fluor without prior separation of mevalonolactone from polar radioactive materials. The nonpolar lactone diffuses into the fluor while polar materials remain in the aqueous phase. For microsomal and solubilized rat liver HMG-CoA reductase, this assay gives results fully comparable to the TLC-based assay of Shapiro et al. (1974) (Philipp & Shapiro 1979). Caution should, however, be exercised when equating nonpolar material with mevalonolactone (Ness & Moffler 1978), and the product should be carefully characterized as mevalonolactone (Young & Rodwell 1977, 1979).

IV. PROPERTIES

A. Optimal pH

HMG-CoA reductase exhibits optimal activity near neutral pH. An exception is that for Pseudomonas HMG-CoA reductase measured for the reverse reaction, oxidative acylation of mevalonate to HMG-CoA (Table IV).

B. Km for (S)-HMG-CoA and for NADPH

Caution should be exercised in the determination of Km values, particularly with crude microsomal preparations. Langdon & Counsell (1976) emphasize that substrate turnover must be minimal, and Ness & Moffler (1979) note that rat liver microsomal preparations contain enzymes that rapidly utilize HMG-CoA.

For the three purified non-mammalian HMG-CoA reductases, Km for (S)-HMG-CoA varies only 2.5 fold - probably not a significant difference. The Km for (S)-HMG-CoA for the chicken and rat liver enzymes are, however, significantly lower (Table IV). This may reflect differences between vertebrate and non-vertebrate HMG-CoA reductase as low values are reported for most microsomal mammalian HMG-CoA reductases.

Km for NADPH for purified rat liver HMG-CoA reductase is comparable to that for yeast or Pseudomonas HMG-CoA reductase and significantly lower than that for the purified chicken liver enzyme. This may reflect a difference between mammalian and avian HMG CoA reductase (e.g., the Km for the microsomal pigeon liver enzyme is also high). However, the relatively high Km values for NADPH reported for microsomal preparations (Table IV) may reflect, in part, microsomal utilization of NADPH by other reactions.

TABLE IV. Km Values and pH Optima for HMG-CoA Reductase

Source	Km (μM)		Optimal pH	Ref.[a]
	(S)-HMG-CoA	NADPH		
PURIFIED				
Pseudomonas	25	32[b]	9.5[c]	(1)
Yeast	12	89		(2)
F. oxysporium[d]	10		7.4	(3)
Chicken liver	1.6	260	6.9	(4)
Rat liver	2.0	67		(5)
MICROSOMAL				
Neurospora	15			(6)
Pea seedling	50		6.8	(7)
Pigeon liver	8.8	680		(8)
Mouse liver	4.2	1900	7	(9)
Rat liver	1.1	26		(10)
Rat intestine	7.0		7.5	(11)
Rat leukocyte	3.5			(12)
Dog liver	10	700	6-8	(13)
Human liver			7.3	(14)
Human lymphocyte	4.5			(15)

[a] References are: (1) Bensch & Rodwell 1970; (2) Kirtley & Rudney 1967; (3) Madhosingh & Orr 1978; (4) Beg *et al.* 1977, 1979; (5) Edwards *et al.* 1979b; (6) Imblum & Rodwell 1974; (7) Brooker & Russell 1975; (8) Hulcher & Oleson 1973; (9) Kandutsch & Saucier 1969; (10) Ness & Moffler 1979, Langdon & Counsell 1976; (11) Shefer *et al.* 1972; (12) Young & Rodwell 1977; (13) Ho *et al.* 1977; (14) Nicolau *et al.* 1974; (15) Laporte *et al.* 1978.

[b] Km for NADH.

[c] For oxidative acylation of mevalonate to HMG-CoA.

[d] Both NADH and NADPH are substrates.

C. Essential Thiol Group

Like many oxidoreductases, HMG-CoA reductase requires reduced thiol groups for optimal activity. Yeast HMG-CoA reductase is inhibited by low concentrations of p-chloromercuribenzoate or iodoacetamide, and dithiothreitol, 2-mercaptoethanol, glutathione and cysteine protect against the action of thiol reagents (Kirtley & Rudney 1967, Durr & Rudney 1960). In addition, purified *Fusarium oxysporium* or chicken liver HMG-CoA reductase require addition of dithiothreitol or similar reagents for optimal activity (Madhosingh & Orr 1978, Beg *et al*. 1979).

D. Molecular Weight

The molecular weights of the bacterial and yeast HMG-CoA reductase are essentially the same. Both the yeast and the rat liver enzymes are tetramers (Table V). Although chicken and rat liver HMG-CoA reductase have subunits of dissimilar molecular weight, antibodies to chicken liver HMG-CoA reductase cross-react with the rat liver enzyme (Beg *et al*. 1978), indicating some degree of homology.

E. Catalytic Activity

Catalytic activity, expressed either as specific activity (pU/mg protein) or as turnover number (moles mevalonate formed per min per mole of subunit), varies over two orders of magnitude. The specific activity and turnover numbers of the yeast and rat liver enzymes are similar (Table V).

TABLE V. Molecular Weight and Specific Activity of Purified HMG-CoA Reductase

Source	M.W. x 10^{-6}		Catalytic Activity		Ref.[b]
	Holo	Subunit	pU/mg x 10^{-6}	Turnover number[a]	
Pseudomonas	280		28[c]		(1)
Yeast	260	60	10[c]	600	(2)
F. oxysporium		11	0.15[d]	2	(3)
Chicken liver		18	3.3[d]	60	(4)
Rat liver	200	47–52	10[e]	500	(5)

[a] Moles mevalonate formed per min per mole of subunit at indicated temperature.

[b] References are: (1) Bensch & Rodwell, 1970; (2) Qureshi *et al.* 1976a; (3) Madhosingh & Orr 1978; (4) Beg *et al.* 1977, 1979; (5) Higgins *et al.* 1974, Brown *et al.* 1974, Tormanen *et al.* 1976, Kleinsek *et al.* 1978, Edwards *et al.* 1979b.

[c] Assayed at 30°C.

[d] Assayed at 37°C.

[e] For enzyme from cholestyramine-fed rats assayed at 37°C.

V. EFFECTORS OF CATALYTIC ACTIVITY

A. Lipids

1. Solubilized HMG-CoA Reductase

Depending on the method employed for purification, the activity of solubilized rat liver HMG-CoA reductase may be affected by addition of lipids. HMG-CoA reductase purified by one method (Heller & Shrewsbury 1976) contains 50 times as much phospholipid as enzyme prepared by alternative techniques (Tormanen *et al.* 1976, Berde *et al.* 1977). Added microsomal lipids have no effect on the phospholipid-rich preparation, but increase the activity of the phospholipid-poor preparation 2.8 fold (Table VI). Phosphatidylcholine and its derivatives give optimal activation of the lipid-poor preparation. Phosphatidylethanolamine has little activating effect and phosphatidylserine is inhibitory (Berde *et al.* 1977) (Table VI).

TABLE VI. Activation of Lipid-Poor, Solubilized Rat Liver HMG-CoA Reductase by Phosphatidylcholine Derivatives[a]

Addition	Fold Activation
None	(1.0)
Microsomal lipids	2.8
Soybean phosphatidylcholine	2.5
Dipalmitoylglycerophosphocholine	2.5
Phosphatidylethanolamine	1.3
Phosphatidylserine	0.3

[a] From Berde *et al.* (1977).

2. Microsomal HMG-CoA Reductase

The choline analogue N-isopropylethanolamine (IPE) is incorporated into the membranes of rat liver cells (Lee et al. 1975) and cultured mouse fibroblast (L-M) cells (Blank et al. 1975) replacing choline and ethanolamine (Lee et al. 1975). IPE has no effect on HMG-CoA reductase of cell extracts. Although without adverse effects on cell growth, IPE added to the medium decreases HMG-CoA reductase activity 75% in cultured glial cells. Choline blocks this decrease. Choline derivatives may thus affect the activity of membrane-bound HMG-CoA reductase (Volpe 1978).

Arrhenius plots of the activity of both microsomal (Mitropolous & Venkatesan 1977, Sipat 1978) and the lipid-rich, soluble HMG-CoA reductase (Heller & Gould 1974) show discontinuities or transition temperatures (T_t) at 28°C and 19°C (Fig. 2).

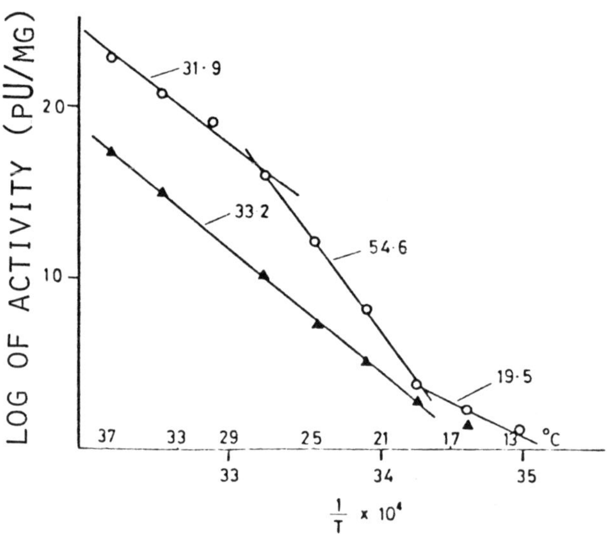

Fig. 2. Arrhenius plots of HMG-CoA reductase activity in liver microsomes of rats fed 2% cholesterol (▲) or a normal diet (O). Numerals indicate Ea values. Reproduced, with permission, from Mitropoulos & Venkatesan (1977).

When rats are fed a diet rich in cholesterol for at least 12 hours, the cholesterol content of the microsomes increases and the T_t at 28°C is abolished (Mitropolous & Venkatesan 1977, Sipat 1978). Although dietary cholesterol rapidly and profoundly decreases HMG-CoA reductase activity (Rodwell et al. 1976), the loss of the 28°C T_t occurs some hours later (Sipat 1978). The Ea's at 37° are, furthermore, similar in normal-fed and cholesterol-fed rats (Fig. 2). The precise physiological significance of this change in the 28°C T_t thus is presently unclear.

B. Substrate Analogs

1. HMG

Purified HMG added to a rat liver postmitochondrial supernatant fraction inhibits conversion of [^{14}C]-acetate to mevalonate without dilution of label incorporation (Fimognari 1964). Beg & Lupien (1972) localized the site of HMG inhibition of cholesterogenesis at HMG-CoA reductase and demonstrated that HMG inhibits HMG-CoA reductase in intact rats. Since HMG forms a portion of the HMG-CoA molecule, inhibition competitive with HMG-CoA might be anticipated. To our knowledge, however, this has only been studied for partially purified chicken liver HMG-CoA reductase (Beg et al. 1979) for which the inhibition kinetics are not strictly competitive (Fig. 3).

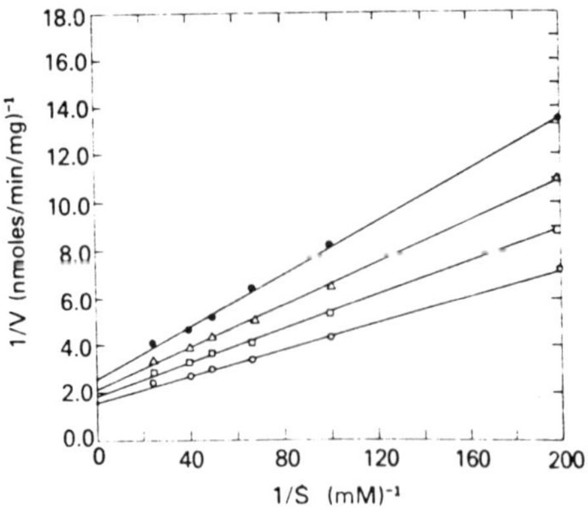

Fig. 3. Inhibition by HMG of the activity of partially purified chicken liver HMG-CoA reductase. [NADPH] = 4 mM. HMG-CoA concentration was varied as shown on the 1/S axis. HMG concentrations are: 0 (O), 2.5 (□), 5 (Δ) and 10 mM (●). Reproduced, with permission, from Beg et al. (1979).

For the oxidation of mevalonate to HMG-CoA by Pseudomonas HMG-CoA reductase, HMG is an inhibitor competitive with mevalonate with a Ki comparable to the Km (Table VII). For this bacterial HMG-CoA reductase other 3-hydroxy-3-methyl-carboxylic acids also inhibit competitively with mevalonate (Table VII).

TABLE VII. Inhibitors Competitive with Mevalonate for *Pseudomonas* HMG-CoA Reductase[a]

Competitive Inhibitor	Ki(mM)
3-Hydroxybutyrate	0.60
DL-3-Hydroxy-3-methylbutyrate	0.30
DL-3-Hydroxy-3-methylglutarate	0.40

[a] Km for mevalonate = 0.35 mM. From Fimognari & Rodwell (1965a).

2. ML-236A and ML-236B

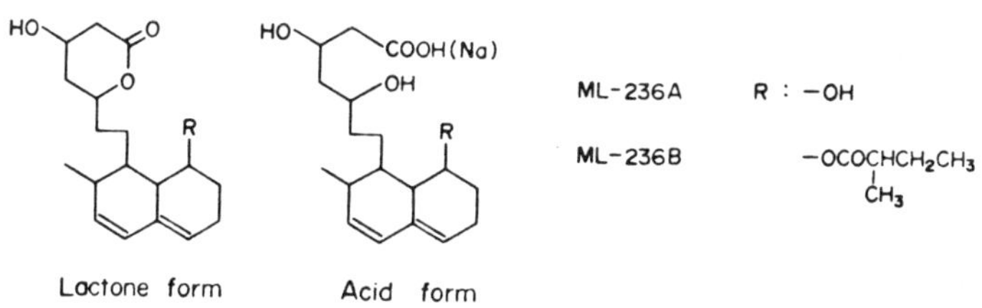

Fig. 4. Structures of ML-236A and ML-236B. Reproduced, with permission, from Endo *et al.* (1976).

The fungal metabolites ML-236A and ML-236B (Fig. 4) are potent inhibitors of HMG-CoA reductase activity competitive with HMG-CoA and noncompetitive with NADPH (Fig. 5). The salts inhibit more effectively than the lactones, and (Na)ML-236B (Ki = 10 nM) is a more potent inhibitor than (Na)ML- 236A (Ki = 220 nM) (Endo *et al.* 1976). For human cell fibroblast reductase, Ki for (Na)ML-236B is 1.1 nM

(Brown *et al.* 1978). When administered to intact animals or cells, ML-236B effectively curtails cholesterogenesis (Kaneko *et al.* 1978, Brown *et al.* 1978) and derepresses synthesis of HMG-CoA reductase protein (Brown *et al.* 1978, Bensch *et al.* 1978).

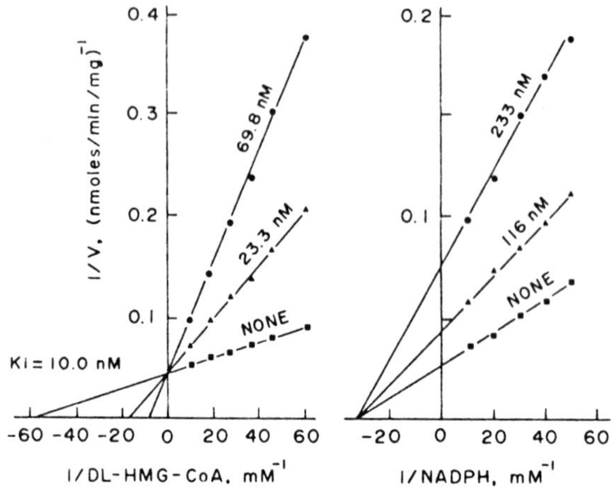

Fig. 5. Inhibition of rat liver HMG-CoA reductase by ML-236B. Reproduced, with permission, from Endo *et al.* (1976).

C. Inactivation

Like other enzymes, HMG-CoA reductase is readily inactivated by heat or extremes of pH. In addition, it is inactivated by cold, coenzyme A, citrinin and certain drugs.

1. Effect of Temperature on Solubilized HMG-CoA Reductase.

Unlike microsomal HMG-CoA reductase, the solubilized enzyme is sensitive to cold. HMG-CoA reductase may be irreversibly inactivated at 4°C following solubilization.

This irreversible inactivation, which is partially blocked by 4 M KCl (Brown *et al.* 1973) or by preincubation at 37°C (Tormanen *et al.* 1977), appears to be caused by a heat-labile protein (Tormanen *et al.* 1977).

Reversible inactivation has been observed for solubilized HMG-CoA reductase from both rat liver (Heller & Gould 1974) (Fig. 6) and dog liver (Ho *et al.* 1977). The mechanism of this inactivation is not known. It may be due to a cold-induced dissociation of subunits since: 1) Reactivation is more rapid at high protein concentrations (Heller & Gould 1975, Ho *et al.* 1977), 2) Initial gel filtration data indicates that the inactive form is smaller than the active form (Heller & Gould 1974), 3) Even at 37°C HMG-CoA reductase is inactivated upon dilution (Nordstrom 1976), 4) Inactivation is prevented by NADPH or $NADP^+$ (Heller & Gould 1975).

Solubilized HMG-CoA reductase is relatively heat-stable. In the presence of salt, it withstands 65°C for 10 min without significant loss of activity (Brown *et al.* 1973).

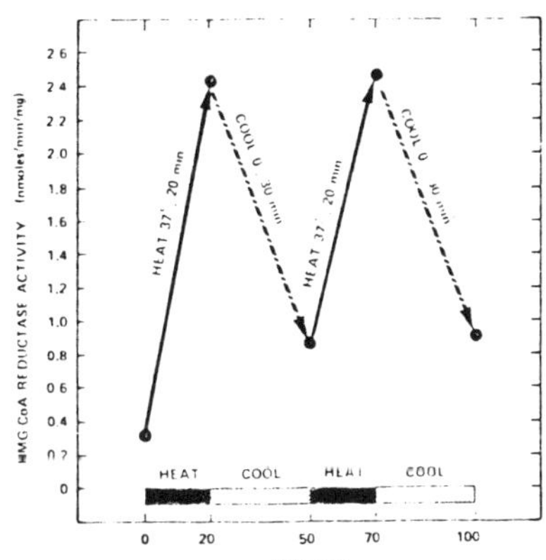

Fig. 6. Reversible cold inactivation of solubilized rat liver HMG-CoA reductase. Reproduced, with permission, from Heller & Gould (1974).

2. Coenzyme A

Mitochondrial yeast HMG-CoA reductase is inactivated in a time-dependent manner by coenzyme A. Inactivation is reversed by dialysis, but not by addition of substrates (Tan-Wilson & Kohlhaw 1978). Rat liver HMG-CoA reductase is also inhibited by CoA (Nordstrom (1976).

3. Citrinin

The fungal antibiotic citrinin is an effective inhibitor of cholesterogenesis in intact animals (Kuroda et al. 1977). Its ability to inhibit HMG-CoA reductase activity is uncompetitive with HMG-CoA and noncompetitive with NADPH (Tanzawa et al. 1977, Ness & Moffler 1978).

4. Drugs

Boots, Guyer and their colleagues have synthesized and evaluated many irreversible inhibitors of yeast and rat liver HMG-CoA reductase. Structurally, these are alkylaryl esters of succinate or glutarate (Boots et al. 1973), half-acid esters of 1-(4-biphenylyl)pentanol (Guyer et al. 1976) or of succinate and glutarate (Boots et al. 1976a) and 4-biphenylalkyl carboxylic acids (Boots et al. 1976b).

D. Phosphorylation/Dephosphorylation

1. Inactivation by MgATP

Addition of MgATP to crude, unwashed rat liver microsomal preparations rapidly and profoundly decreases HMG-CoA reductase activity (Fig. 7). Both Mg^{2+} and ATP are required. This represents a decrease in the catalytic efficiency of HMG-CoA reductase. Removal of MgATP by centrifugation does not restore activity (Beg et al. 1973, Nordstrom et al. 1977), suggesting that MgATP is not an allosteric effector of HMG-CoA reductase activity.

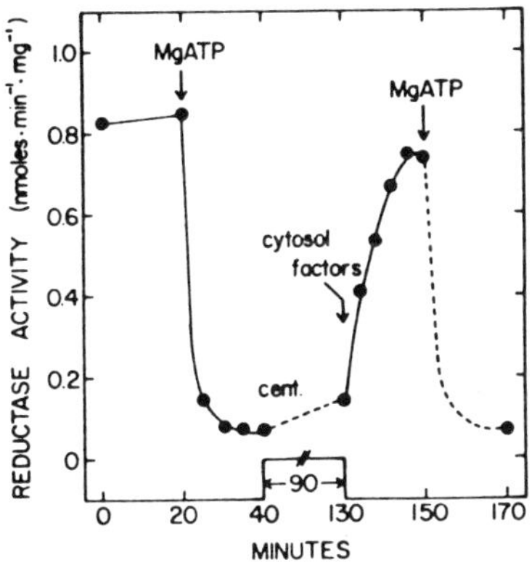

Fig. 7. Reversible modulation of rat liver microsomal HMG-CoA reductase activity. Reproduced, with permission, from Nordstrom *et al.* (1977). Similar data has been obtained for liver microsomal preparations of several other vertebrates (C. F. Hunter & V. W. Rodwell, unpublished observations).

The MgATP-dependent decrease in activity requires a protein, reductase kinase, present in liver cytosol and on unwashed microsomal preparations (Beg *et al.* 1973, Brown *et al.* 1975, Nordstrom *et al.* 1977, Ingebritsen *et al.* 1978) (Fig. 8).

ADP may also be required for inactivation since an ATP-regenerating system blocks inactivation of crude microsomal preparations (Brown *et al.* 1975) and inactivation of solubilized HMG-CoA reductase is enhanced by ADP (Nordstrom *et al.* 1977, Ursini *et al.* 1977).

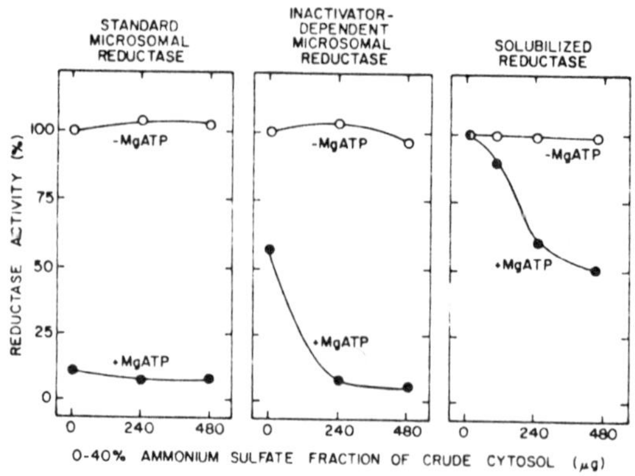

Fig. 8. Dependence of inactivation of rat liver HMG-CoA reductase upon a cytosol protein, reductase kinase. LEFT: Inactivation of HMG-CoA reductase of crude microsomal preparations is independent of added reductase kinase. CENTER: Inactivation of washed microsomal HMG-CoA reductase is partially dependent on added reductase kinase. RIGHT: Inactivation of solubilized, partially purified HMG-CoA reductase is totally dependent on added reductase kinase. Reproduced, with permission, from Nordstrom et al. (1977).

2. Reactivation

Activity is restored to MgATP-inactivated HMG-CoA reductase by a different cytosol protein, reductase phosphatase, (Fig. 7) (Beg et al. 1973, Brown et al. 1975, Nordstrom et al. 1977) or by certain phosphatase preparations

(Ingebritsen et al. 1978, Beg et al. 1978, Philipp & Shapiro 1979). Reductase phosphatase has an apparent molecular weight of 30,000 (Nordstrom et al. 1977, Ingebritsen et al. 1978). The activity copurifies with phosphorylase phosphatase (Ingebritsen et al. 1978) and is effectively inhibited by 50 mM NaF (Nordstrom et al. 1977). Activation of HMG-CoA reductase in a postmitochondrial supernatant fraction is inhibited by cAMP + MgATP or by NaF (Goodwin & Margolis 1978).

3. Mechanism

Active and inactive HMG-CoA reductase have the same Km for (S)-HMG-CoA and for NADPH (Brown et al. 1975, Nordstrom et al. 1977) and approximately the same molecular weight (Nordstrom et al. 1977). Inactivation thus does not appear to involve altered substrate affinity or subunit dissociation/association.

When washed microsomal preparations are incubated with [γ-^{32}P]MgATP, decreased HMG-CoA reductase activity is paralleled by an increase in acid-precipitable ^{32}P (Beg et al. 1978, Bové & Hegardt 1978). HMG-CoA reductase activity is restored and acid-soluble ^{32}P is released following incubation with cytosol (Bové & Hegardt 1978) or with partially purified phosphorylase phosphatase (Beg et al. 1978). No radioactivity is incorporated into acid-precipitable material when microsomes are inactivated with [^{14}C]ATP or with [α-^{32}P]ATP. Immunoprecipitation of partially purified, inactive rat liver HMG-CoA reductase with antibodies to the chicken liver enzyme precipitates ^{32}P counts. For SDS gels of immunoprecipitated material, >90% of the ^{32}P counts migrate where HMG-CoA reductase should migrate (Beg et al. 1978).

4. Inactivation/Activation of Reductase Kinase

Reductase kinase appears also to be an interconvertible enzyme. The active form of reductase kinase appears to be phosphorylated and the inactive form dephosphorylated (Fig. 9). While direct evidence of covalent modification is lacking, reductase kinase activity is activated by MgATP plus an appropriate protein kinase and is inactivated by phosphorylase phosphatase (Ingebritsen et al. 1978, Gibson & Ingebritsen 1978).

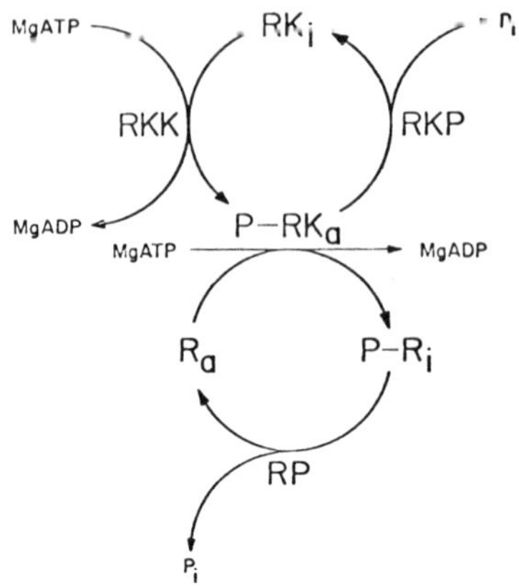

Fig. 9. Proposed model for regulation of the catalytic activity of HMG-CoA reductase. Interconversion of vertebrate liver HMG-CoA reductase between a catalytically active form (R_a) and a phosphorylated, less active form ($P-R_i$) is catalyzed by reductase kinase ($P-RK_a$) and reductase phosphatase (RP). Reductase kinase also exists in an active form ($P-RK_a$) and a less active form (RK_i). Their interconversion requires reductase kinase kinase (RKK) and reductase kinase phosphatase (RKP). RKK is distinct from RK. Evidence that $P-RK_a$ is phosphorylated is at present inferential.

E. Activation by Sulfite

Mouse liver microsomal HMG-CoA reductase is activated three-fold by $MgCl_2$ plus Na_2SO_3, but not Na_2SO_4 (Berndt *et al.* 1976). Similarly, rat liver glycogen synthase is converted from the D form to the I (non-phosphorylated) form when crude glycogen pellets are incubated with $MgCl_2$ and Na_2SO_3. This latter conversion requires a heat-labile factor present in microsomes (Hizukuri & Larner 1964). Activation of HMG-CoA reductase by sulfite resembles the activation observed when microsomes are incubated with cytosol. Both processes are inhibited by NaF (Berndt *et al.* 1976).

VI. MECHANISM OF ACTION

A. Stereospecificity of the Hydride Transfer

For both NADPH molecules, the hydride comes from the A side of NADPH (Fig. 10) for both yeast (Dugan & Porter 1971) and rat liver HMG-CoA reductase (Beedle *et al.* 1972).

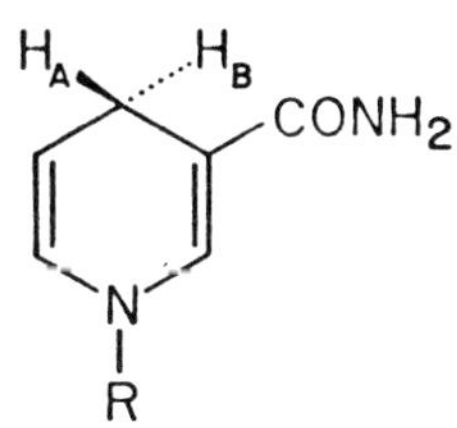

Fig. 10. A and B hydrogens of NADPH.

B. Mevaldate as an Intermediate

Enzyme-bound mevaldate is a likely intermediate in the reductive deacylation of HMG-CoA (Durr & Rudney 1960, Bensch & Rodwell 1970). Mevaldate is both reduced to mevalonate and oxidized to HMG-CoA by yeast (Qureshi et al. 1976b) and bacterial HMG-CoA reductase (Bensch 1970, Bensch & Rodwell 1970). Mevaldate has not, however, been detected as a free intermediate. In the presence of a trapping pool, no isotope is incorporated into the mevaldate from either [^{14}C]HMG-CoA (Durr & Rudney 1960, Bensch & Rodwell 1970) or from [^{14}C]mevalonate (Bensch & Rodwell 1970).

C. Model of the Reaction

Reductive deacylation of HMG-CoA to mevalonate probably occurs via a random sequential mechanism (Qureshi et al. 1976b). The model (Fig. 11) incorporates a dissociable X-H group that may be the essential thiol group (see IV, C).

Protonation of (I) (Reaction A) is followed by random binding of HMG-CoA and NADPH (Reaction B) forming complex (III) in which HMG-CoA is bound both by CoA and by its R group. Internal reduction (Reaction C) forms enzyme-bound mevaldate thiohemiacetal (IV). Uptake of NADPH and H^+ and release of $NADP^+$ (Reaction D) forms enzyme-bound mevaldate and bound CoA (V). Internal reduction (Reaction E) yields enzyme-bound mevalonate (VI). CoA and mevalonate are then released in random order (Reaction F) forming (VII). $NADP^+$ then dissociates (Reaction G) reforming (I). Kinetic constants are shown in Table VIII. Essential features of the model include:

1. The presence at the catalytic site of a group X that undergoes rapid protonic equilibrium.
2. HMG-CoA is bound both via coenzyme A and by other portions of the molecule.
3. Slow dissociation of the enzyme-NADPH and enzyme-$NADP^+$ complexes.

The charged state of a group X present at the catalytic site, implicated for other oxidoreductases, depends on which form of the coenzyme is bound:

(A)	(B)
NADPH | E-X-H	$NADP^+$ | $E-X^-$

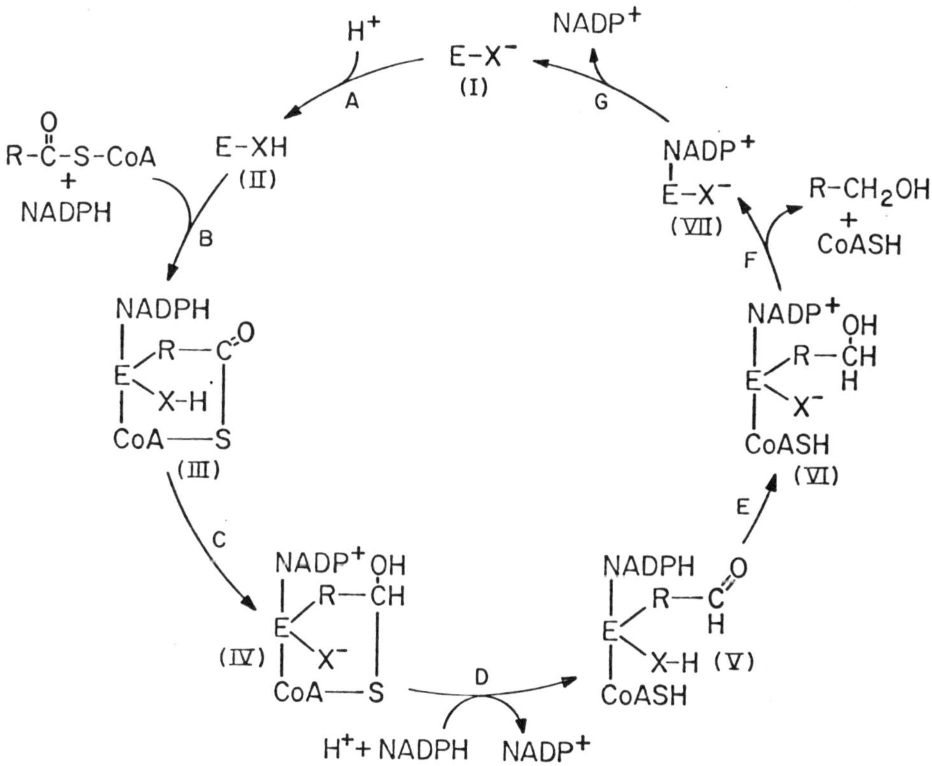

Fig. 11. Presumed intermediates for reductive deacylation of HMG-CoA.

TABLE VIII. Kinetic Constants for Reduction of HMG-CoA or of Mevaldate

Enzyme Source	Km (μM)					Ref.[a]
	Oxidant		Reductant		CoA	
	HMG-CoA	Mevaldate	NADPH	NADH		
Yeast	1.2		89			(1)
Pseudomonas	25			32		(2)
Pseudomonas[b]		9000		360		(2)
Yeast[b]		8000	24			(3)
Yeast[c]		400			7	(3)
Yeast[d]		500	16			(3)

[a] References are: (1) Kirtley & Rudney 1967; (2) Bensch & Rodwell 1970; (3) Qureshi *et al*. 1976b.

[b] No CoA present.

[c] CoA present; [NADPH] = 160 μM.

[d] [CoA] = 200 μM.

Reductions by pyridine nucleotides are facilitated by this X-H group hydrogen bonding to the carbonyl oxygen of the substrate. This generates sufficient carbonium ion character to allow hydride transfer.

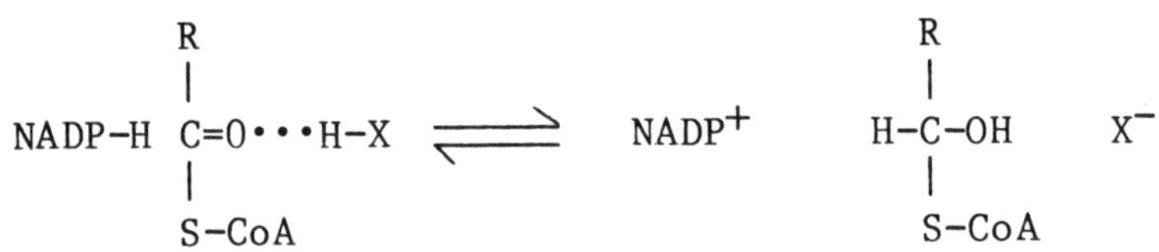

The enzyme-NADPH complex (A) should have affinity for HMG-CoA and mevaldate and the enzyme-$NADP^+$ complex (B) for mevalonate and mevaldate thiohemiacetal.

Double-reciprocal plots of initial velocity data for <u>Pseudomonas</u> HMG-CoA reductase give intersecting lines when both NADH and HMG-CoA are varied (Fig. 12). A sequential mechanism is proposed (Bensch & Rodwell 1970).

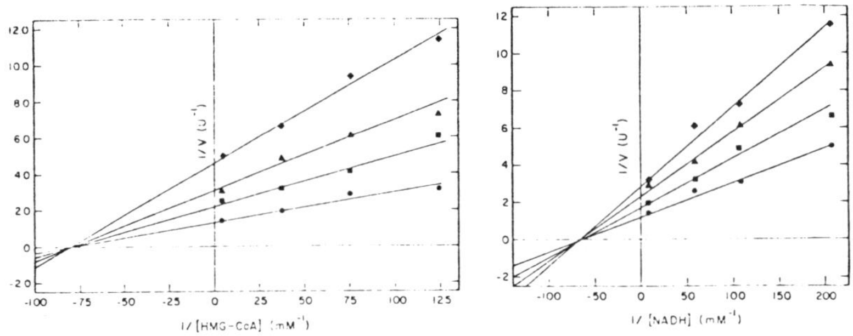

Fig. 12. Initial velocity data for HMG-CoA and NADH for <u>Pseudomonas</u> HMG-CoA reductase. LEFT: micromolar concentrations of NADH are 4.8 (◆), 9.3 (▲), 17.1 (■) and 112 (●). RIGHT: micromolar concentrations of HMG-CoA are 8.0 (◆), 13.1 (▲), 26.6 (■) and 229 (●). Reproduced, with permission, from Bensch & Rodwell (1970).

For yeast HMG-CoA reductase, comparable experiments yield parallel lines (Fig. 13), and a ping-pong mechanism was initially proposed (Kirtley & Rudney 1967).

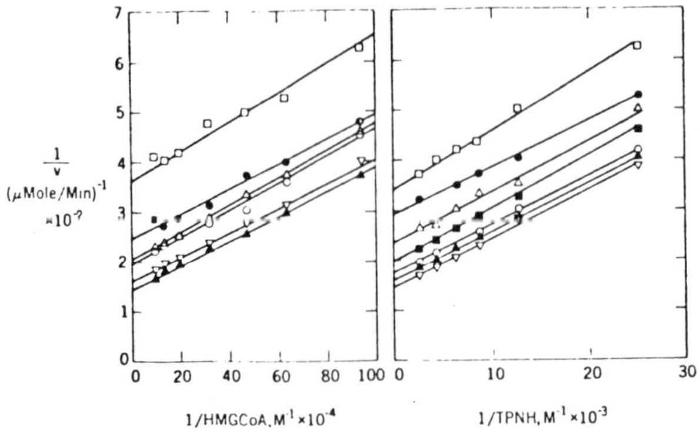

Fig. 13. Initial velocity data for HMG-CoA and NADPH for yeast HMG-CoA reductase. LEFT: micromolar concentrations of NADPH are 40 (□), 79 (●), 119 (Δ), 159 (0), 237 (∇) and 396 (▲). RIGHT: micromolar concentrations of HMG-CoA are 1.1 (□), 1.6 (●), 2.1 (Δ), 3.2 (■), 5.3 (0), 7.4 (▲) and 10.6 (∇). Reproduced, with permission, from Kirtley & Rudney (1967).

Parallel lines can be obtained in a random sequential reaction if the release of substrates (HMG-CoA and NADPH) is slow relative to V_{max} (Bar-Tana & Cleland 1975). The rapid equilibrium assumption then does not apply, and the lines may appear parallel. Definitive evidence for a ping-pong mechanism should include parallel lines in the reverse reaction and isotopic exchange for the half reactions (Piszkiewicz 1977). Rétey et al. (1972) observed no isotope exchange for HMG-CoA and CoA in the presence or absence of $NADP^+$. Intersecting initial velocity plots were obtained for the half reaction mevaldate to HMG-CoA for yeast HMG-CoA reductase at low concentrations of mevaldate (Qureshi et al. 1976b). These and other data (see below) thus support a sequential mechanism for both the bacterial and yeast enzymes.

D. Reduction of Mevaldate

Reduction of mevaldate (Fig. 14) probably occurs by reactions analogous to those of reactions D–G (Fig. 11). Since NADPH is present, the X group is protonated (II) and the enzyme has an affinity for the aldehyde. The mechanism appears to be random sequential. Double reciprocal plots of initial velocity data give intersecting lines for the bacterial (Bensch and Rodwell 1970) (Fig. 15) and the yeast enzymes (Qureshi *et al.* 1976b). Table IX lists kinetic constants. Although reduction of mevaldate occurs in the absence of CoA, CoA decreases the Km for mevaldate 20 fold, consistent with binding sites for the CoA and HMG portions of HMG-CoA.

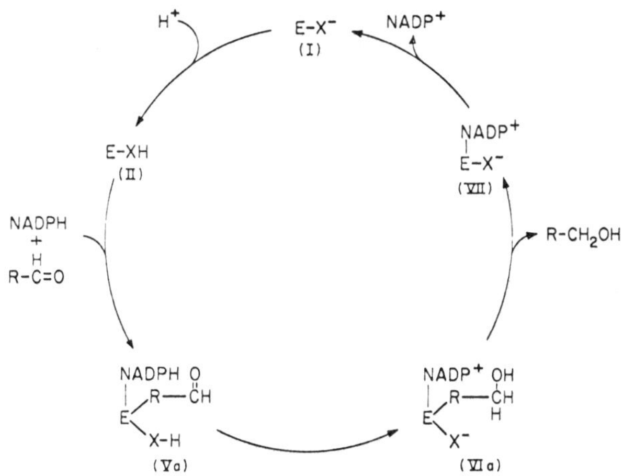

Fig. 14. Presumed intermediates for reduction of mevaldate.

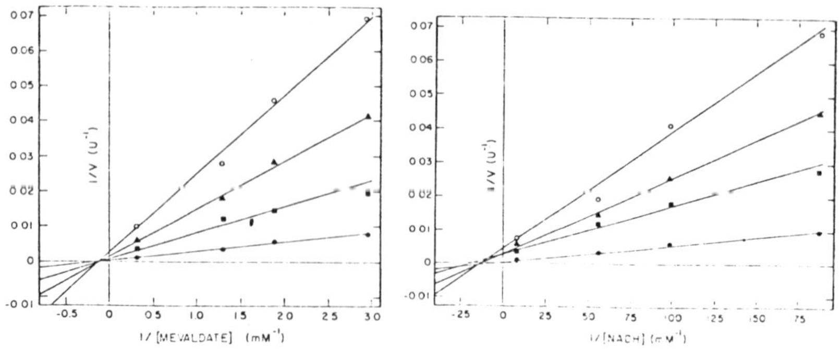

Fig. 15. Initial velocity data for mevaldate and NADH for the half reaction: mevaldate + NADH + $H^+ \rightleftharpoons$ mevalonate + NAD^+ catalyzed by Pseudomonas HMG-CoA reductase. LEFT: micromolar concentrations of NADH are 5.3 (O), 10.1 (▲), 17.9 (■) and 121 (●). RIGHT: micromolar concentrations of mevaldate are 335 (O), 533 (▲), 771 (■) and 3,240 (●). Reproduced, with permission, from Bensch & Rodwell (1970).

E. Oxidation of Mevaldate

Oxidation of mevaldate may occur by an ordered sequential mechanism (Fig. 16). Table IX lists kinetic constants. CoA may be necessary for mevaldate binding, since in the presence of $NADP^+$ the unprotenated X^- groups lacks affinity for mevaldate. The X^- group has affinity for hydroxyl groups such as the 5-hydroxy group of mevalonate or the presumed thiohemiacetal intermediate. If CoA is bound, then the thiohemiacetal can be formed and the reaction will proceed.

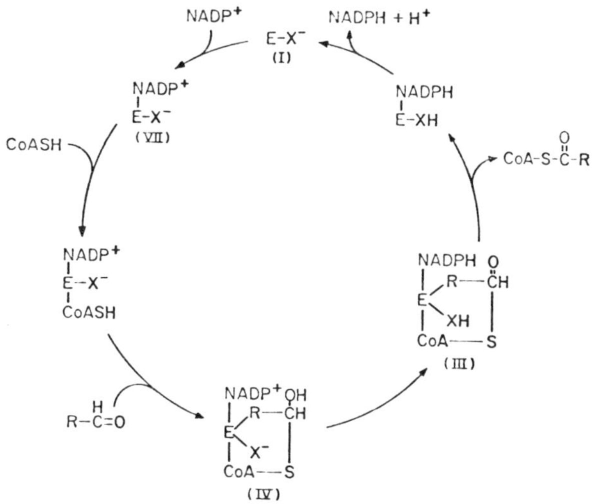

Fig. 16. Presumed intermediates for oxidation of mevaldate.

Double reciprocal plots of initial velocity data are intersecting when CoA and mevaldate or $NADP^+$ and mevaldate are the variable substrates. When $NADP^+$ and CoA are varied, intersecting lines are obtained at low mevaldate concentrations. Parallel lines are obtained when mevaldate is saturating (Fig. 17). The change from intersecting to parallel lines is consistent with an ordered sequential mechanism with mevaldate binding second. The binding of the third substrate may be random (Qureshi _et al._ 1976b).

TABLE IX. Kinetic Constants for Oxidation of Mevalonate or Mevaldate

Enzyme Source	Km (μM)					
	Reductant		Oxidant		CoA	Ref.[a]
	Mevalonate	Mevaldate	$NADP^+$	NAD^+		
Pseudomonas	300			270	39	(1)
Pseudomonas		610		60	46	(1)
Yeast[b]	230				7	(2)
Yeast[c]		70	33			(2)
Yeast[d]			1.0		22	(2)
Yeast[e]			0.25		13	(2)

[a] References are: (1) Bensch & Rodwell 1970; (2) Qureshi *et al*. 1976b.

[b] [$NADP^+$] = 8,000 μM.

[c] [CoA] = 200 μM.

[d] [Mevaldate] = 10,000 μM.

[e] [Mevaldate] = 500 μM.

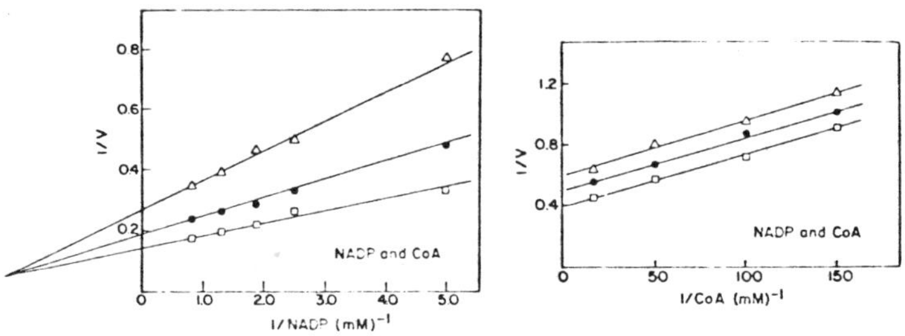

Fig. 17. Initial velocity data for oxidation of mevaldate to HMG-CoA by yeast HMG-CoA reductase. LEFT: micromolar concentrations of CoA are 10 (Δ), 20 (●) and 50 (□). [Mevaldate] = 0.5 mM. RIGHT: micromolar concentrations of $NADP^+$ are 400 (Δ), 530 (●) and 800 (□). [Mevaldate] = 10 mM. Reproduced, with permission, from Qureshi et al. (1976b).

F. Inhibition by $NADP^+$ and Mevalonate

For reduction of HMG-CoA (Kirtley & Rudney 1967) or mevaldate (Bensch 1970, Qureshi et al. 1976b), NAD^+/$NADP^+$ inhibit competitively with NADH/NADPH and noncompetitively with mevaldate (Table X). For reduction of high concentrations of HMG-CoA (Kirtley & Rudney 1967) or of mevaldate (Qureshi et al. 1976b), mevalonate inhibits uncompetitively with NADPH. However, at low mevaldate concentration, inhibition by mevalonate is noncompetitive with NADPH (Table X). The change from uncompetitive to noncompetitive inhibition is consistent with an ordered sequential mechanism or with a random sequential mechanism in which one reactant dissociates slowly. In the proposed model (Fig. 11), loss of $NADP^+$ from complex (VII) is envisaged as rate-limiting (see below).

For mevaldate reduction mevalonate inhibits non-competitively with mevaldate, presumably due to the ability of the X-H group of the E-NADPH complex to bind mevaldate but not mevalonate.

G. Rate-Limiting Step

V_{max} for the overall reaction is the same as V_{max} for mevaldate reduction. The rate-limiting step may thus occur in the second half reaction. Since reduction occurs in the absence of CoA, release of CoA probably is not rate-limiting. Release of $NADP^+$ may be the rate-limiting step in catalysis by HMG-CoA reductase (Qureshi *et al*. 1976b).

TABLE X. Substrate and Product Inhibition of Reduction of HMG-CoA or of Mevaldate[a]

Inhibitor	Variable substrate	Fixed substrate: Substrate	Fixed substrate: Concentration relative to Km	Type of inhibition	Ref.[b]
$NADP^+$	NADPH	HMG-CoA	140	C	(1)
$NADP^+$	NADPH	Mevaldate	2.5	C	(2)
$NADP^+$	NADPH	Mevaldate	50	C	(2)
		+ CoA	29		
NAD^+	NADH	Mevaldate	0.7	C	(3)
$NADP^+$	Mevaldate	NADPH	6.7	NC	(2)
Mevalonate	NADPH	HMG-CoA	140	UC	(1)
Mevalonate	NADPH	Mevaldate	50	UC	(2)
		+ CoA	29		
Mevalonate	NADPH	Mevaldate	2.5	NC	(2)
Mevalonate	NADPH	Mevaldate	1.2	NC	(2)
		+ CoA	29		
Mevalonate	Mevaldate	NADPH	6.7	NC	(2)

[a] With the exception of line 5, which is for *Pseudomonas* HMG-CoA reductase, all data are for yeast HMG-CoA reductase. Abbreviations are: C = Competitive, UC = uncompetitive, NC = noncompetitive.

[b] References are: (1) Kirtley & Rudney 1967; (2) Qureshi *et al.* 1976b; (3) Bensch 1970.

REFERENCES

Ackerman, M. E., Redd, W. L., Tormanen, C. D., Hardgrove, J. E. & Scallen, T. J. (1977) J. Lipid Res. 18, 408-413

Baqir, Y. A. & Booth, R. (1977) Biochem. J. 164, 501-508

Bar-Tana, J. & Cleland, W. W. (1974) J. Biol. Chem. 249, 1271-1276

Beg, Z. H. & Lupien, P. J. (1972) Biochim. Biophys. Acta 260, 439-448

Beg, Z. H., Allmann, D. W. & Gibson, D. W. (1973) Biochem. Biophys. Res. Commun. 54, 1362-1369

Beg, Z. H., Stonik, J. A. & Brewer, H. B., Jr. (1977) FEBS Lett. 80, 123-129

Beg, Z. H., Stonik, J. A. & Brewer, H. B., Jr. (1978) Proc. Nat. Acad. Sci. U.S.A. 75, 3678-3682

Beg, Z. H., Stonik, J. A. & Brewer, H. B., Jr. (1979) Biochim. Biophys. Acta 572, 83-94

Beedle, A. S., Munday, K. A. & Wilton, D. C. (1972) Eur. J. Biochem. 28, 151-155

Bensch, W. R. (1970) M.S. Thesis, Purdue University, West Lafayette, IN

Bensch, W. R. & Rodwell, V. W. (1970) J. Biol. Chem. 245, 3755-3762

Bensch, W. R., Ingebritsen, T. S. & Diller, E. R. (1978) Biochem. Biophys. Res. Commun. 82, 247-254

Berde, C. B., Heller, R. A. & Simoni, R. D. (1977) Biochim. Biophys. Acta 488, 112-120

Berndt, J. & Gaumert, R. (1971) Hoppe-Seyler's Z. Physiol. Chem. 352, 345-350

Berndt, J. & Gaumert, R. (1974) Hoppe-Seyler's Z. Physiol. Chem. 355, 905-910

Berndt, J., Hegardt, F. G., Bové, J., Gaumert, R., Still, J. & Cardó, M-T. (1976) Hoppe-Seyler's Z. Physiol. Chem. 357, 1277-1282

Blank, M. L., Piantadosi, C., Ishaq, K. S. & Snyder, F. (1975) Biochem. Biophys. Res. Commun. 62, 983-988

Boots, M. R., Boots, S. G., Noble, C. M. & Guyer, K. E. (1973) J. Pharm. Sci. 62, 952-957

Boots, M. R., Marecki, P. E., Boots, S. G. & Guyer, K. E. (1976a) J. Pharm. Sci. 65, 724-727
Boots, S. G., Boots, M. R., Guyer, K. E. & Marecki, P. E. (1976b) J. Pharm. Sci. 65, 1374-1380
Bové, J. & Hegardt, F. G. (1978) FEBS Lett. 90, 198-202
Brooker, J. D. & Russell, D. W. (1975) Arch. Biochem. Biophys. 167, 723-729
Brown, M. S., Dana, S. E., Dietschy, J. M. & Siperstein, M. D. (1973) J. Biol. Chem. 248, 4731-4738
Brown, M. S., Dana, S. E. & Siperstein, M. D. (1974) J. Biol. Chem. 249, 6585-6589
Brown, M. S., Brunschede, G. Y. & Goldstein, J. L. (1975) J. Biol. Chem. 250, 2502-2509
Brown, M. S., Faust, J. R., Goldstein, J. L., Kaneko, I. & Endo, A. (1978) J. Biol. Chem. 253, 1121-1128
Dempsey, M. E. (1974) Annu. Rev. Biochem. 43, 967-990
Dugan, R. E. & Porter, J. W. (1971) J. Biol. Chem. 246, 5361-5364
Durr, I. F. & Rudney, H. (1960) J. Biol. Chem. 235, 2572-2578
Edwards, P. A., Lemangello, D. & Fogelman A. M. (1979a) J. Lipid Res. 20, 40-46
Edwards, P. A., Lemangello, D. & Fogelman, A. M. (1979b) Biochim. Biophys. Acta in the press.
Endo, A., Kuroda, M. & Tanzawa, K. (1976) FEBS Lett. 72, 323-326
Fimognari, G. M. (1964) Ph.D. Thesis, Univ. of California, San Francisco Medical Center, San Francisco, CA
Fimognari, G. M. & Rodwell, V. W. (1965a) Biochemistry 4, 2086-2090
Fimognari, G. M. & Rodwell, V. W. (1965b) Science 147, 1038
Gibson, D. M. & Ingebritsen, T. S. (1978) Life Sci. 23, 2649-2664
Goldfarb, S. & Pitot, H. C. (1971) J. Lipid Res. 12, 512-515
Goodwin, C. D. & Margolis, S. (1978) J. Lipid Res. 19, 747-756
Guyer, K. E., Boots, S. G., Marecki, P. E. & Boots, M. R. (1976) J. Pharm. Sci. 65, 548-552

Hamprecht, B. & Lynen, F. (1970) Eur. J. Biochem. 14, 323-336
Heller, R. A. & Gould, R. G. (1974) J. Biol. Chem. 249, 5254-5260
Heller, R. A. & Gould, R. G. (1975) Biochim. Biophys. Acta 388, 254-259
Heller, R. A. & Shrewsbury, M. A. (1976) J. Biol. Chem. 251, 3815-3822
Higgins, M. J. P., Brady, D. & Rudney, H. (1974) Arch. Biochem. Biophys. 163, 271-282
Hizukuri, S. & Larner, J. (1964) Biochemistry 3, 1783-1788
Ho, P. P. K., Esterman, M. A., Towner, R. D. & Turner, R. K. (1977) Enzyme 22, 242-248
Huber, J., Latzin, S. & Hamprecht, B. (1973) Hoppe-Seyler's Z. Physiol. Chem. 354, 1645-1647
Hulcher, F. H. & Oleson, W. H. (1973) J. Lipid Res. 14, 625-631
Imblum, R. L. & Rodwell, V. W. (1974) J. Lipid Res. 15, 211-222
Ingebritsen, T. S., Lee, H-S., Parker, R. A. & Gibson, D. M. (1978) Biochem. Biophys. Res. Commun. 81, 1268-1277
Kandutsch, A. A. & Saucier, S. E. (1969) Arch. Biochem. Biophys. 135, 201-208
Kaneko, I., Hazama-Shimada, Y. & Endo, A. (1978) Eur. J. Biochem. 87, 313-321
Kawashi, T. & Rudney, H. (1970) Biochemistry 9, 1700-1705
Kirtley, M. E. & Rudney, H. (1967) Biochemistry 6, 230-238
Kleinsek, D. A., Ranganathan, S. & Porter, J. W. (1977) Proc. Natl. Acad. Sci. U.S.A. 74, 1431-1435
Kuroda, M., Hazama-Shimada, Y. & Endo, A. (1977) Biochim. Biophys. Acta 486, 254-259
Langdon, R. B. & Counsell, R. E. (1976) J. Biol. Chem. 251, 5820-5823
Laporte, M., Astruc, M., Tabacik, C., Decomps, B. & Crastes de Paulet, A. (1978) FEBS Lett. 86, 225-229
Lee, T-C., Blank, M. L., Piantadosi, C., Ishaq, K. S. & Snyder, F. (1975) Biochim. Biophys. Acta 409, 218-224
Madhosingh, C. & Orr, W. (1978) Biochim. Biophys. Acta 523, 283-296
Mitropoulos, K. A. & Venkatesan, S. (1977) Biochim. Biophys. Acta 489, 126-142

McNamara, D. J. & Rodwell, V. W. (1972) in Biochemical Regulatory Mechanisms in Eukaryotic Cells (Kun, E. & Grisolia, S. eds), pp. 205-243, John Wiley & Sons, Inc., London, New York, Sydney and Toronto

Ness, G. C. & Moffler, M. H. (1978) Federation Proc. 37, 1708

Ness, G. C. & Moffler, M. H. (1979) Biochim. Biophys. Acta 572, 333-344

Nicolau, G., Shefer, S., Salen, G. & Mosbach, E. H. (1974) J. Lipid Res. 15, 94-98

Nordstrom, J. L. (1976) Ph.D. Thesis, Purdue University, West Lafayette, IN

Nordstrom, J. L., Rodwell, V. W. & Mitschelen, J. J. (1977) J. Biol. Chem. 252, 8924-8934

Philipp, B. W. & Shapiro, D. J. (1979) Federation Proc. 38, 1328

Piszkiewicz, D. (1977) Kinetics of Chemical and Enzyme-catalyzed Reactions, pp. 119-124, Oxford University Press, New York

Qureshi, N., Dugan, R. E., Nimmannit, S., Wu, W. H. & Porter, J. W. (1976a) Biochemistry 15, 4185-4190

Qureshi, N., Dugan, R. E., Cleland, W. W. & Porter, J. W. (1976b) Biochemistry 15, 4191-4197

Rétey, J., von Stetten, E., Coy, U. & Lynen, F. (1970) Eur. J. Biochem. 15, 72-76

Rodwell, V. W., McNamara, D. J. & Shapiro, D. J. (1973) Advan. Enzymol. 38, 373-412

Rodwell, V. W., Nordstrom, J. L. & Mitschelen, J. J. (1976) Advan. Lipid Res. 14, 1-74

Sabine, J. R. (1977) Cholesterol, Marcel Dekker, Inc., New York and Basel

Shapiro, D. J., Nordstrom, J. L., Mitschelen, J. J., Rodwell, V. W. & Schimke, R. T. (1974) Biochim. Biophys. Acta 370, 369-377

Shefer, S., Hauser, S., Lepar, V. & Mosbach, E. H. (1972) J. Lipid Res. 13, 402-412

Sipat, A. B. (1978) Ph.D. Thesis, Univ. of Adelaide, South Australia

Siperstein, M. D. (1970) Curr. Top. Cell. Regul. 2, 65-100

Tan-Wilson, A. & Kohlhaw, G. B. (1978) Biochem. Biophys. Res. Commun. 85, 70-76

Tanzawa, K. Kuroda, M. & Endo, A. (1977) Biochim. Biophys. Acta 488, 97-101

Tormanen, C. D., Redd, W. L., Srikantaiah, M. V. & Scallen, T. J. (1976) Biochem. Biophys. Res. Commun. 68, 754-762

Tormanen, C. D., Srikantaiah, M. V., Hardgrave, J. E. & Scallen, T. J. (1977) J. Biol. Chem. 252, 1561-1565

Ursini, F., Valente, M., Ferri, L. & Gregolin C. (1977) FEBS Lett. 82, 97-101

Volpe, J. J. (1978) Biochem. Biophys. Res. Commun. 80, 1000-1006

Young, N. L. & Rodwell, V. W. (1977) J. Lipid Res. 18, 572-581

Young, N. L. & Rodwell, V. W. (1979) Clinical and Environmental Applications of Quantitative TLC, (Touchstone, J. C. & Rogers, D., eds) Wiley-Interscience, New York, in the press

SUBJECT INDEX